for ArcGIS 10

GISTUTORIAL

Basic Workbook

1

Wilpen L. Gorr
Kristen S. Kurland

ESRI Press
REDLANDS, CALIFORNIA

Esri Press, 380 New York Street, Redlands, California 92373-8100

Originally published as *GISTutorial: Workbook for ArcView 9*

GIS Tutorial 1: Basic Workbook

15 14 13 12   5 6 7 8 9 10 11

Printed in the United States of America

Ask for Esri Press titles at your local bookstore or order by calling 800-447-9778, or shop online at www.esri.com/esripress. Outside the United States, contact your local Esri distributor or shop online at www.eurospanbookstore.com/Esri.

Esri Press titles are distributed to the trade by the following:

In North America:
Ingram Publisher Services
Toll-free telephone: 800-648-3104
Toll-free fax: 800-838-1149
E-mail: customerservice@ingrampublisherservices.com

In the United Kingdom, Europe, Middle East and Africa, Asia, and Australia:
Eurospan Group
3 Henrietta Street
London WC2E 8LU
United Kingdom
Telephone: 44(0) 1767 604972
Fax: 44(0) 1767 601640
E-mail: eurospan@turpin-distribution.com

Contents

Part 2 Working with spatial data

Preface

GIS Tutorial 1: Basic Workbook is the direct result of the authors' experiences teaching GIS to high school students in a summer program at Carnegie Mellon University, undergraduate and graduate students in several departments and disciplines at Carnegie Mellon University, as well as working professionals. *GIS Tutorial 1* is a hands-on workbook with step-by-step exercises that take the reader from the basics of using ArcGIS Desktop interfaces through performing advanced spatial analyses.

Instructors can use this book for the lab portion of a GIS course, or individuals can use it for self-study. You can learn a lot about GIS concepts and principles by "doing" and we provide many short notes on a "just-in-time" basis to help this kind of learning.

The book has three parts. Part 1, "Using and making maps," is essential for all beginning students. Then come the chapters of part 2, "Working with spatial data," and part 3, "Learning advanced GIS applications." These are largely independent of each other, and you can use them in the order that best fits your needs or your class's needs.

In chapter 1, readers learn the basics of working with existing GIS data and maps. In chapters 2 and 3, they learn how to build maps from GIS data. The exercises in chapter 4 teach readers how to create geodatabases and import data into them.

Chapter 5 explores the basic data types used within GIS and then shows readers how to use the Internet to download GIS data. Editing spatial data is a large part of GIS work, and chapter 6 teaches how to digitize vector data and transform data to match real-world coordinates. In chapter 7, students learn how to map address data as points through the geocoding process. Chapters 8 and 9 cover spatial analysis using geoprocessing tools and analysis workflow models.

Chapters 10 and 11 provide instructions on two ArcGIS extensions. Chapter 10 introduces ArcGIS 3D Analyst, allowing students to create 3D scenes, conduct fly-through animations, and conduct line-of-sight studies. Finally, chapter 11 introduces ArcGIS Spatial Analyst for creating and analyzing raster maps, including hillshades, density maps, site suitability surfaces, and risk index surfaces.

To reinforce the skills learned in the step-by-step exercises and to provoke critical problem-solving skills, there are short Your Turn assignments throughout each chapter and advanced assignments at the end of each chapter. The quickest way to increase GIS skills is to follow up step-by-step instructions with independent work, and the assignments provide these important learning components.

This book comes with a DVD containing exercise and assignment data and a DVD containing a trial version of ArcGIS Desktop 10, ArcEditor license. You will need to install the software and data in order to perform the exercises and assignments in this book. (If you have an earlier version of ArcView, ArcEditor, or ArcInfo installed, you will need to uninstall it.) The ArcGIS Desktop 10 DVD provided with this book will work for instructors and basic-level students in exercise labs that previously used an ArcView license of ArcGIS Desktop. Instructions for installing the data and software that come with this book are included in appendix D.

For teacher resources and updates related to this book, go to `www.esri.com/esripress`.

Acknowledgments

We would like to thank all who made this book possible.

We have taught GIS courses at Carnegie Mellon University since the late 1980s, always with lab materials that we had written. With the feedback and encouragement of students, teaching assistants, and colleagues, we eventually wrote a book that became this book. We are forever grateful for the encouragement and feedback we received.

Faculty at other universities who have taught GIS using *GIS Tutorial Workbook for ArcView 9* have also provided valuable feedback. They include Don Dixon of California State University, Sacramento; Mike Rock of Columbus State Community College; Piyusha Singh of State University of New York at Albany; An Lewis of the University of Pittsburgh; and George Tita at the University of California, Irvine.

We are very grateful to the many public servants and vendors who have generously supplied us with interesting GIS applications and data, including Kevin Ford of Facilities Management Services, Carnegie Mellon University; Barb Kviz of the Green Practices Program, Carnegie Mellon University; Susan Golomb and Mike Homa of the City Planning Department, City of Pittsburgh; Richard Chapin of infoUSA Inc.; Pat Clark and Traci Jackson of Jackson Clark Partners, Pennsylvania Resources Council; Commander Kathleen McNeely, Sergeant Mona Wallace, and John Shultie of the Pittsburgh Bureau of Police; Mayor Robert Duffy of Rochester, New York; Lieutenant Todd Baxter, Lieutenant Michael Wood, and Jeff Cheal of the Rochester, New York, Police Department; Kirk Brethauer of Southwestern Pennsylvania Commission (www.spcregion.org); and Tele Atlas for use of its U.S. datasets contained within the ESRI Data & Maps Media Kit.

Finally, thanks to the great team at ESRI Press who tested, edited, designed, and produced this book, including Claudia Naber, Michael Schwartz, Riley Peake, David Boyles, and the entire production team.

Part 1
Using and making maps

1 Introduction

This first chapter familiarizes you with some of the basic features of ArcGIS and illustrates some fundamentals of GIS. You will work with map layers and underlying attribute data tables for U.S. states, cities, counties, and streets. All layers you will use are made up of spatial features consisting of points, lines, and polygons. Each geographic feature has a corresponding data record, and you will work with both features and their data records.

Learning objectives

- *Open and save a map document*
- *Work with map layers*
- *Navigate in a map document*
- *Measure distances*
- *Work with feature attributes*
- *Select features*
- *Work with attribute tables*
- *Label features*

Tutorial 1-1

Open and save a map document

ArcMap is the primary mapping component of ArcGIS Desktop software from ESRI. ESRI offers three licensing levels of ArcGIS Desktop, each with increasing capabilities: ArcView, ArcEditor, and ArcInfo. Together, ArcMap, ArcCatalog, ArcScene, and ArcGlobe—all of which you will use in this book—make up ArcGIS Desktop, the world's most popular GIS software.

Launch ArcMap

1 **From the Windows taskbar, click Start, All Programs, ArcGIS, ArcMap 10.** Depending on your operating system and how ArcGIS and ArcMap have been installed, you may have a different navigation menu.

2 **In the resulting ArcMap - Getting Started window, click Existing Maps and Browse for more.**

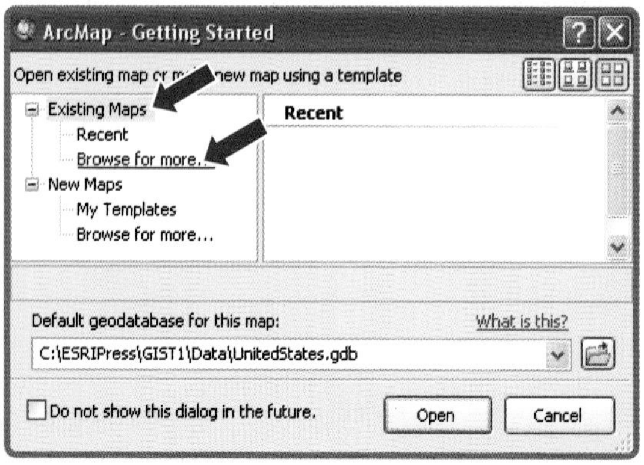

Open an existing map document

1 **Browse to the drive that has the \ESRIPress\GIST1\Maps\ folder installed (e.g., C:\ESRIPress\GIST1\Maps\).**

2 **Click the Tutorial1-1.mxd (or Tutorial1-1) icon and click Open.**

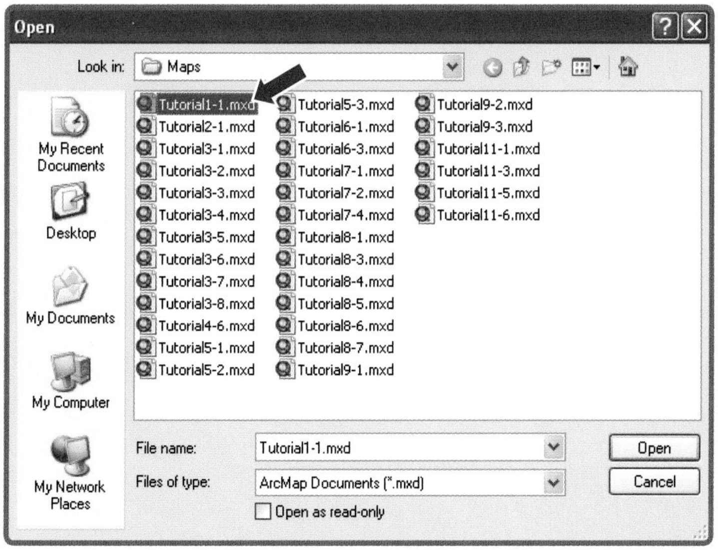

The Tutorial1-1.mxd map document opens in ArcMap, showing a map consisting of the US States layer (with boundaries of the lower 48 contiguous states). The US Cities layer (not yet turned on) is the subset of cities with population greater than 300,000. The left panel of the ArcMap window is the table of contents (TOC). It serves as a legend for the map—plus has several other uses you will learn about in this chapter. Note that the Tools toolbar, which is floating on the right side of the screen on the next page, may be docked somewhere in the interface. If you wish, you can anchor it by clicking in its top area, dragging it to a side or top of the map display window and releasing when you see a thin rectangle materialize. If you do not see the Tools toolbar at all, click Customize, Toolbars, Tools to make it visible. You will learn to use many of the tools in this toolbar in this chapter.

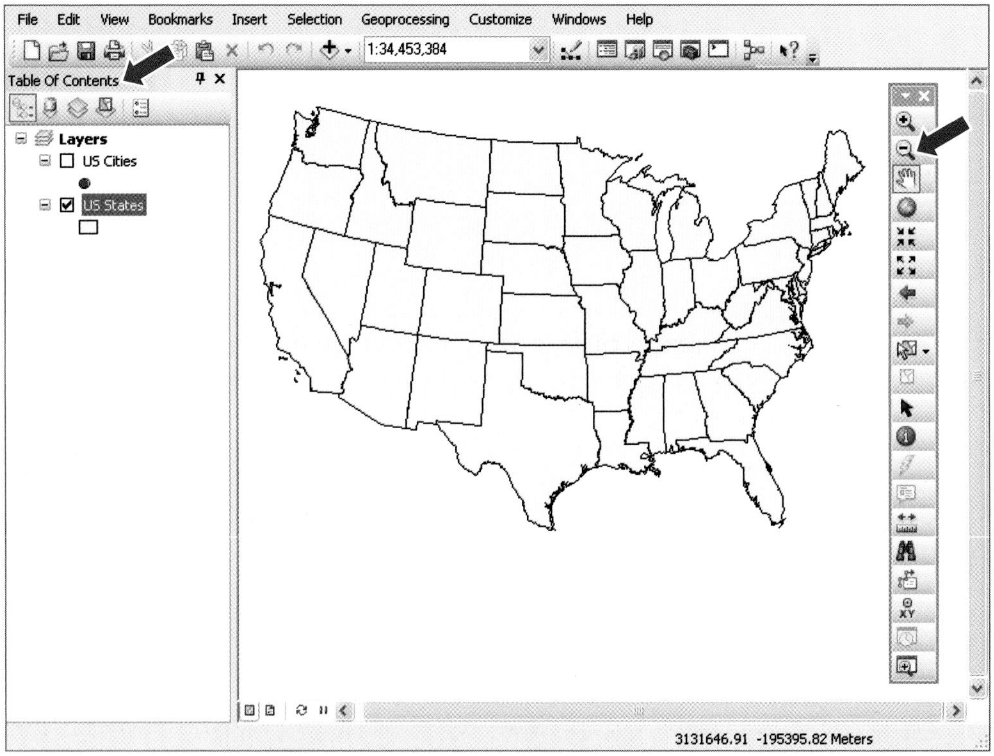

Save the map document to a new location

You will save all files that you modify or create while working through the tutorials in this book in the MyExercises folder.

1. Click File, Save As.
2. Navigate to the \ESRIPress\GIST1\MyExercises\Chapter1\ folder and save the map as **Tutorial1-1.mxd**.
3. Click Save.

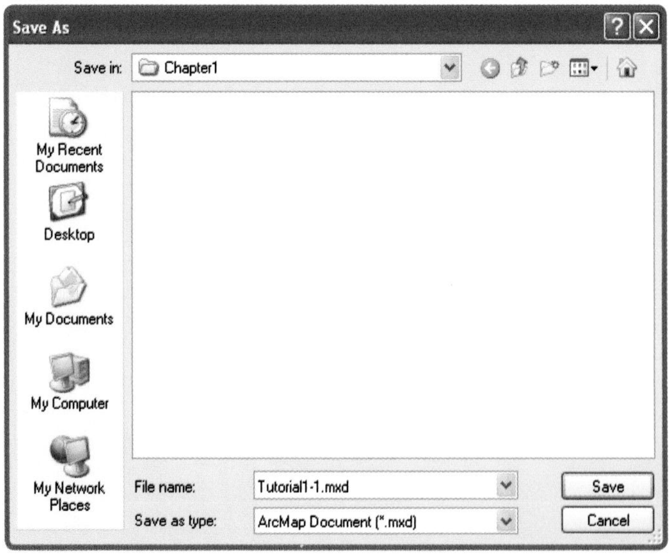

Tutorial 1-2

Work with map layers

Map layers are references to data sources such as point, line, and polygon shapefiles, geodatabase feature classes, raster images, and so forth representing spatial features that can be displayed on a map. ArcMap displays map layers from a map document such as Tutorial1-1.mxd, but the map document does not contain copies of the map layers. The map layer files remain external to the map document wherever they exist on computer storage media. Next, you will use the map document's table of contents (TOC) for the map layers in the document.

Turn a layer on and off

Before GIS existed, mapmakers drew separate layers on clear plastic sheets and then carefully stacked the sheets to make a map composition. Now with GIS, working with layers is much easier.

1 **Click the small check box to the left of the US Cities layer in the TOC to turn that layer on.** The TOC is the panel on the left side of the view window. A check mark appears if the layer is turned on. If the TOC accidentally closes, click Windows, Table of Contents to reopen it.

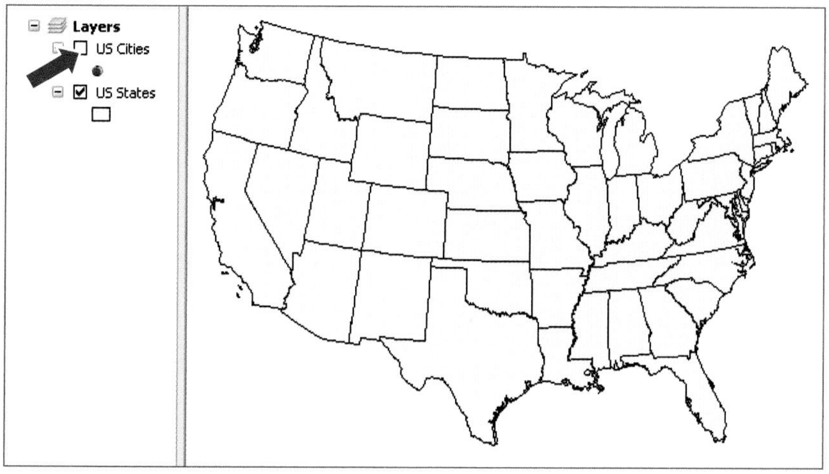

2 **Click the check box to the left of the US Cities layer in the TOC again to turn the layer off.**

Add and remove map layers

You can add map layers to the TOC from their storage locations.

1 **Click the Add Data button** ✚.

2 **In the Add Data browser, click the Connect to Folder button** .

3 **Click the drive on which the \ESRIPress\GIST1\ folder is installed, browse to and click the Data folder, and click OK.** After this, you will always be able to connect directly to the Data folder when searching for or saving data map layers and data tables.

4 **In the Add Data window, double-click the \ESRIPress\GIST1\Data\ folder icon, double-click UnitedStates.gdb, and click COCounties.** ArcMap randomly picks a color for the Colorado counties layer. You will learn how to change the color and other layer symbols later.

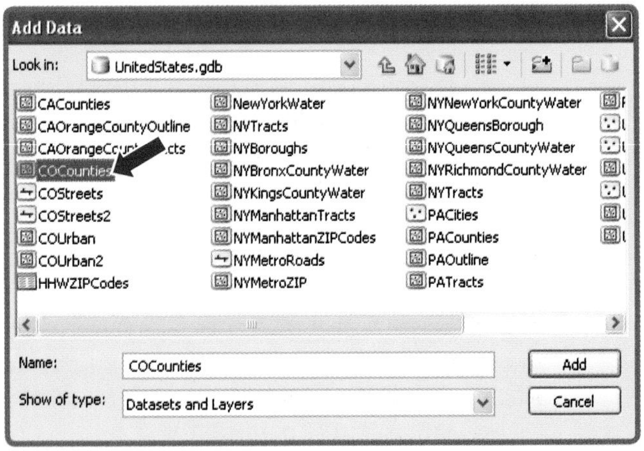

5 **Click Add.** ArcMap places the new layer with Colorado counties correctly over the state of Colorado because all map layers have coordinates tied to specific locations on the earth's surface.

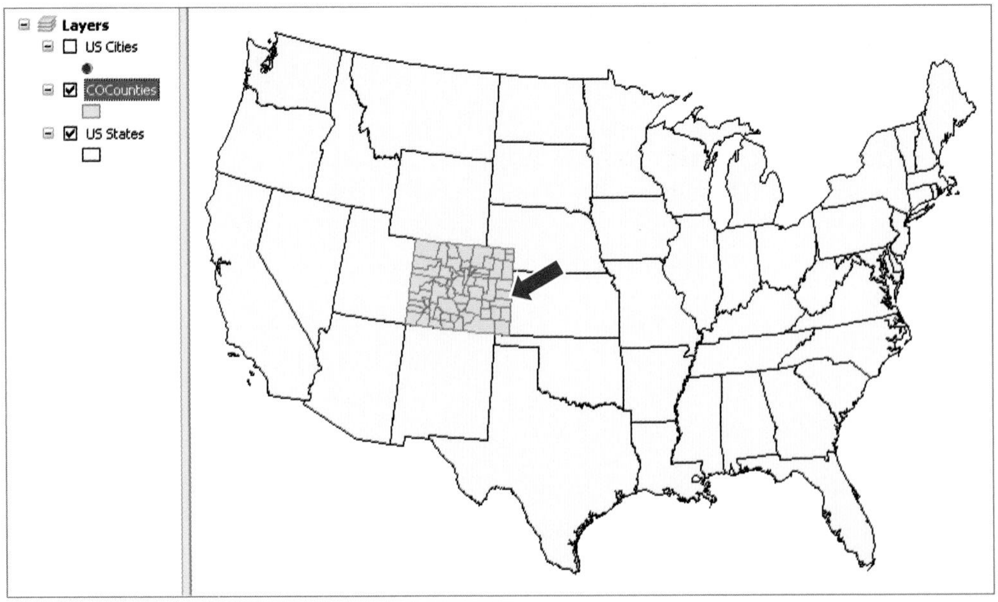

6 **Right-click COCounties in the TOC and click Remove.** This action removes the map layer from the map document but does not delete it from its storage location.

Using relative paths

When you add a layer to a map, ArcMap stores the paths in the map document. When you open a map, ArcMap locates the layer data it needs using these stored paths. If ArcMap cannot find the data for a layer, the layer will still appear in the ArcMap TOC, but of course it will not appear on the map. Instead, ArcMap places a red exclamation mark (!) next to the layer name to indicate that its path needs repair. You can view information about the data source for a layer and repair it by clicking the Source tab in the Layers Properties window.

Paths can be absolute or relative. An example of an absolute path is C:\ESRIPress\GIST1\Data\UnitedStates.gdb\USCities. To share map documents saved with absolute paths, everyone who uses the map must have exactly the same paths to map layers on his or her computer. Instead, the relative path option is favored.

Relative paths in a map specify the location of the layers relative to the current location on disk of the map document (.mxd file). Because relative paths do not contain drive letter names, they enable the map and its associated data to point to the same directory structure regardless of the drive or folder in which the map resides. If a project is moved to a new drive, ArcMap will still be able to find the maps and their data by traversing the relative paths.

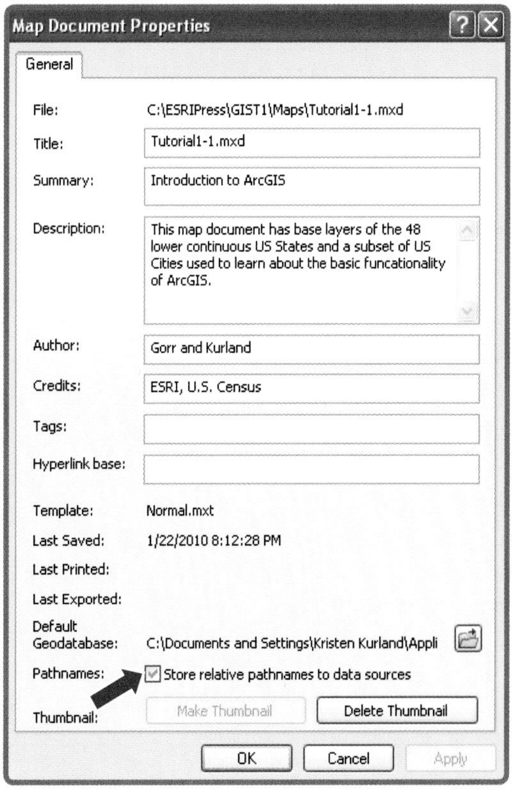

1 **Click File, Map Document Properties.** Notice the option is set to Store relative pathnames to data sources.

2 **Click OK.**

3 **Save your map document.**

Drag and drop a layer from the Catalog window

The Catalog window allows you to explore, maintain, and use GIS data with its many ArcCatalog utility functions. From Catalog, you will drag and drop a map layer into the TOC as an alternative method of adding data.

1 Click Windows, Catalog.

2 In the Catalog window, navigate to \ESRIPress\GIST1\Data\UnitedStates.gdb.

3 Drag and drop COCounties into the top of the TOC window.

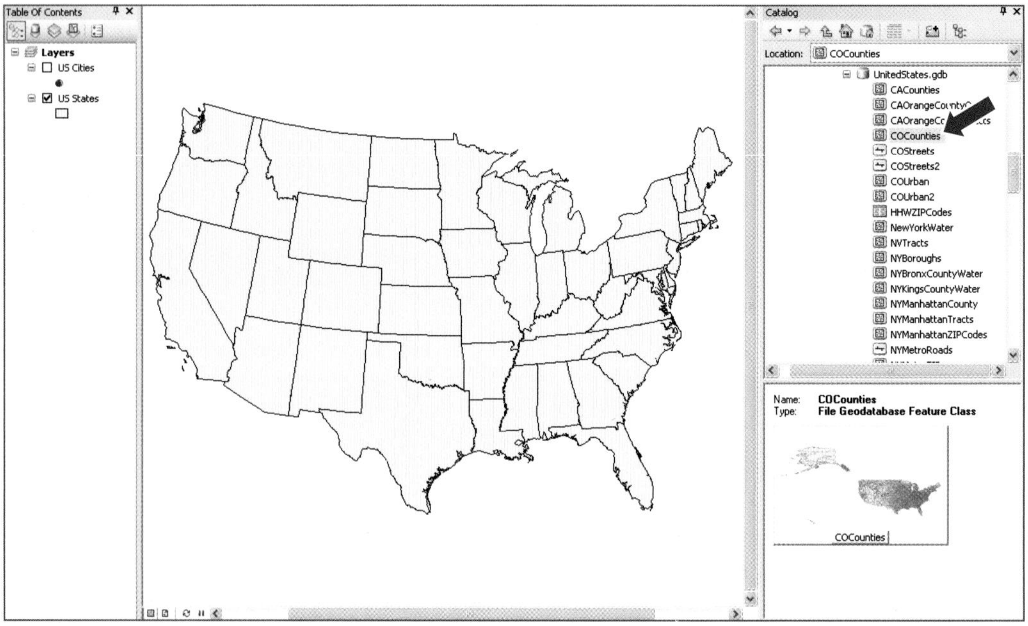

The map layers in the TOC draw in order from the bottom up, so if you dropped COCounties below US States, the states will cover COCounties. If COCounties is covered, remove it and drag and drop it again from Catalog, this time above US States.

Use Auto Hide for the Catalog window

Notice that when you opened the Catalog window, it opened in pinned-open mode, which keeps the window open and handy for use, but covers part of your map. The Auto Hide feature of this application window along with other application windows (such as the TOC and Search window) keeps the windows available for immediate use, but hides them in between uses so that you have more room for your map.

1. **Click the Auto Hide button on top of the Catalog window.** The window closes but leaves a Catalog button on the right side of the ArcMap window Catalog.

2. **Click the Catalog button.** The Catalog window opens. Next, you will simulate having completed a Catalog task by clicking the map document. The window will auto hide.

3. **Click any place on the map or TOC.** You can pin the window open again, which you will do next.

4. **Click the Catalog button and click the Unpinned Auto Hide button.** That pins the Catalog window open until you click the pin again to auto hide or close the window. Try clicking the map or TOC to see that the Catalog window remains open.

5. **Close the Catalog window.**

YOUR TURN

Use the Add Data or Catalog button to add COStreets, also found in \ESRIPress\GIST1\Data\UnitedStates.gdb. These are street centerlines for Jefferson County, Colorado. You may have difficulty seeing the streets because they occupy a small area of the map (look carefully above the center of Colorado). Later in these exercises you will learn how to zoom in for a closer look at small features such as the streets.

Change a layer's display order

Next, you will change the drawing order of layers, but you must have the List By Drawing Order button selected to enable such changes.

1 **Make sure that the List By Drawing Order button is selected in the TOC and turn on the US Cities layer.**

2 **Drag the US Cities layer to the bottom of the TOC and drop it.** Because ArcMap draws the US Cities layer first now, the US States and Counties layers cover its point markers.

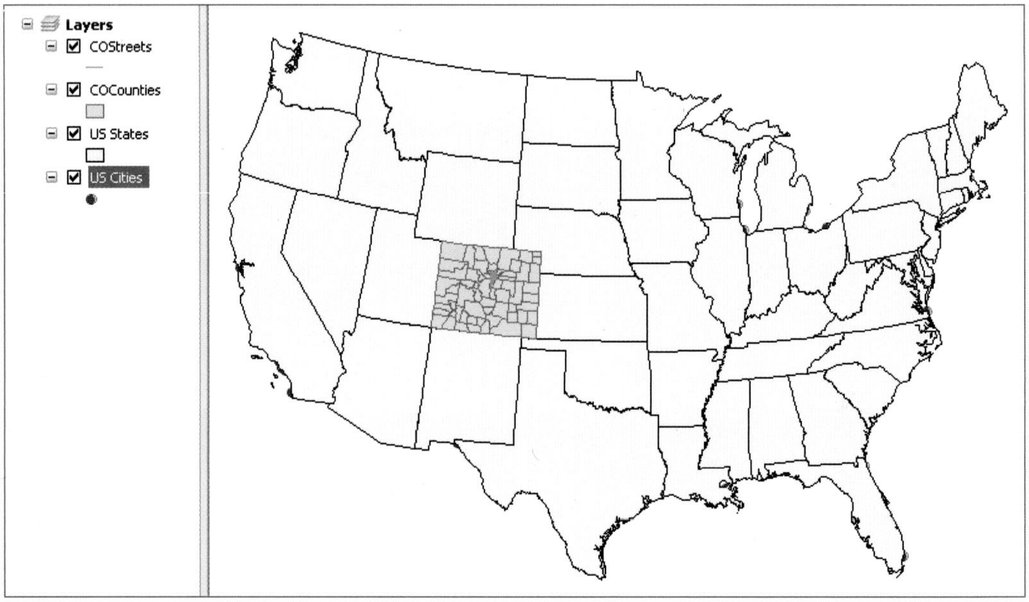

3 **Drag the US Cities layer to the top of the TOC and drop it.** ArcMap now draws the US Cities last, so you can see its points again.

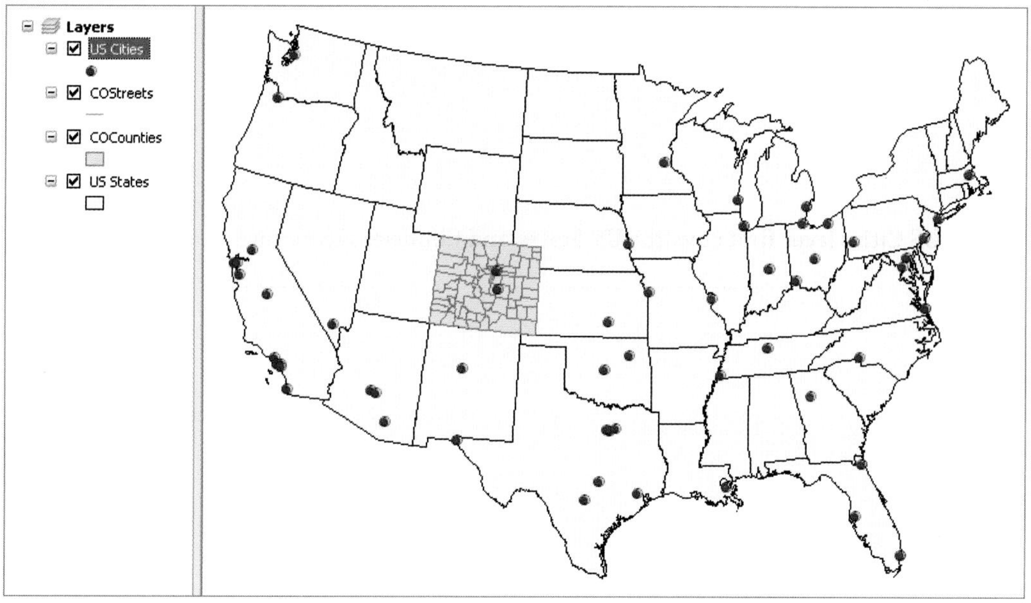

Change a layer's color

One of the nicest capabilities of ArcGIS is how easy it is to change colors and other symbols of layers. First you will change the color fill of a layer's polygons.

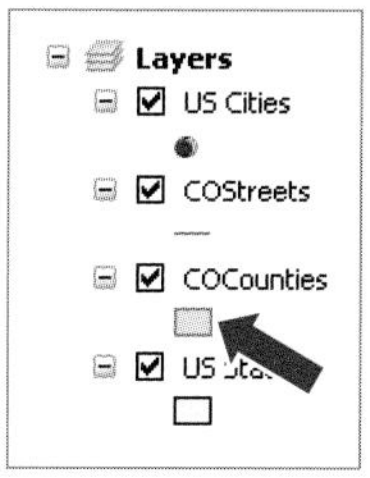

1 **Click the COCounties layer's legend symbol in the TOC.** The legend symbol is the rectangle below the layer name in the TOC.

2 **Click the Fill Color button in the Current Symbol section of the Symbol Selector window.**

3 **Click the Tarragon Green tile in the Color Palette.**

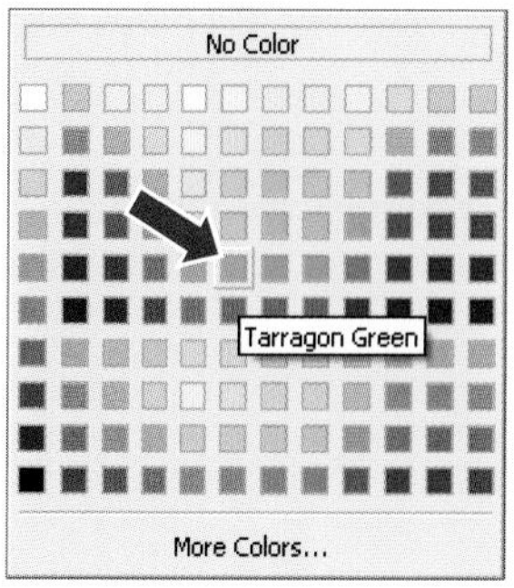

4 **Click OK.** The layer's color changes to Tarragon Green on the map.

Change a layer's outline color

Now you will change the outline color of a layer's polygons.

1. **Click the COCounties layer's legend symbol.**
2. **Click the Outline Color button in the Current Symbol section of the Symbol Selector window.**
3. **Click the Black tile in the Color Palette.**
4. **Click OK.**
5. **Click File and Save to save your map document.**

YOUR TURN

Change the color of the COStreets layer, choosing a medium shade of gray. You will see the results later in the exercise.

Tutorial 1-3

Navigate in a map document

When you open a map document, you see the entire map, a view called the full extent. You can zoom in to any area of the map resulting in that area filling the map window, giving you a close-up view. The current view of the map is its current extent. You can zoom out, pan, and use several additional means of moving about in your map document. These include the Magnifier window for close-up views without zooming in, the Overview window that shows where you are on the full map when zoomed in, and spatial bookmarks for saving a map extent for future use.

Zoom In

1 Click the Zoom In button on the Tools toolbar.

2 Click and hold down the mouse button on a point above and to the left of the state of Florida.

3 Drag the mouse down to the bottom and to the right of the state of Florida and release.

The process you performed in steps 2 and 3 is sometimes called "dragging a rectangle."

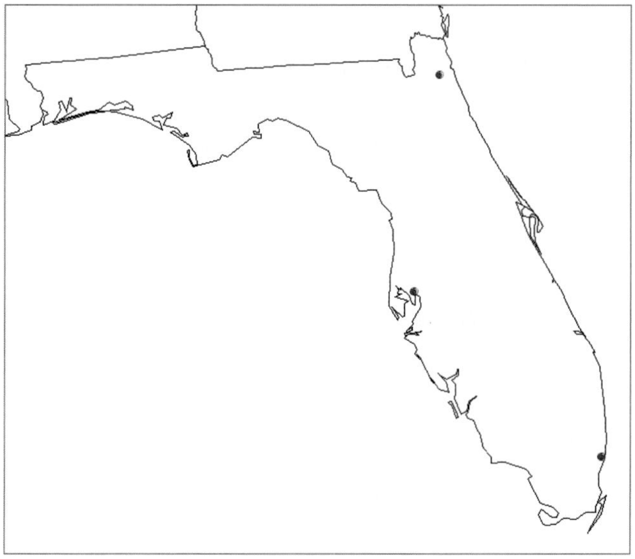

Fixed Zoom In and Zoom Out

This is an alternative for zooming in by fixed amounts.

1. Click the Fixed Zoom In button. This zooms in a fixed distance on the center of the current display.
2. Click the map to zoom in centered on the point you pick.
3. Click the Fixed Zoom Out button. This zooms out a fixed distance from the center of the current zoomed display.

Pan

Panning shifts the current display in any direction without changing the current scale.

1. Click the Pan button.
2. Move the cursor anywhere onto the map display.
3. Hold down the left mouse button and drag the mouse in any direction.
4. Release the mouse button.

Full, previous, and next extent

The following steps introduce tools that navigate through views you have already created.

1. Click the Full Extent button. This zooms to a full display of all layers, regardless of whether they are turned on or turned off.
2. Click the Go Back to Previous Extent button. This returns the map display to its previous extent.
3. Continue to click this button to step back through all of the views.
4. Click the Go to Next Extent button. This moves forward through the sequence of zoomed extents you have viewed.
5. Continue to click this button until you reach full extent.

YOUR TURN

Zoom to the county polygons in Colorado, and then zoom and pan so the streets in Jefferson County, Colorado, are in the center of the display. Leave your map zoomed in to the streets.

Open the Magnifier window

The Magnifier window adjusts the map display to see more detail or get an overview of an area. This window works like a magnifying glass. As you pass the window over the map display, you see a magnified view of the location under the window. Moving the window does not affect the current map extent.

1 **Click Windows, Magnifier.**

2 **Drag the Magnifier over an area of the map to see crosshairs for area selection, and then release to see the zoomed details.**

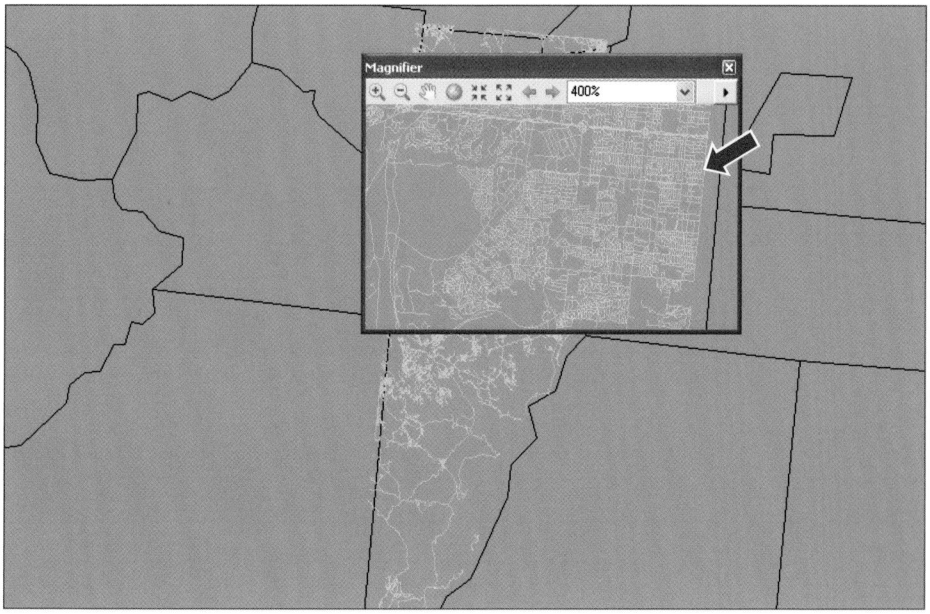

3 Drag the Magnifier window to a new area to see another detail on the map.

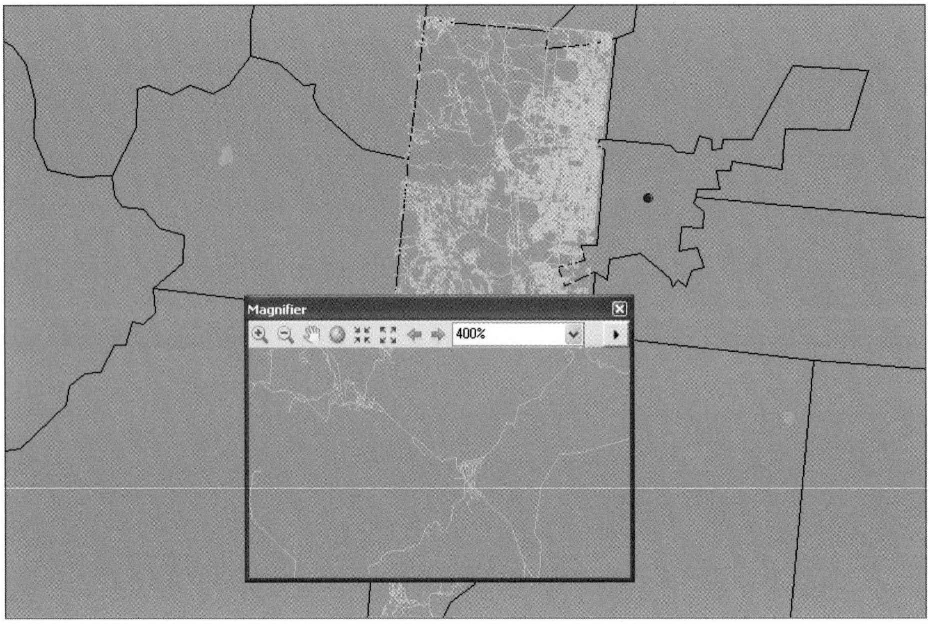

Magnifier properties

You will change the magnification property of the Magnifier window.

1 Right-click the title bar of the Magnifier window and click Properties.

2 Change the Magnify By percentage to 800% if it is not already at that power, and then click OK.

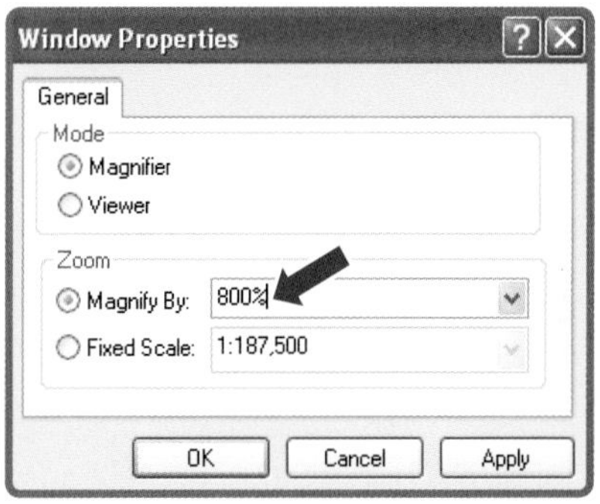

3 Drag the Magnifier window to a different location and see the resulting view.

4 Close the Magnifier window.

Use the Overview window

The Overview window shows the full extent of the layers in a map. A box shows the currently zoomed area. You can move the box to pan the map display. You can also make the box smaller or larger to zoom the map display in or out.

1 **Click the Zoom to Full Extent button**  **.**

2 **Zoom to a small area of the map in the northwest corner of the United States (two or three complete states).**

3 **Click Windows, Overview.** The Overview window shows the current extent of the map highlighted with a red rectangle.

4 **Move the cursor to the center of the red box, click and drag to move it to a new location, and release.** The extent of the map display updates to reflect the changes made in the Layers Overview window. Note that if you right-click the top of the Layers Overview window and click Properties, you can modify the display.

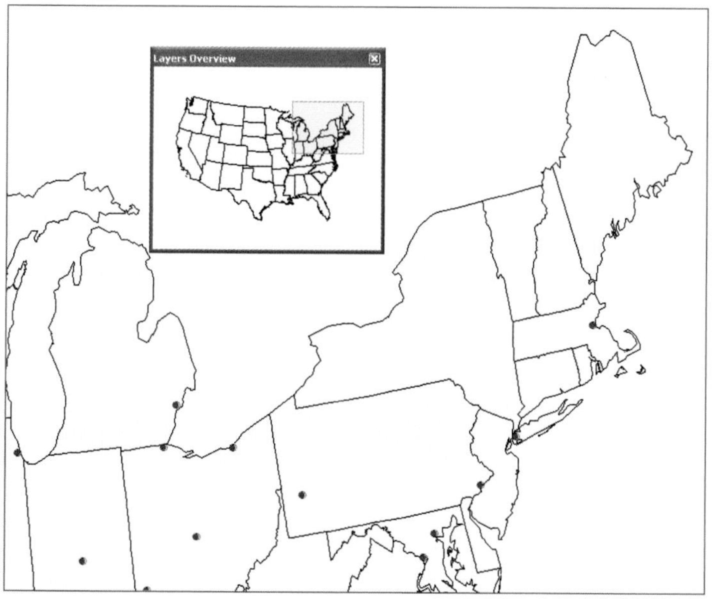

5 **Close the Layers Overview window.**

Create spatial bookmarks

Spatial bookmarks save the extent of a map display or geographic location so you can return to it later without having to use Pan and Zoom tools.

1. **Click the Zoom to Full Extent button .**

2. **Zoom to the state of Florida.**

3. **Click Bookmarks, Create, and type Florida in the Bookmark Name field.**

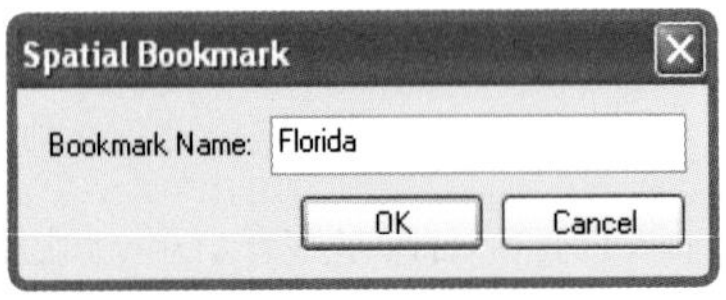

4. **Click OK.**

5. **Click the Zoom to Full Extent button .**

6. **Click Bookmarks, Florida.** ArcMap zooms to the saved bookmark of Florida.

7. **Save your map document.**

YOUR TURN

Create spatial bookmarks for the states of California, New York, and Texas. Try out your bookmarks. Use Bookmarks, Manage to remove the California bookmark.

Tutorial 1-4

Measure distances

Maps have coordinates enabling you to measure distances along paths that you choose with your mouse and cursor.

Change measurement units

While a map's coordinates are in specific units such as feet or meters, you can set the measurement tool to gauge distances in any relevant units.

1 **Zoom to the full extent, then zoom to the state of Washington (uppermost western state).**

2 **On the Tools toolbar, click the Measure button.** The Measure window opens with the Measure Line tool enabled. The current map units are meters, but miles are more familiar in the United States, so you will change the units to Miles.

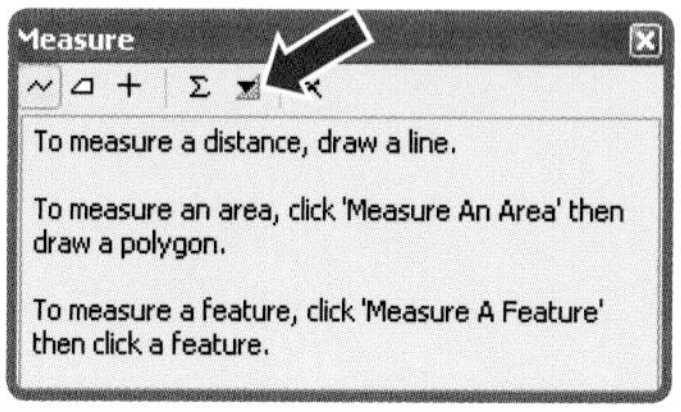

3 **In the Measure window, click the Units drop-down button.**

4 **Click Distance and Miles, and leave the Measure window open.**

Measure the width of Washington state

1 **Move the mouse to the westernmost boundary of the state of Washington and click it.** You do not need to match the selections made below. Any measurement will demonstrate the procedure.

2 **Move the mouse in a straight line to the eastern boundary of Washington until you reach its eastern edge, then double-click the edge.** The distance should be around 300 miles.

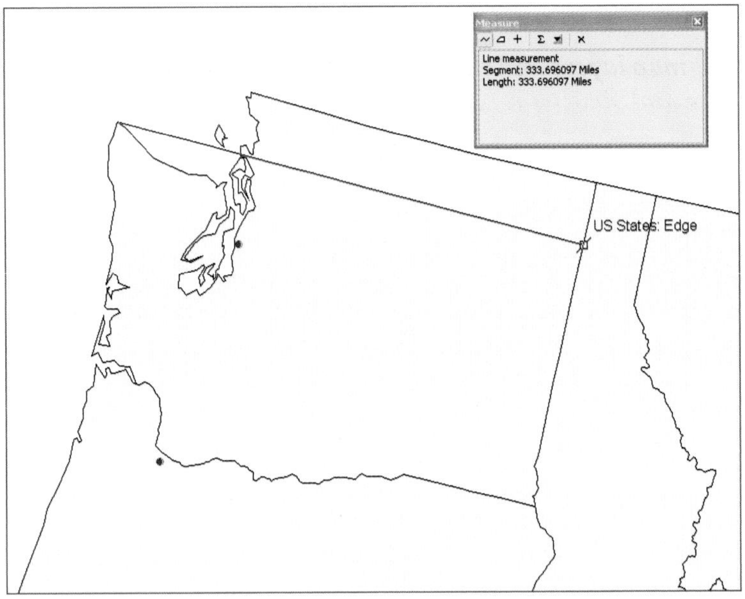

3 **Close the Measure window.**

YOUR TURN

Measure the north-south distance (top to bottom) of Washington. The distance should be roughly 250 miles. Zoom to full extent and measure the north-south distance of the continental United States from the southern tip of Texas to the northern edge of North Dakota. This distance is approximately 1,600 miles. Measure the east-west distance of the continental United States from Washington to Maine. This distance should be approximately 2,500 miles. Close the Measure window when finished.

Tutorial 1-5

Work with feature attributes

Graphic features of map layers and their data records are connected, so you can start with a feature and view its record. You can also find features on a map using feature attributes.

Use the Identify tool

To display the data attributes of a map feature, you can click the feature with the Identify tool. This tool is the easiest way to learn something about a location on a map.

1 **Zoom to the full extent of the map.**

2 **On the Tools toolbar, click the Identify button. Click anywhere on the map.**

3 **From the Identify window, click the Identify from drop-down list and click US States.**

4 **Click inside the state of Texas.** The state temporarily flashes and its attributes appear in the Identify dialog box. Note one of the field values for the state, such as Hispanic population. Next, you will use the Identify tool's options to control which features it will process.

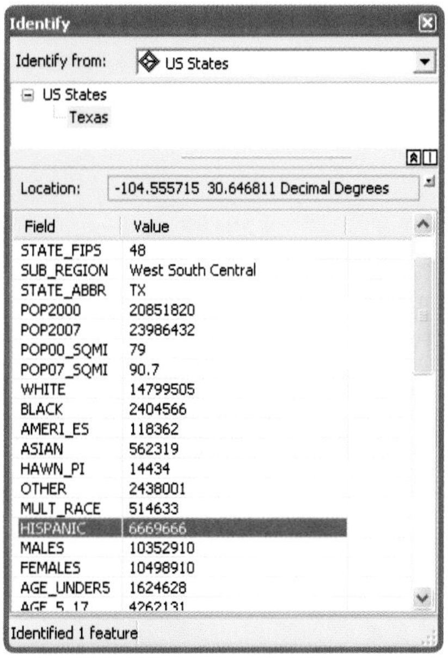

5 In the Identify window, click the Identify from drop-down list and click US Cities.

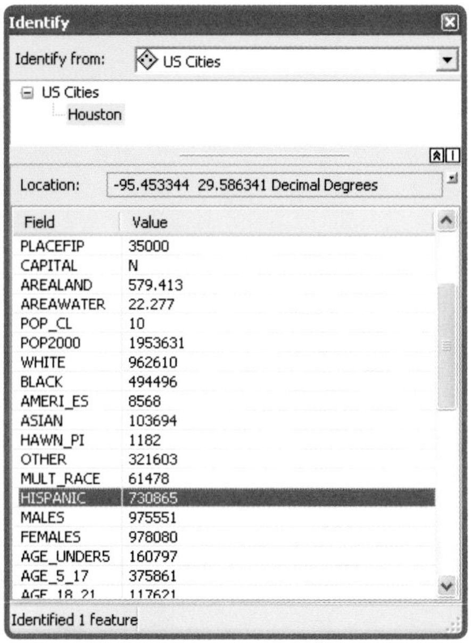

6 Click the red circular point marker for Houston (at the southeastern side of Texas).

7 Make sure the point of the arrow is inside the circle when you click the mouse button. Notice which feature flashes—that is the feature for which you get information.

8 Continue clicking a few other cities to see the identify results. Hold down the Shift key to retain information for more than one city. Then click the name of a city in the top panel of the Identify window to view that city's information.

Use advanced Identify tool capabilities

You can use the Identify tool to navigate and create spatial bookmarks.

1 Without holding down the Shift key, click Houston with the Identify tool.

2 Right-click the name Houston in the Identify window and click Flash. This flashes Houston's point marker.

3 Right-click the name Houston again and click Zoom To. The map display zooms to Houston, Texas. ArcMap identifies the US Cities only because its layer is set in the dialog box.

4 Right-click the name Houston once again and click Create Bookmark.

5 Close the Identify window.

6 Click the Full Extent button.

7 Click Bookmarks, Houston.

YOUR TURN

Restrict the Identify results to the COCounties layer and identify Colorado counties. Practice making bookmarks for various counties using the Identify tool. Close the Identify window.

Find features

Use the Find tool to locate features in a layer or layers based on their attribute values. You can also use this tool to select, flash, zoom, bookmark, identify, or unselect the feature in question.

1 From the Tools toolbar, click the Find button.

2 Click the Features tab.

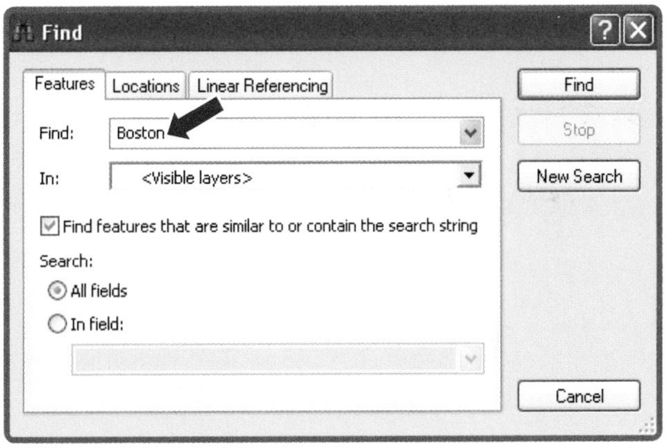

3 Type **Boston** as the feature to find.

4 Click Find. The results appear in the bottom section of the Find window.

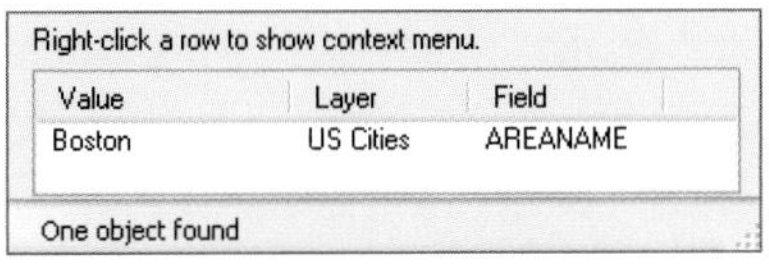

5 Right-click the city name Boston and click Zoom To.

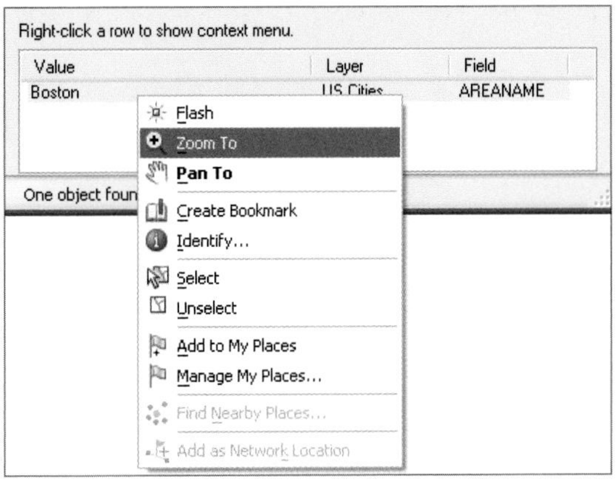

The extent zooms to the city of Boston, Massachusetts.

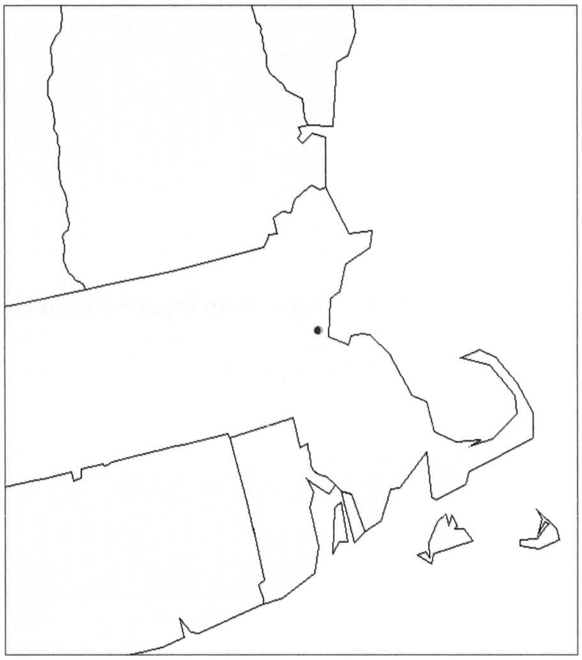

YOUR TURN

Find other U.S. cities and practice showing them using other find options such as Flash Features, Identify Feature(s), and Set Bookmark. When finished, clear any selected features, and zoom to the full extent.

Tutorial 1-6

Select features

You can work with a subset of one or more features in a map layer by selecting them. For example, before you move, delete, or copy a feature (as you will learn about in future chapters), you must select it. Selected features appear highlighted in the layer's attribute table and on the map.

Use the Select Features tool

1 Zoom to the full extent of the map.

2 Turn off the COStreets and COCounties layers.

3 From the Tools toolbar, click the Select Features button.

4 Click inside Texas. This action selects Texas and highlights it with a blue outline.

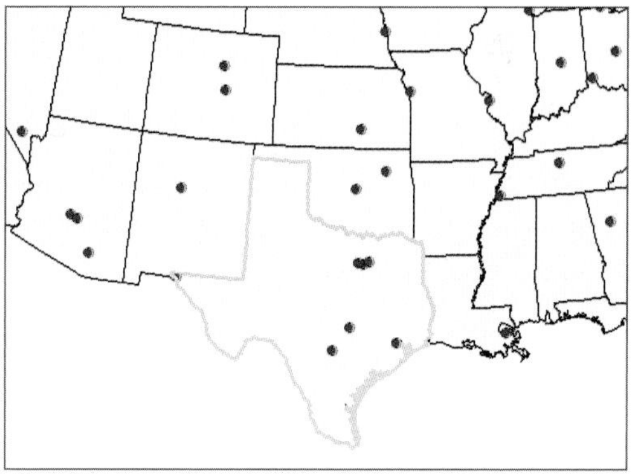

5 Hold down the Shift key and click inside the four states adjacent to Texas.

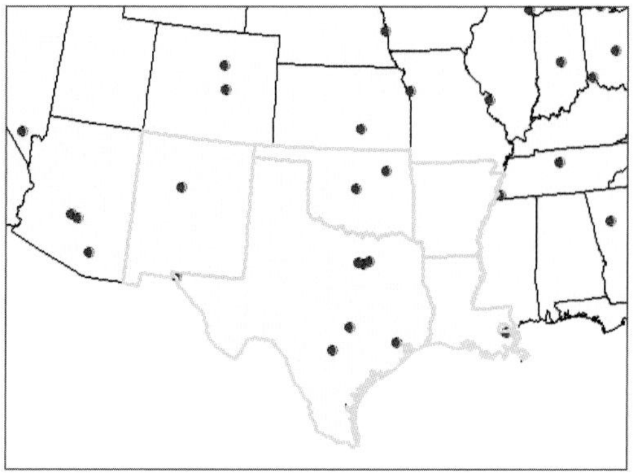

Change the selection color and clear a selection

Sometimes you will want to produce a map with certain features selected. Then it is desirable to be able to change the selection color for purposes at hand.

1 **Click Selection, Selection Options.**

2 **Click the color box in the Selection Tools Settings frame.**

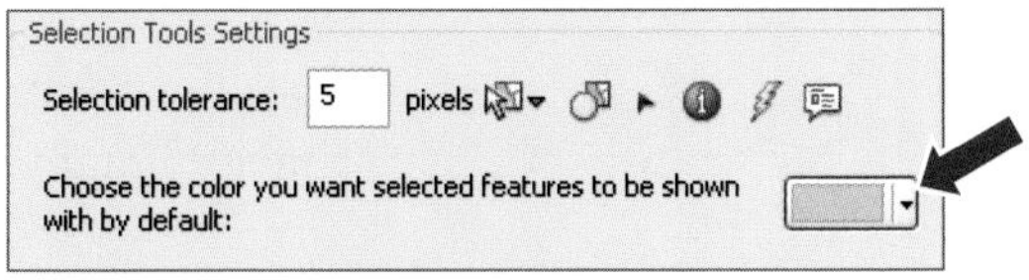

3 **Pick Amethyst as the new selection color and click OK.** The selection color for map features will now be amethyst.

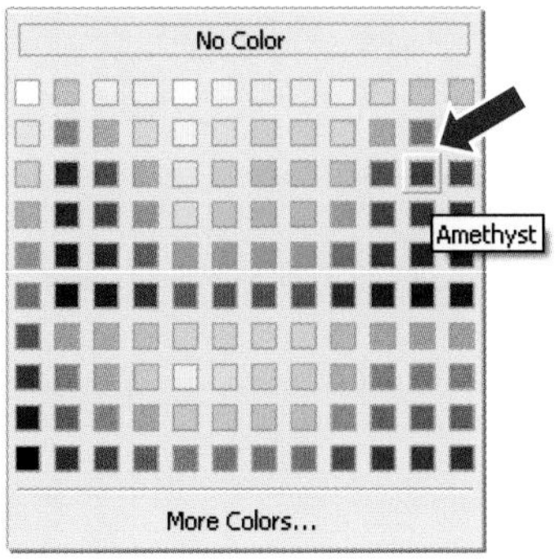

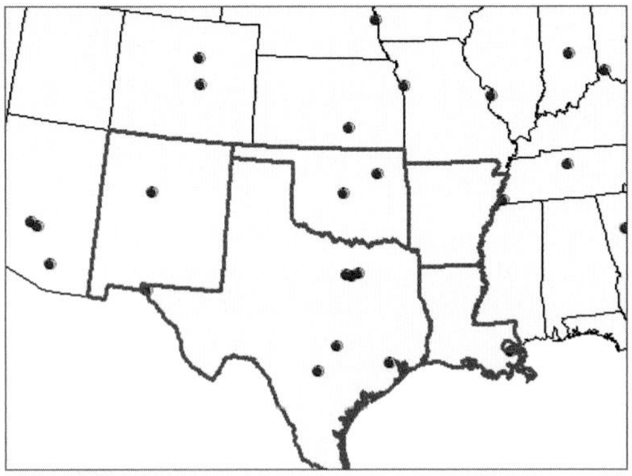

4 **Click Selection, Clear Selected Features.**

Change selection symbol

In addition to changing the color of selected features, you can change the symbol for the entire map or for individual layers.

1 Right-click the US Cities layer in the TOC.

2 Click Properties. The resulting tabbed Layer Properties window is one that you will use often. It allows you to modify many properties of a map layer.

3 Click the Selection tab and Symbol button.

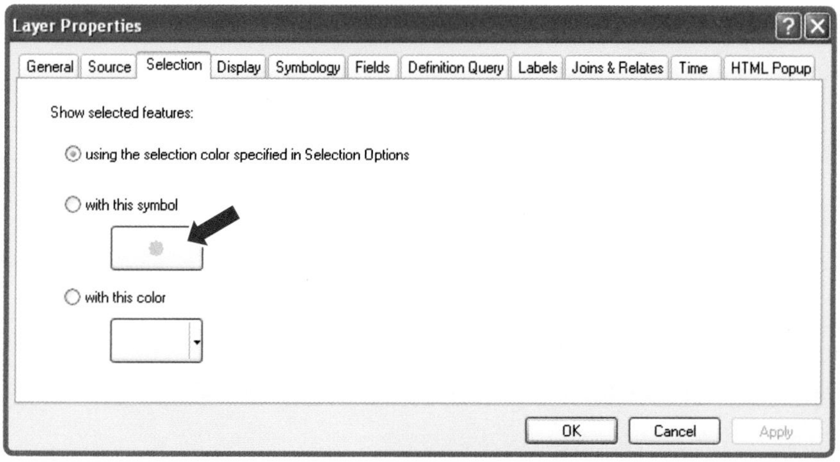

4 Pick a new symbol and/or color for point features, and click OK twice.

5 Click a city feature to see the new selection symbol.

6 Clear the selected features.

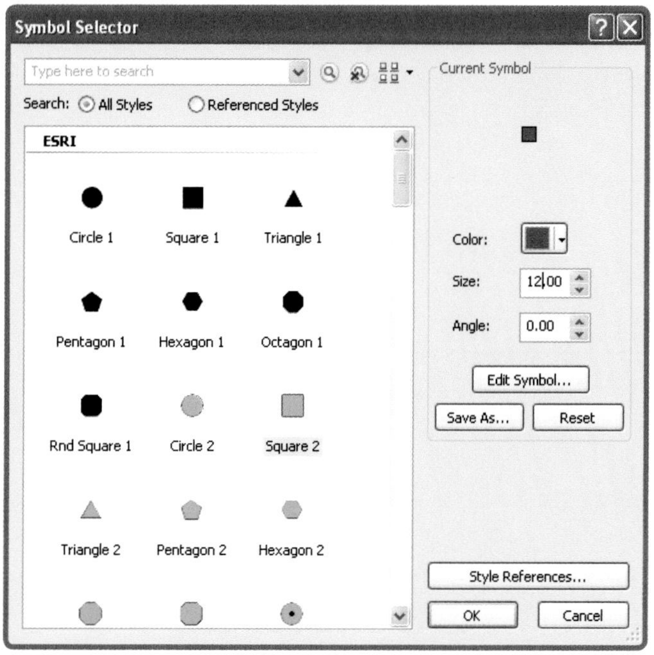

Set selectable layers

When there are many layers in a map document, you may want to restrict which ones are selectable. That simplifies the selection process.

1 **Click the List By Selection button in the TOC.**

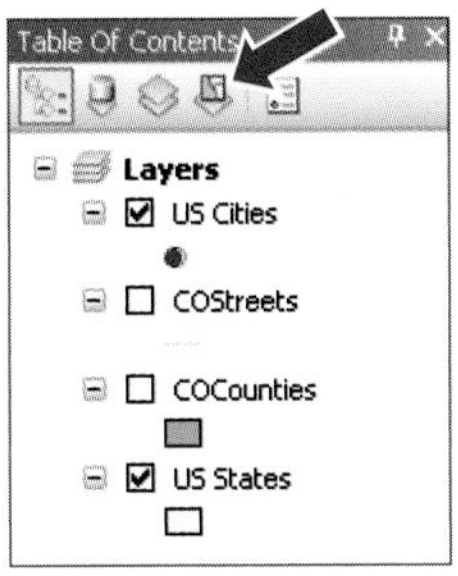

2 **Click off the selection boxes (third to last graphic on each line) for Streets, Counties, and US States to make only US Cities selectable.**

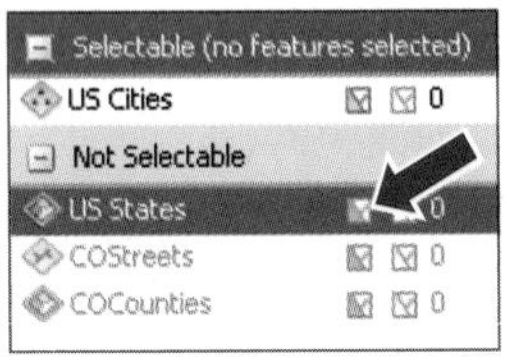

3 **Click the Select Features button ▾ and click a city.** The selected city gets the selection symbol and color that you chose on the previous page and is listed in the TOC.

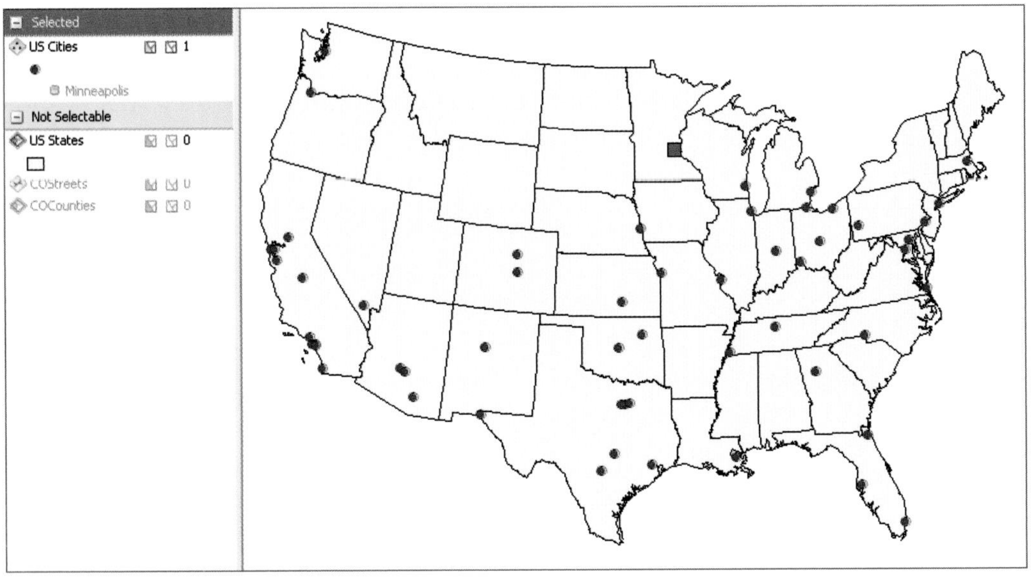

4 **Clear the selected features.**

Select by graphic

Selecting features by using graphics is a shortcut to select multiple features.

1 **Click the drop-down list of the Select Features button and click Select by Circle.**

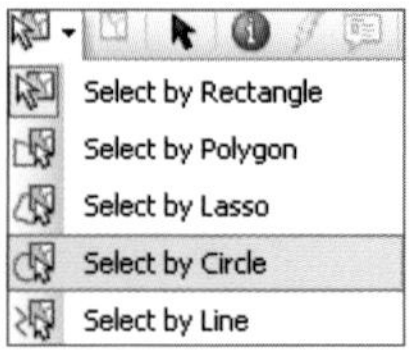

2 **Click inside the state of Florida and drag to draw a circle that includes the three cities in Florida.**

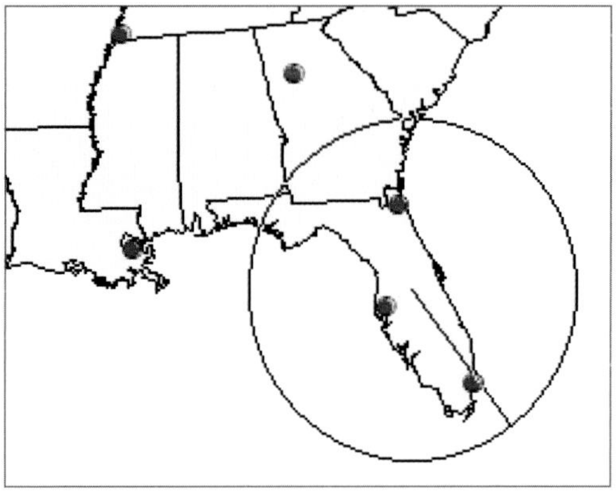

The resulting map will show multiple cities selected and the resulting names in the TOC.

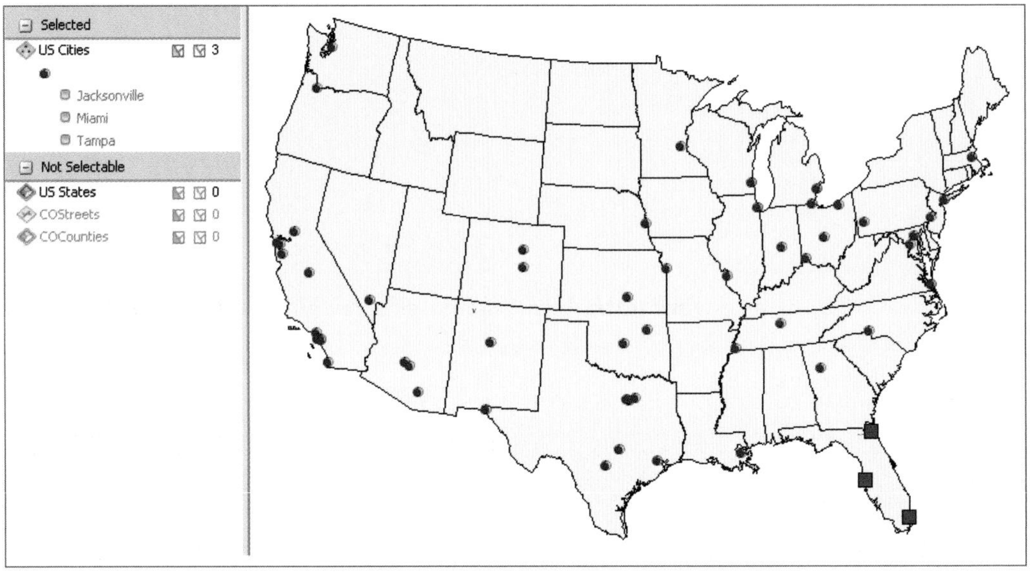

YOUR TURN

Create a new layer from selected features by selecting all cities in Texas using the Select by Lasso graphic. After the cities are selected, right-click the US Cities layer in the TOC, and click Create Layer from Selected Features. Name the new layer **Texas Cities**. Clear the selected features. Turn the new layer off.

Tutorial 1-7

Work with attribute tables

You can view and work with data associated with map features in the layer's attribute table.

Open the attribute table of the US Cities layer and select a record

To explore the attributes of a layer on a map, open its attribute table and select a feature.

1 **Right-click the US Cities layer in the TOC.**

2 **Click Open Attribute Table. The table opens, containing one record for each US City point feature.** Every layer has an attribute table with one record per feature.

3 **Scroll down in the table until you find Chicago and click the record selector (gray cells on the left side of the table) for Chicago to select that record.** If a feature is selected in the attribute table, it also is selected on the map.

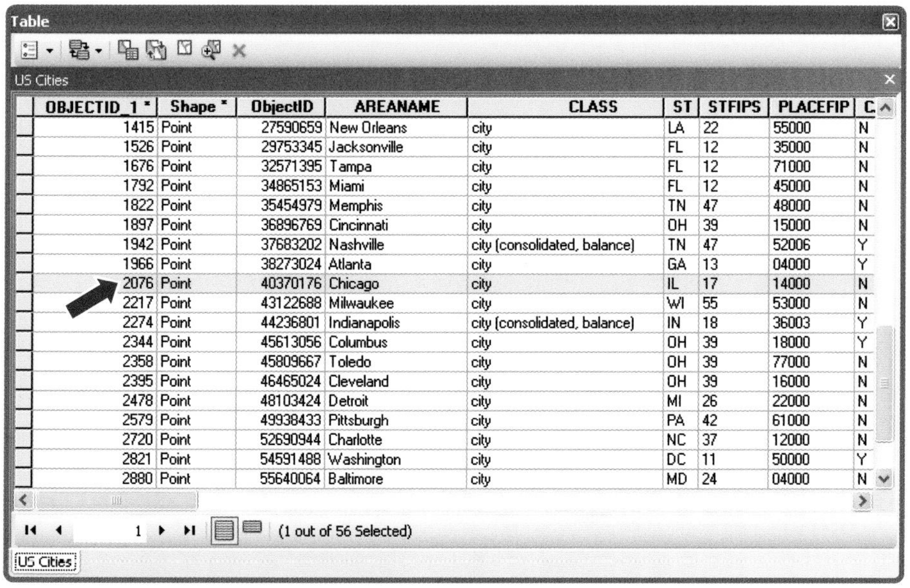

OBJECTID_1 *	Shape *	ObjectID	AREANAME	CLASS	ST	STFIPS	PLACEFIP	C
1415	Point	27590659	New Orleans	city	LA	22	55000	N
1526	Point	29753345	Jacksonville	city	FL	12	35000	N
1676	Point	32571395	Tampa	city	FL	12	71000	N
1792	Point	34865153	Miami	city	FL	12	45000	N
1822	Point	35454979	Memphis	city	TN	47	48000	N
1897	Point	36896769	Cincinnati	city	OH	39	15000	N
1942	Point	37683202	Nashville	city (consolidated, balance)	TN	47	52006	Y
1966	Point	38273024	Atlanta	city	GA	13	04000	Y
2076	Point	40370176	Chicago	city	IL	17	14000	N
2217	Point	43122688	Milwaukee	city	WI	55	53000	N
2274	Point	44236801	Indianapolis	city (consolidated, balance)	IN	18	36003	Y
2344	Point	45613056	Columbus	city	OH	39	18000	Y
2358	Point	45809667	Toledo	city	OH	39	77000	N
2395	Point	46465024	Cleveland	city	OH	39	16000	N
2478	Point	48103424	Detroit	city	MI	26	22000	N
2579	Point	49938433	Pittsburgh	city	PA	42	61000	N
2720	Point	52690944	Charlotte	city	NC	37	12000	N
2821	Point	54591488	Washington	city	DC	11	50000	Y
2880	Point	55640064	Baltimore	city	MD	24	04000	N

4 In the table, click the Clear Selected Features button.

Select features on the map and see selected records

Selecting features also selects their records.

1 Resize the US Cities table to see both the map and table on the screen.

2 Click the Select Features button, hold down the Shift key, and select all cities in Texas on the map.

3 In the US Cities table, click the Show Selected Records button. This shows only the records for the features selected in the map: the cities in Texas.

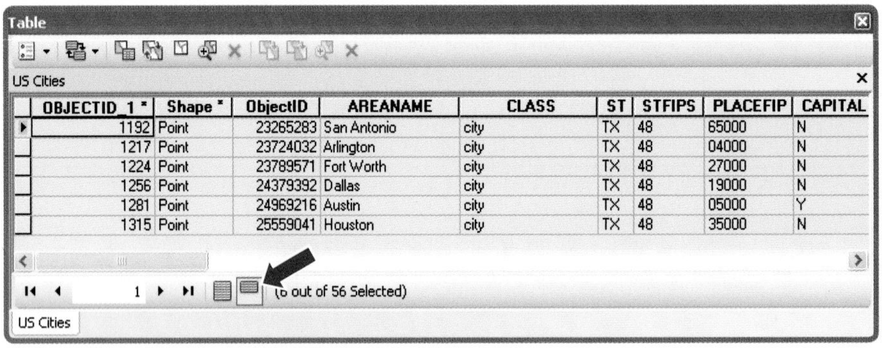

OBJECTID_1 *	Shape *	ObjectID	AREANAME	CLASS	ST	STFIPS	PLACEFIP	CAPITAL
1192	Point	23265283	San Antonio	city	TX	48	65000	N
1217	Point	23724032	Arlington	city	TX	48	04000	N
1224	Point	23789571	Fort Worth	city	TX	48	27000	N
1256	Point	24379392	Dallas	city	TX	48	19000	N
1281	Point	24969216	Austin	city	TX	48	05000	Y
1315	Point	25559041	Houston	city	TX	48	35000	N

4 Click the Show All Records button to show all records again.

5 In the US Cities table, click the Clear Selected Features button.

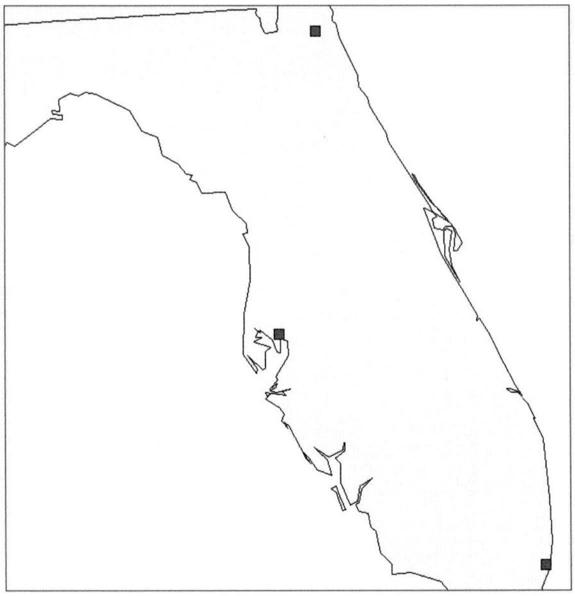

Switch selections

You can select most records in a layer by first selecting the few not to be selected and then reversing the selection.

1 On the map select all of the cities in Florida.

2 Click Selection, Zoom to Selected Features.

3 Click the drop-down option of the Table Options button.

4 Click Switch Selection.

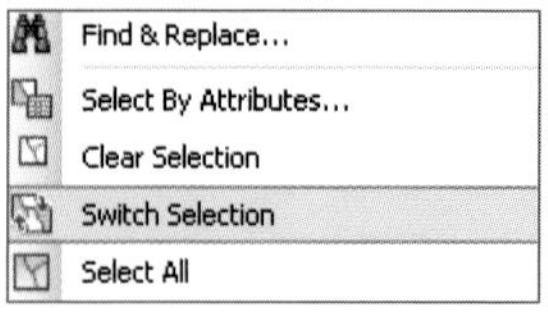

This reverses the selection. It selects all of those that were not selected and deselects those that were selected.

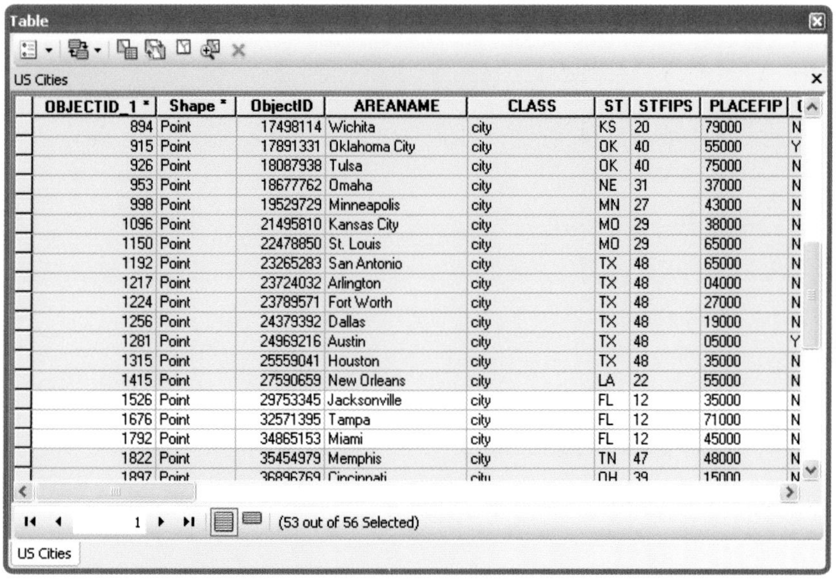

OBJECTID_1 *	Shape *	ObjectID	AREANAME	CLASS	ST	STFIPS	PLACEFIP	(
894	Point	17498114	Wichita	city	KS	20	79000	N
915	Point	17891331	Oklahoma City	city	OK	40	55000	Y
926	Point	18087938	Tulsa	city	OK	40	75000	N
953	Point	18677762	Omaha	city	NE	31	37000	N
998	Point	19529729	Minneapolis	city	MN	27	43000	N
1096	Point	21495810	Kansas City	city	MO	29	38000	N
1150	Point	22478850	St. Louis	city	MO	29	65000	N
1192	Point	23265283	San Antonio	city	TX	48	65000	N
1217	Point	23724032	Arlington	city	TX	48	04000	N
1224	Point	23789571	Fort Worth	city	TX	48	27000	N
1256	Point	24379392	Dallas	city	TX	48	19000	N
1281	Point	24969216	Austin	city	TX	48	05000	Y
1315	Point	25559041	Houston	city	TX	48	35000	N
1415	Point	27590659	New Orleans	city	LA	22	55000	N
1526	Point	29753345	Jacksonville	city	FL	12	35000	N
1676	Point	32571395	Tampa	city	FL	12	71000	N
1792	Point	34865153	Miami	city	FL	12	45000	N
1822	Point	35454979	Memphis	city	TN	47	48000	N
1897	Point	36896769	Cincinnati	city	OH	39	15000	N

5 In the US Cities table, click the drop-down option of the Table Options button.

6 Click Clear Selection.

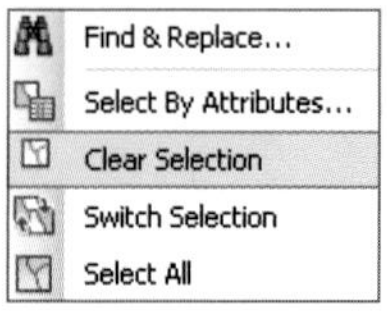

Move a field

1 Click the gray title of the POP2000 field in the US Cities table.

2 Click, drag, and release the POP2000 field to the right of the AREANAME field.

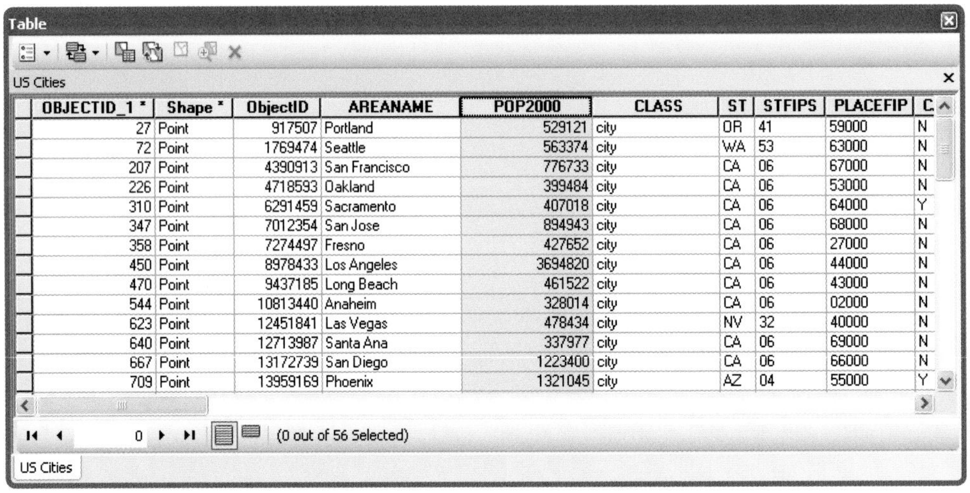

OBJECTID_1 *	Shape *	ObjectID	AREANAME	POP2000	CLASS	ST	STFIPS	PLACEFIP	C
27	Point	917507	Portland	529121	city	OR	41	59000	N
72	Point	1769474	Seattle	563374	city	WA	53	63000	N
207	Point	4390913	San Francisco	776733	city	CA	06	67000	N
226	Point	4718593	Oakland	399484	city	CA	06	53000	N
310	Point	6291459	Sacramento	407018	city	CA	06	64000	Y
347	Point	7012354	San Jose	894943	city	CA	06	68000	N
358	Point	7274497	Fresno	427652	city	CA	06	27000	N
450	Point	8978433	Los Angeles	3694820	city	CA	06	44000	N
470	Point	9437185	Long Beach	461522	city	CA	06	43000	N
544	Point	10813440	Anaheim	328014	city	CA	06	02000	N
623	Point	12451841	Las Vegas	478434	city	NV	32	40000	N
640	Point	12713987	Santa Ana	337977	city	CA	06	69000	N
667	Point	13172739	San Diego	1223400	city	CA	06	66000	N
709	Point	13959169	Phoenix	1321045	city	AZ	04	55000	Y

Sort a field

1 In the US Cities table, right-click the AREANAME field name.

2 Click Sort Ascending Sort Ascending . This sorts the table from A to Z by the name of each U.S. city.

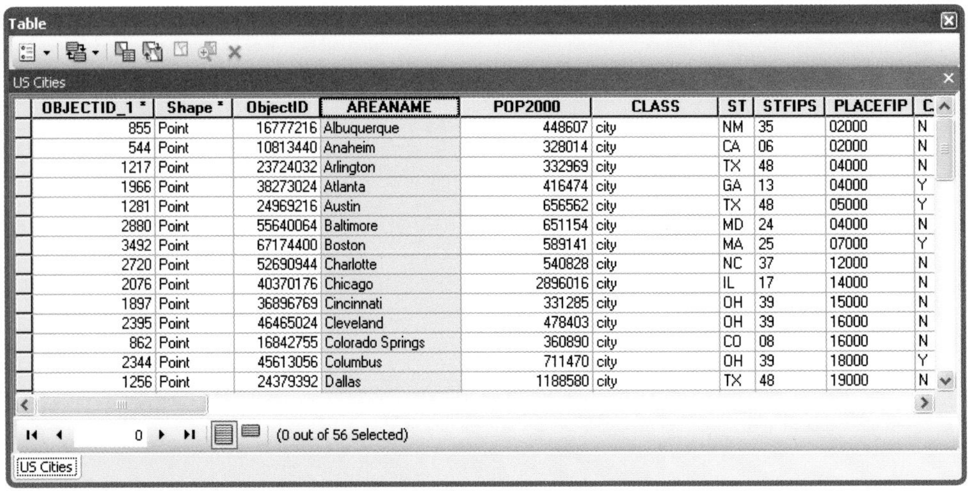

OBJECTID_1 *	Shape *	ObjectID	AREANAME	POP2000	CLASS	ST	STFIPS	PLACEFIP	C
855	Point	16777216	Albuquerque	448607	city	NM	35	02000	N
544	Point	10813440	Anaheim	328014	city	CA	06	02000	N
1217	Point	23724032	Arlington	332969	city	TX	48	04000	N
1966	Point	38273024	Atlanta	416474	city	GA	13	04000	Y
1281	Point	24969216	Austin	656562	city	TX	48	05000	Y
2880	Point	55640064	Baltimore	651154	city	MD	24	04000	N
3492	Point	67174400	Boston	589141	city	MA	25	07000	Y
2720	Point	52690944	Charlotte	540828	city	NC	37	12000	N
2076	Point	40370176	Chicago	2896016	city	IL	17	14000	N
1897	Point	36896769	Cincinnati	331285	city	OH	39	15000	N
2395	Point	46465024	Cleveland	478403	city	OH	39	16000	N
862	Point	16842755	Colorado Springs	360890	city	CO	08	16000	N
2344	Point	45613056	Columbus	711470	city	OH	39	18000	Y
1256	Point	24379392	Dallas	1188580	city	TX	48	19000	N

3 Right-click the POP2000 field name.

4 Click the Sort Descending button Sort Descending. This sorts the field from the highest populated city to the lowest populated city.

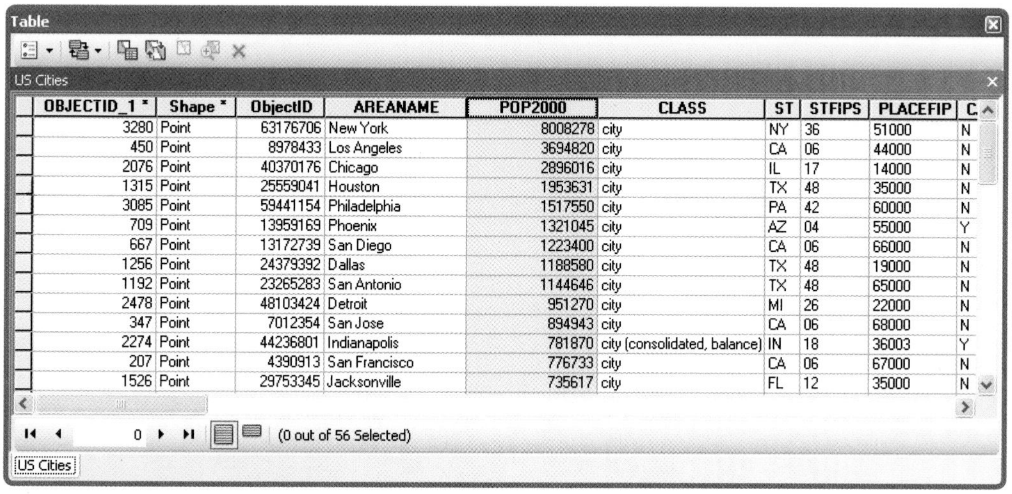

OBJECTID_1 *	Shape *	ObjectID	AREANAME	POP2000	CLASS	ST	STFIPS	PLACEFIP	C
3280	Point	63176706	New York	8008278	city	NY	36	51000	N
450	Point	8978433	Los Angeles	3694820	city	CA	06	44000	N
2076	Point	40370176	Chicago	2896016	city	IL	17	14000	N
1315	Point	25559041	Houston	1953631	city	TX	48	35000	N
3085	Point	59441154	Philadelphia	1517550	city	PA	42	60000	N
709	Point	13959169	Phoenix	1321045	city	AZ	04	55000	Y
667	Point	13172739	San Diego	1223400	city	CA	06	66000	N
1256	Point	24379392	Dallas	1188580	city	TX	48	19000	N
1192	Point	23265283	San Antonio	1144646	city	TX	48	65000	N
2478	Point	48103424	Detroit	951270	city	MI	26	22000	N
347	Point	7012354	San Jose	894943	city	CA	06	68000	N
2274	Point	44236801	Indianapolis	781870	city (consolidated, balance)	IN	18	36003	Y
207	Point	4390913	San Francisco	776733	city	CA	06	67000	N
1526	Point	29753345	Jacksonville	735617	city	FL	12	35000	N

Use Advanced Sorting

1 In the US Cities table, move the ST (State) field to the right of the POP2000 field.

2 Right-click the ST field and click Advanced Sorting Advanced Sorting... .

3 Make selections as follow.

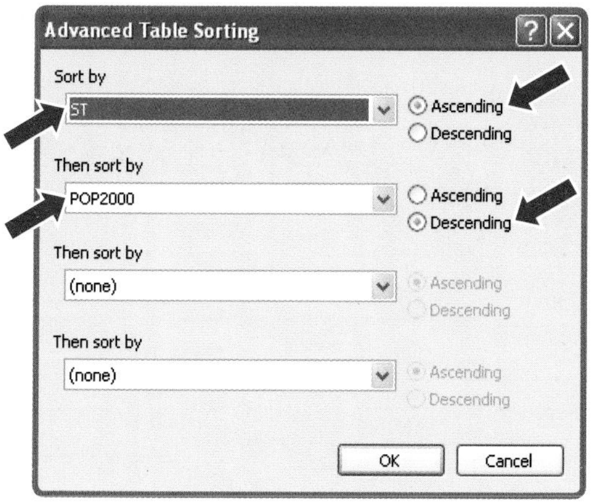

4 Click OK. This sorts the table first by state and then by population of each U.S. city for that state.

Table

US Cities

OBJECTID_1 *	Shape *	ObjectID	AREANAME	POP2000	ST	CLASS	STFIPS	PLACEFIF
709	Point	13959169	Phoenix	1321045	AZ	city	04	55000
726	Point	14221315	Tucson	486699	AZ	city	04	77000
715	Point	14024707	Mesa	396375	AZ	city	04	46000
450	Point	8978433	Los Angeles	3694820	CA	city	06	44000
667	Point	13172739	San Diego	1223400	CA	city	06	66000
347	Point	7012354	San Jose	894943	CA	city	06	68000
207	Point	4390913	San Francisco	776733	CA	city	06	67000
470	Point	9437185	Long Beach	461522	CA	city	06	43000
358	Point	7274497	Fresno	427652	CA	city	06	27000
310	Point	6291459	Sacramento	407018	CA	city	06	64000
226	Point	4718593	Oakland	399484	CA	city	06	53000
640	Point	12713987	Santa Ana	337977	CA	city	06	69000
544	Point	10813440	Anaheim	328014	CA	city	06	02000
827	Point	16187393	Denver	554636	CO	city	08	20000

0 (0 out of 56 Selected)

US Cities

YOUR TURN

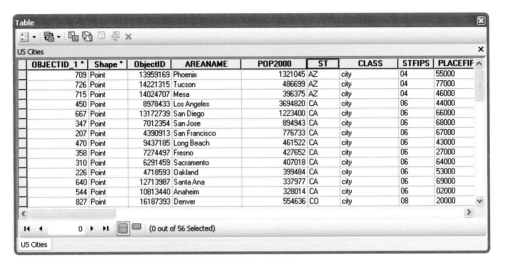

Move and sort by other field names. Try sorting by other multiple fields. For example, you could sort US Cities alphabetically or by whether or not they are state capitals.

Get statistics

You can get descriptive statistics, such as the mean and maximum value of an attribute, in ArcMap by opening a map layer's attribute table using the Statistics function.

1 Zoom to the full extent.

2 Right-click US States in the TOC, click Selection, and click Make This The Only Selectable Layer.

3 Hold down the Shift key and use the Select Features tool to select the state of Texas and the four states adjacent to it.

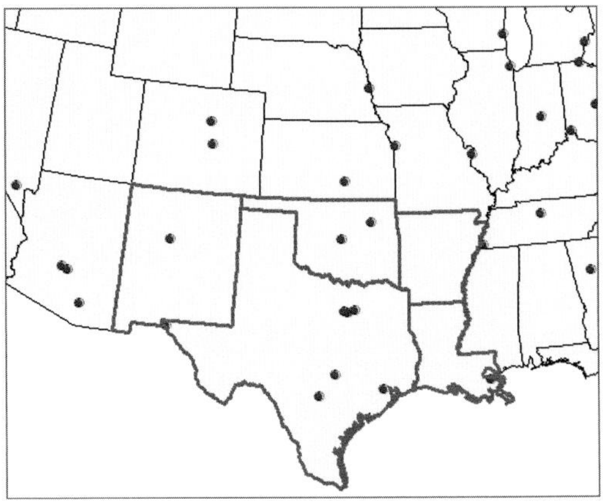

4 In the TOC, right-click US States.

5 Click Open Attribute Table.

6 Right-click the column heading for the POP2000 attribute.

7 Click Statistics. The resulting window has statistics for the five selected states; for example, the mean 2000 population is 6,652,779.

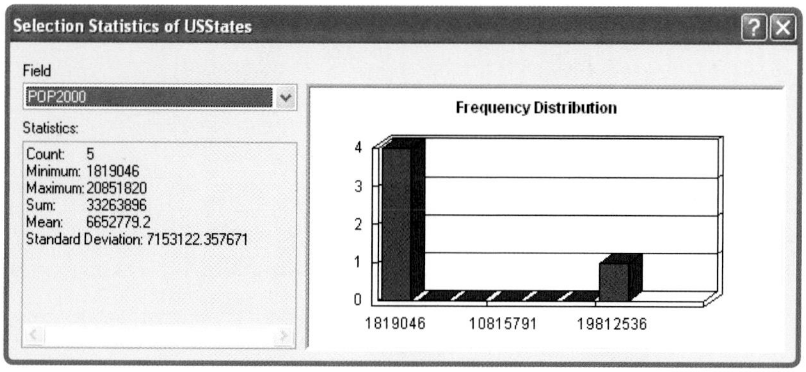

YOUR TURN

Get statistics for a new selection of states and attribute of your choice.

Tutorial 1-8

Label features

Labels are text items on the map derived from one or more feature attributes that ArcMap places dynamically depending on map scale.

Set label properties

There are many label properties that you can set. Here you get started by setting the data value source.

1 Click Bookmarks, Florida to zoom to the state of Florida.

2 Right-click the US Cities layer in the TOC, click Properties, and click the Labels tab.

3 Click the Label Field drop-down arrow and click AREANAME if it is not already selected.

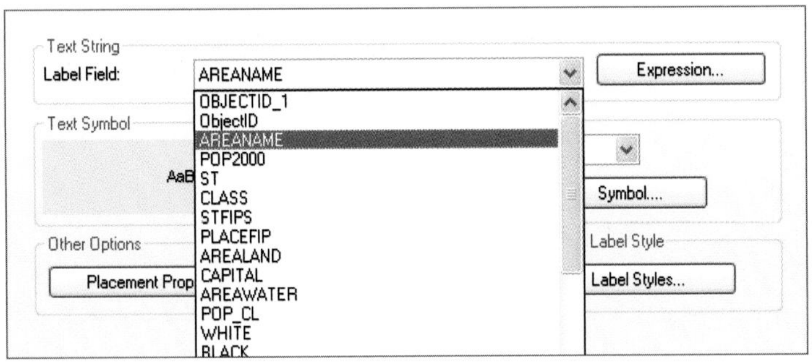

4 Click OK.

Label features

1 Right-click the US Cities layer in the TOC.

2 Click Label Features.

3 Zoom out to see additional states.

Turn labels off

1 Right-click the US Cities layer in the TOC.

2 Click Label Features again.

3 Labels in the map toggle off. Click Label Features again to turn them back on.

Convert labels to annotation

You can convert labels to graphics in order to edit them individually. You can convert all labels, only labels in a zoomed window, or labels from selected features only.

1 Click Bookmarks, Florida.

2 Right-click the US Cities layer in the TOC.

3 Click Convert Labels to Annotation.

4 Make selections as shown in the graphic to label features in the state of Florida only.

5 Click Convert.

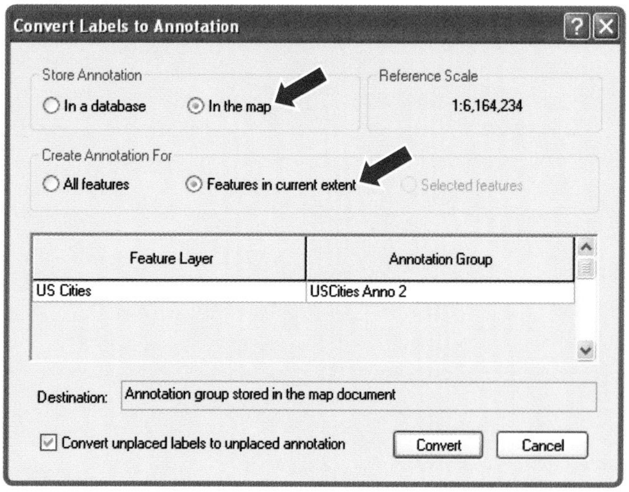

Edit a label graphic

Once labels become graphics, you can move, scale, and otherwise change them individually.

1 Click the Select Elements button.

2 Click the text label for Miami and move it into the state of Florida.

3 Similarly move the label for Jacksonville.

4 Save your map document and exit ArcMap.

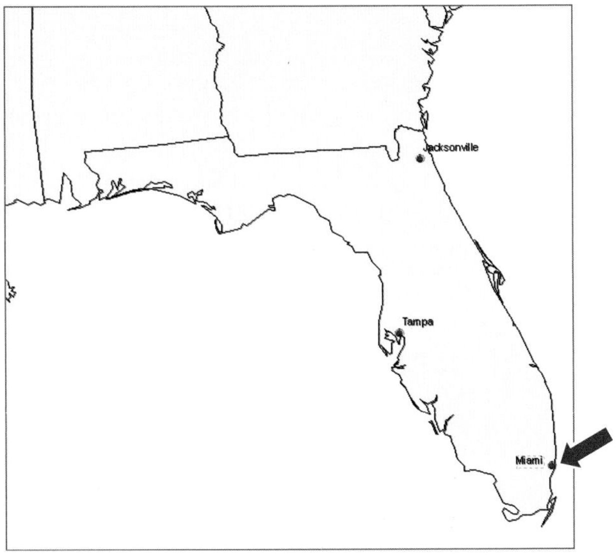

Assignment 1-1

Evaluate U.S. housing statistics

In this assignment, you will compare statistics for U.S. states on the number of housing units, number of renter- and owner-occupied units, and highest number of vacant units.

Start with the following:

- \ESRIPress\GIST1\Data\UnitedStates.gdb\USStates—polygon layer of U.S. states with Census 2000 data
- Attributes of States table—attribute table for U.S. states that includes the following fields needed for the assignment:

    STATE_ABBR—two-letter state abbreviation

    HSE_UNITS—number of housing units per state

    RENTER_OCC—number of renter-occupied units per state

    OWNER_OCC—number of owner-occupied units per state

    VACANT—number of vacant housing units per state

Create a map document and get statistics

Create a new blank map document with path and name **\ESRIPress\GIST1\MyAssignments\Chapter1\Assignment1-1YourName.mxd** using relative paths. Add the above layer of the U.S. States and symbolize it with a hollow fill and a medium gray outline. Using the States attribute table and a bright red selection color, select the five states having the highest number of vacant housing units. Label every state (including unselected states) with its abbreviation.

Create a Word document

Create a Microsoft Word document with path and name **\ESRIPress\GIST1\MyAssignments\Chapter1\Assignment1-1YourName.doc**. In the Word file, create a table with statistics as shown for the five states with the highest number of vacant units only.

Attribute	Mean	Minimum	Maximum
HSE_Units			
Renter_Occ			
Owner_Occ			
Vacant			

With your finished map document, in ArcMap, click File and Export Map, and browse to your \ESRIPress\GIST1\MyAssignments\Chapter1\ folder to save **Assignment1-1YourName.jpg** there. In your Word document, place the insertion point after your table and click Insert, Picture, From File. Then browse to \ESRIPress\GIST1\MyAssignments\Chapter1\Assignment1-1YourName.jpg and insert the map image.

Hint

Select a statistic in the Statistics output table, press Ctrl+C to copy the statistic, click in the appropriate cell of your Word table, and press Ctrl+V to paste it.

WHAT TO TURN IN

If your work is to be graded, turn in the following files:

ArcMap document: \ESRIPress\GIST1\MyAssignments\Chapter1\Assignment1-1YourName.mxd

Image file: \ESRIPress\GIST1\MyAssignments\Chapter1\Assignment1-1YourName.jpg

Word document: \ESRIPress\GIST1\MyAssignments\Chapter1\Assignment1-1YourName.doc

If instructed to do so, instead of the above individual files, turn in a compressed file, **Assignment1-1YourName.zip**, with all files included. Do not include path information in the compressed file.

Assignment 1-2

Facilitate the Erin Street crime watch

Crime prevention depends to a large extent on "informal guardianship," meaning that neighborhood residents keep an eye on suspicious behavior and intervene in some fashion, including calling the police. Neighborhood associations called crime watch groups enhance guardianship, so police departments actively promote and support them and keep them informed on crime trends. Suppose that a police officer of a precinct has a notebook computer and a portable color projector for use at crime watch meetings. Your job is to get the officer ready for a meeting with the 100 Block Erin Street crime watch group of the Middle Hill neighborhood of Pittsburgh.

Start with the following:

- \ESRIPress\GIST1\Data\Pittsburgh\Midhill.gdb\Streets—line layer for street centerlines in the Middle Hill neighborhood of Pittsburgh
- Attributes of Streets—table for streets in the Middle Hill neighborhood that includes the following fields needed for the assignment:

  FNAME—street name

  Address ranges

  LEFTADD1—beginning house number on the left side of the street

  LEFTADD2—ending house number on the left side of the street

  RGTADD1—beginning house number on the right side of the street

  RGTADD2—ending house number on the right side of the street
- \ESRIPress\GIST1\Data\Pittsburgh\Midhill.gdb\Buildings—polygon layer for buildings in the Middle Hill neighborhood of Pittsburgh
- \ESRIPress\GIST1\Data\Pittsburgh\Midhill.gdb\Curbs—line layer for curbs in the Middle Hill neighborhood of Pittsburgh
- \ESRIPress\GIST1\Data\Pittsburgh\Midhill.gdb\CADCalls—point layer for 911 computer-aided dispatch police calls in the Middle Hill neighborhood of Pittsburgh
- Attributes of CADCalls table—table for CADCalls points that includes the following attributes needed for the assignment:

  NATURE_COD—call type

  CALLDATE—date of incident

  ADDRESS—addresses of incident location

Change the map and get statistics

Create a new blank map and add the above layers. Save the map document as **\ESRIPress\GIST1\MyAssignments\Chapter1\Assignment1-2YourName.mxd** with relative paths that includes a final zoomed view of the "Erin block" streets (see "Hints") selected and labeled with street names. Display streets, curbs, and buildings as medium-light gray, and CADCalls as bright red circles.

Create a spatial bookmark of the zoomed area called **Erin Street.**

Create a table of addresses, dates of calls, and call types for crimes in the 100 block of Erin Street (see "Hints"). The street names include Davenport, Erin, and Trent. Create a Microsoft Word document called **\ESRIPress\GIST1\MyAssignments\Chapter1\Assignment1-2YourName.doc** and paste the table into it as directed in the hints below.

Address	Date	Call Type

Hints

- The 100 block of Erin Street is the segment of Erin Street whose addresses range from 100 to 199 and is perpendicular to Webster and Wylie streets. The crime reports are prepared for the blocks on either side of Erin Street in this range. Use both the attribute table and Identify tool to find and label these streets.
- Although it appears that there are only six incident points, there are actually 13 total because multiple incidents are at the same locations. Use the Select Features button and information in the attribute table to get the data on all relevant calls.
- In the attributes of the CADCalls table, click the Table Options button and "Export." Save the selected records to a dBASE (.dbf) file. Open the dBASE file in Microsoft Excel, edit the records, and paste from there into Assignment1-2YourName.doc. When opening the dBASE in Excel, choose the Files of Type drop-down menu and choose All Files (*.*). This will allow you to choose the dBASE file. Otherwise only the XML file associated with the dBASE file will appear.

WHAT TO TURN IN

If your work is to be graded, turn in the following files:

ArcMap document: \ESRIPress\GIST1\MyAssignments\Chapter1\Assignment1-2YourName.mxd

Word document: \ESRIPress\GIST1\MyAssignments\Chapter1\Assignment1-2YourName.doc

If instructed to do so, instead of the above individual files, turn in a compressed file, **Assignment1-2YourName.zip**, with all files included. Do not include path information in the compressed file.

2

Map design

In this chapter you will learn all steps necessary to compose common maps from available map layers. One type of map that you will create is a choropleth map that color-codes polygons to convey information about areas. The second is a "point feature map" that uses point markers to display spatial patterns in point data. You will use U.S. states and counties, plus census tracts and detailed census data for Pennsylvania.

Learning objectives

- *Create choropleth maps*
- *Create group layers*
- *Set threshold scales for dynamic display*
- *Create choropleth maps using custom attribute scales*
- *Create point maps*
- *Create a point feature map based on a definition query*
- *Create hyperlinks*
- *Create MapTips*

Tutorial 2-1

Create choropleth maps

A choropleth map is a map in which polygon areas are colored or shaded to represent attribute values. In this tutorial, you will use U.S. Census population data to create choropleth maps for states, counties, and census tracts.

Open a map document

1 On your desktop, click Start, All Programs, ArcGIS, ArcMap 10.

2 In the ArcMap – Getting Started window, click Existing Maps and Browse for more. Browse to the drive and folder where you installed \ESRIPress\GIST1\Maps\, and double-click Tutorial2-1.mxd. ArcMap opens a map with no layers added. You will add the layers needed for the exercises next.

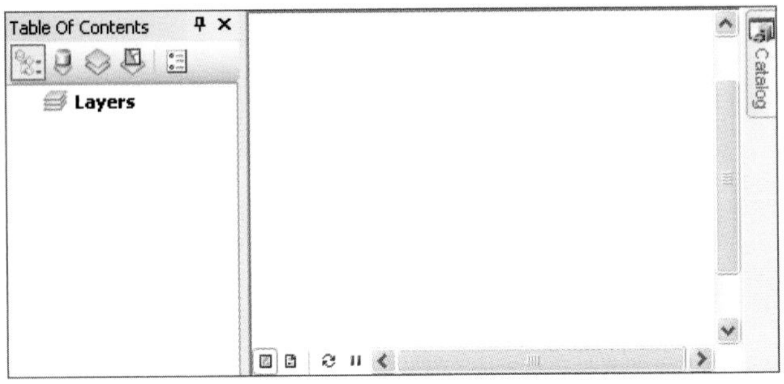

Add a layer

1 Click the Add Data button .

2 Navigate to \ESRIPress\GIST1\Data\ and double-click UnitedStates.gdb. Click USStates and Add.

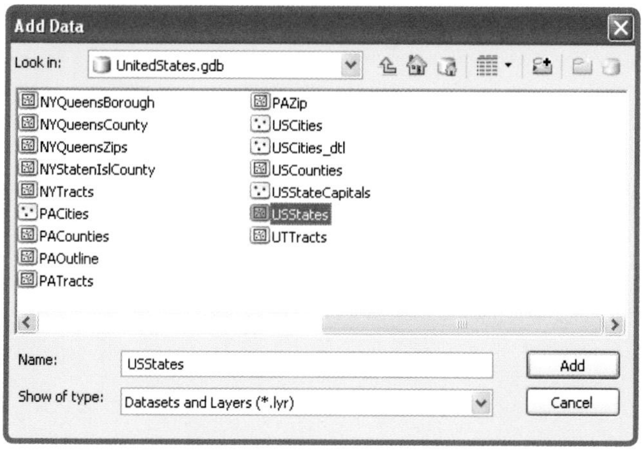

ArcMap draws the 48 contiguous states of the United States using a random color. You will change the colors later in the exercises.

Change a layer's name

1 Right-click the USStates layer in the TOC.

2 Click Properties.

3 Click the General tab. Notice that the current layer name is "USStates."

4 Type **Population By State** as the new layer name and click OK.

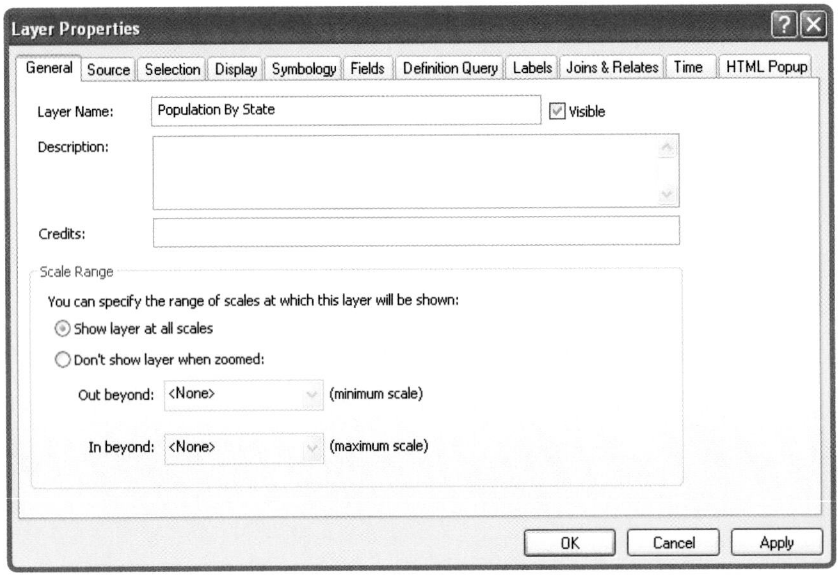

Select a census attribute to display state population

1 Right-click the Population By State layer in the TOC.

2 Click Properties.

3 Click the Symbology tab.

4 In the Show box, click Quantities and Graduated colors.

5 In the Fields box, click the Value drop-down list and POP2007.

6 Click the Color Ramp drop-down list, scroll down, and click the yellow-to-green-to-blue color ramp.

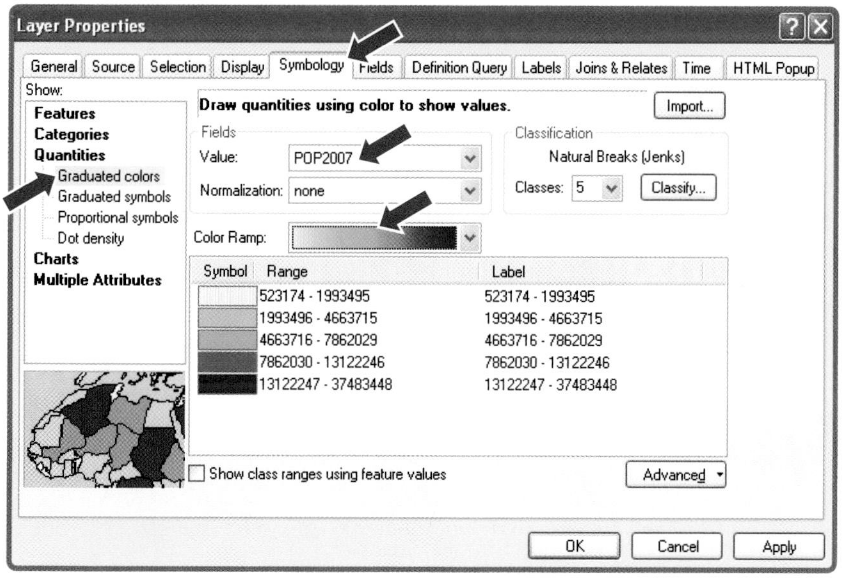

7 Click OK. The result is a classification consisting of five value intervals ranging from lowest to highest of 2007 population with darker colors for higher population. By default, ArcMap uses a method called natural breaks to construct the classification intervals. You will learn how to change classifications later.

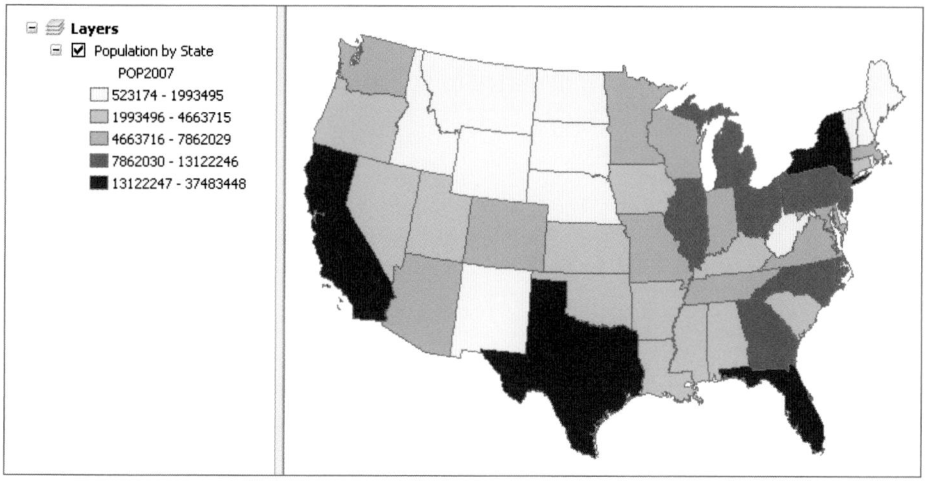

Tutorial 2-2

Create group layers

Group layers contain other layers, allowing for better organization of the layers in your map. Group layers have behavior similar to other layers in the TOC. Turning off the visibility of a group layer turns off the visibility of all its component layers.

Add a group layer to the map

1 Right-click Layers in the TOC.

2 Click New Group Layer.

3 Right-click the resulting New Group Layer and click Properties.

4 Click the General tab.

5 Type **Population By County** as the group layer name (but do not click OK).

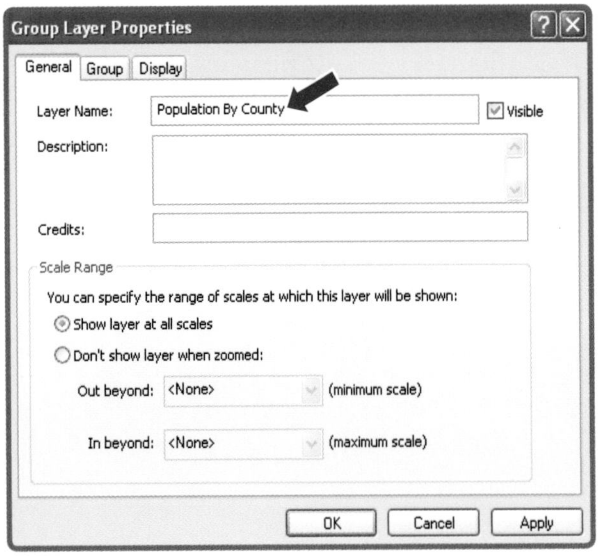

Add a layer to the group

1 Click the Group tab in the Group Layer Properties window.

2 Click the Add button and navigate to \ESRIPress\GIST1\Data\UnitedStates.gdb.

3 Hold down the Ctrl key.

4 Click USStates and USCounties. Release the Ctrl key, then click Add and OK.

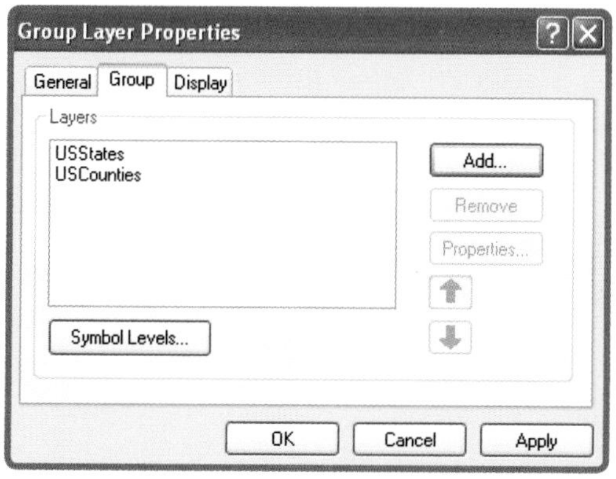

ArcMap displays the U.S. counties with a random color.

Change the symbology for states

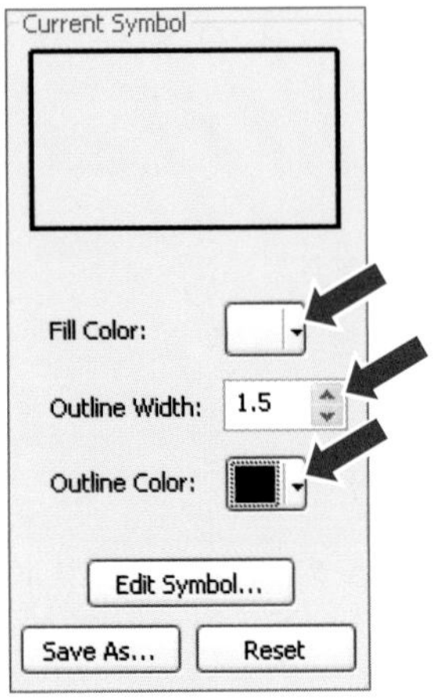

1 **Within the Population By County group layer, if it is not already the top layer, click the USStates layer and drag it above the USCounties layer.** If by mistake you drag the States layer outside of the Population By County group, just drag it back inside. This is another way to add layers to a group—simply add them to the TOC and then drag them inside a group.

2 **Click the legend symbol below the USStates layer name in the group layer.**

3 **In the Current Symbol panel, change the Fill Color to No Color, type an Outline Width of 1.5, change Outline Color to Black, and click OK.**

You will see the USStates' symbology change in the Population by County layer group.

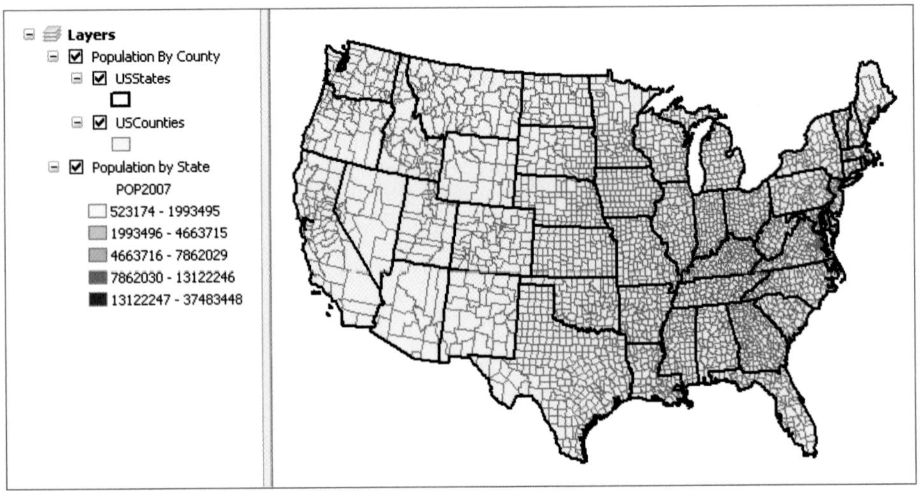

Select a census attribute to display county population

1 **Right-click the USCounties layer in the group layer and click Properties.**

2 **Click the Symbology tab.** The current symbol for the counties layer is Single symbol.

3 **In the Show box, click Quantities, Graduated colors.**

4 **In the Fields panel, click the Value drop-down list and click POP2007. Click OK.**

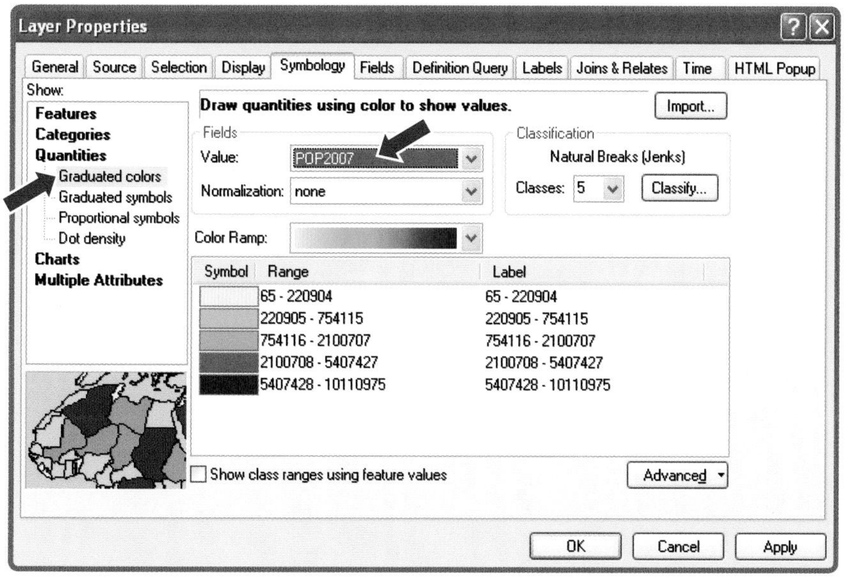

The result is a classification of the U.S. counties into five value ranges of 2007 population.

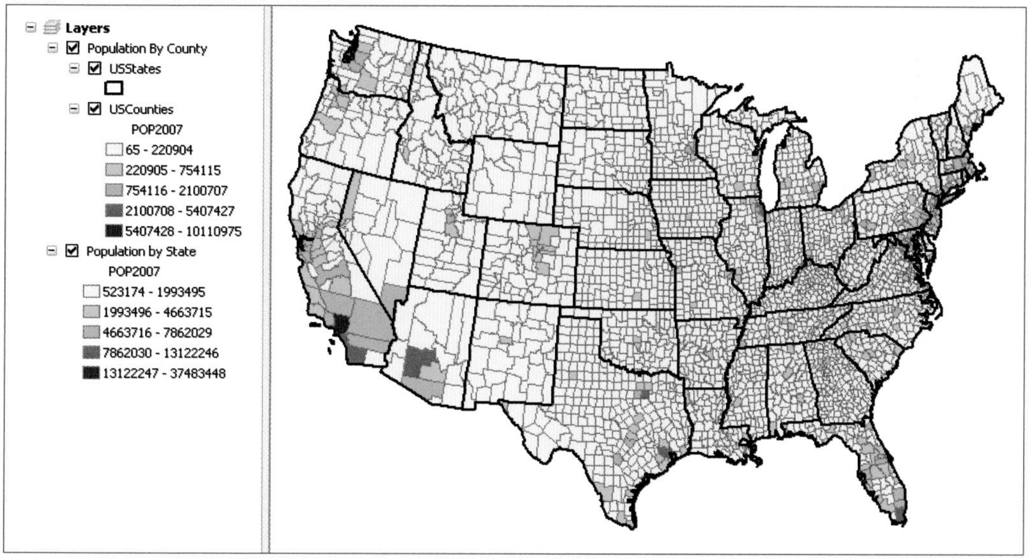

5 **Collapse the tree structures in the TOC by clicking the boxes that have minus signs (-) for Population By County and Population By State.** You can reverse this process by clicking the boxes again, which would again have plus signs (+) indicating that they can be expanded.

YOUR TURN

Note: You are required to complete this Your Turn exercise because the layer group that you create is used in a later exercise.

Turn off the Population By County group layer and the Population By State layer. Create a new group layer called **Population By Census Tract**. Add the census tract layers for Utah (UTTracts) and Nevada (NVTracts) and the USStates layer to the Population By Census Tract group layer. The census tract layers are located in \ESRIPress\GIST1\Data\UnitedStates.gdb. Classify the census tracts using graduated colors based on the POP2007 field. Choose No Color for Fill Color and a black 1.5-width line for the USStates layer.

Notice that the resulting classification intervals for the two states differ. Later in these exercises you will learn how to build custom numerical scales. Then you could create a single scale for both states.

Saving layer files

You can use layer files (.lyr) to quickly add a map layer that you previously classified or otherwise symbolized to a map document. You can save layer files and add them to your map document the same way you add other data.

1 **Turn off the layer groups Population By Census Tract and Population By State, then turn on only the layer group Population By County.**

2 **Right-click the Population By County layer group and click Save As Layer File.**

3 **Navigate to the \ESRIPress\GIST1\MyExercises\Chapter2\ folder.**

4 **Type PopulationByCounty.lyr in the Name field.**

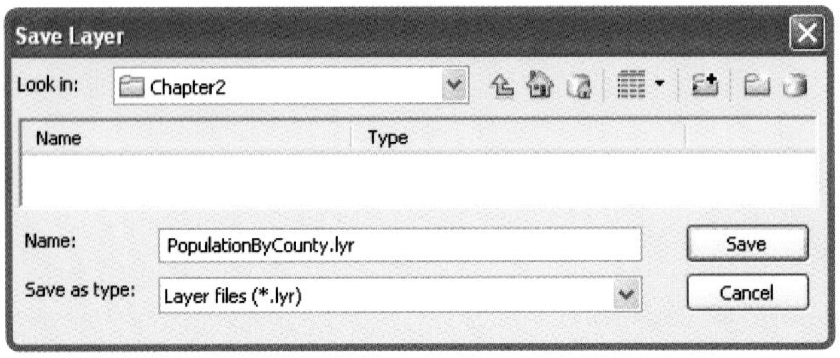

5 **Click Save.** Now you can add the saved layer to any map that you create. Note that you also can save ungrouped layers, such as Population By State, as a layer file for reuse.

Add group layers from the Catalog

Next, you will use an additional way to add data to a map document, using ArcGIS's utility application, Catalog.

1 Click Windows, Catalog.

2 Navigate to \ESRIPress\GIST1\MyExercises\Chapter2\.

3 Click PopulationByCounty.lyr.

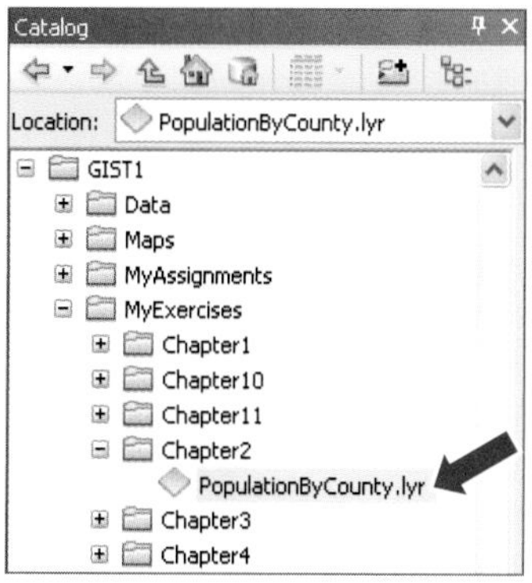

4 Drag and drop this layer into the TOC. You now have a second copy of the group layer in the TOC that is classified the same as the original layer group.

Remove group layers

1 Right-click the duplicate layer group that was just added.

2 Click Remove.

Tutorial 2-3

Set threshold scales for dynamic display

If a layer is turned on in the TOC, ArcMap will draw it, regardless of the map scale (that is, how far you are zoomed in or out). To automatically display layers at an appropriate map scale, you can set a layer's visible scale range to specify the range at which ArcMap draws the layer. Note that the further zoomed in you are, the larger the map scale. For example, 1:24,000 is a common map scale for which 1 inch on your computer screen is 24,000 inches on the ground while 1:1 is real-life scale.

Set a visible scale based on the current scale

1. **Zoom to states in the northeastern part of the country as shown at right.**

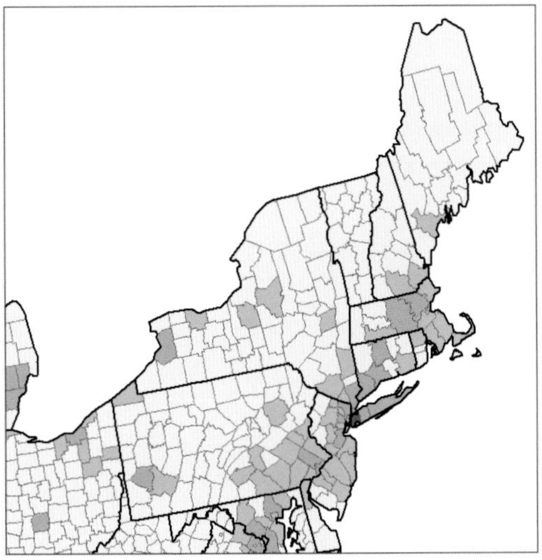

2. **Click the plus (+) sign to expand the Population By County group layer, then right-click the USCounties layer in the Population By County group layer.**

3. **Click Visible Scale Range, Set Minimum Scale.** ArcMap sets the scale to display this layer when zoomed in this close or closer. Zooming out any further will turn off the polygons for this layer.

4. **Click the Full Extent button.**

Now the county polygons will not display, and you will see only the outline for USStates.

Set a maximum scale based on the current scale

1 **Go back to the previous extent so that the county polygons display again.**

2 **Right-click the USStates layer in the Population By County group layer.**

3 **Click Visible Scale Range, Set Maximum Scale.**

4 **Zoom in a little closer.** ArcMap does not display the black outline polygons for the states when zoomed in beyond the maximum scale just set. Zooming out enough will turn on the state polygons again. The layer's check box is gray if the layer is not displayed.

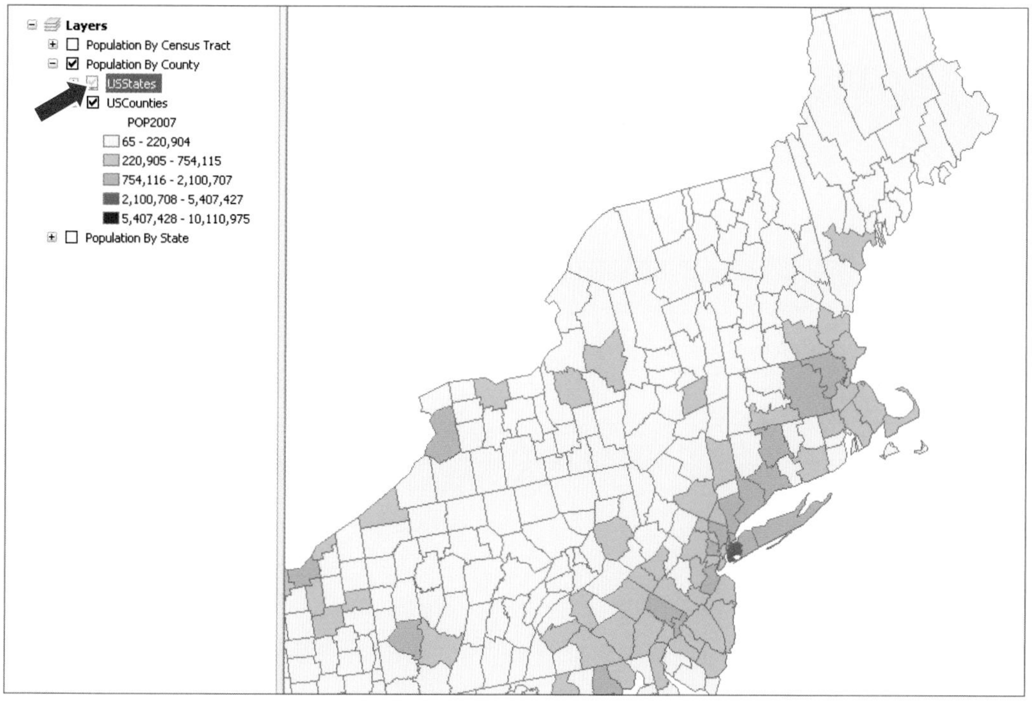

Clear a layer's visible scale

1 Right-click the USStates layer in the Population By County layer group.

2 Click Visible Scale Range, Clear Scale Range. ArcMap again displays the outline polygons for the states when zoomed to this scale.

Set a minimum visible scale for a specific layer

Instead of setting a map extent by its current scale, you can set it by the layer properties.

1 Turn the Population By County layer group off and Population By Census Tract group layer back on, then expand it.

2 Zoom to the full extent of the map.

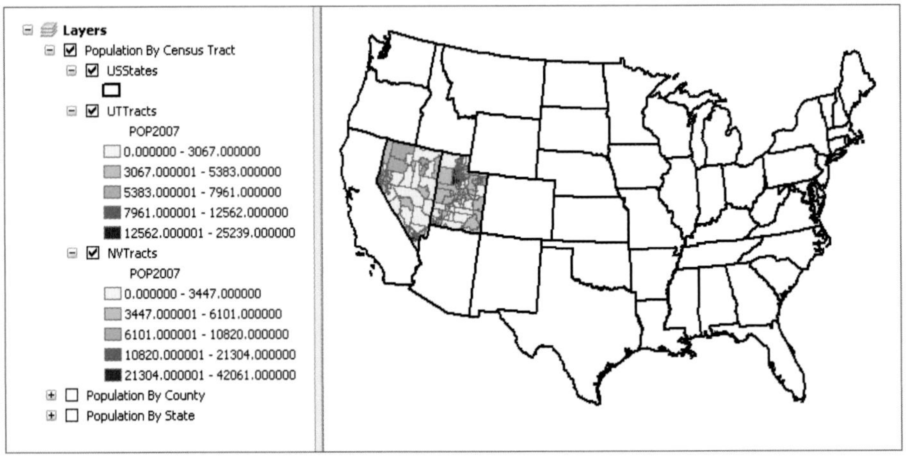

3 In the TOC, right-click the NVTracts layer and click Properties.

4 Click the General tab.

5 Click the Don't show layer when zoomed radio button.

6 Type **8,000,000** in the Out beyond field and click OK. If you zoom out beyond this scale, the layer will not be visible.

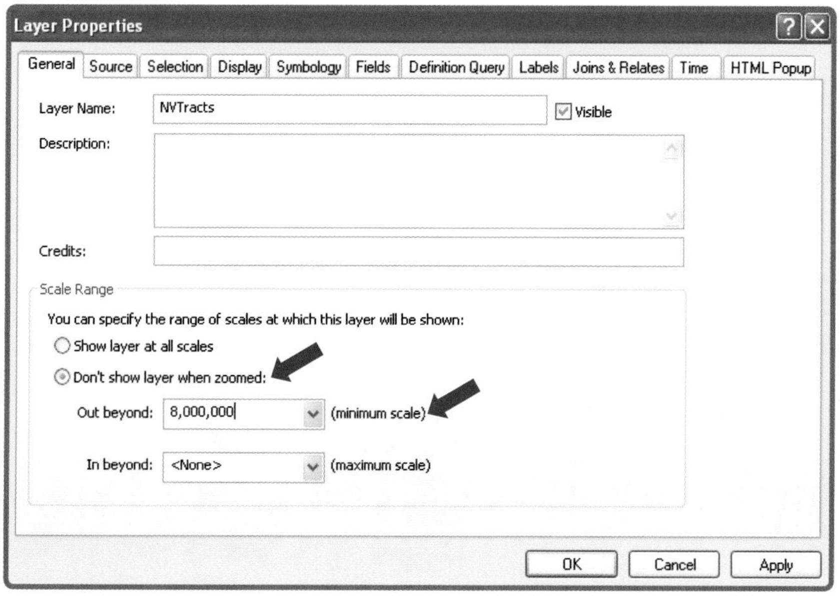

The census tracts for Nevada disappear in the map display. ArcMap does not show the Nevada census tract polygons when zoomed out past a scale of 1:8,000,000 as shown below with a scale of 1:34,453,384.

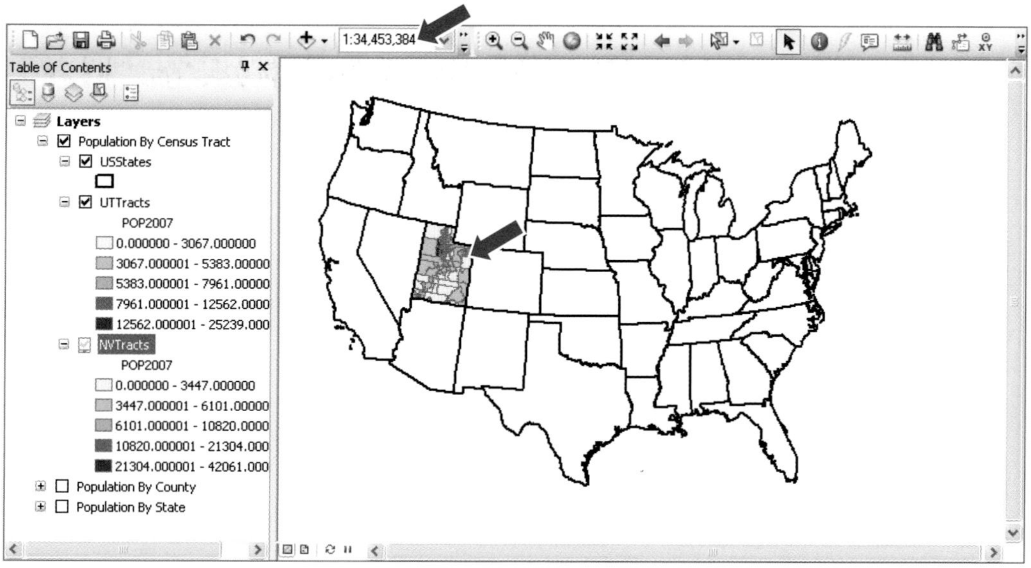

YOUR TURN

Use layer properties to set the Utah Tracts minimum scale to the same as Nevada Tracts (1:8,000,000). Utah tracts will disappear when zoomed to full extent.

Type a specific scale

1 **From the Standard Toolbar, click inside the scale box and type 1:7,900,000 as the scale and press Enter.**

2 **Pan to the Utah and Nevada tracts.** The tracts for these states are now visible because the scale is less than the minimum scale.

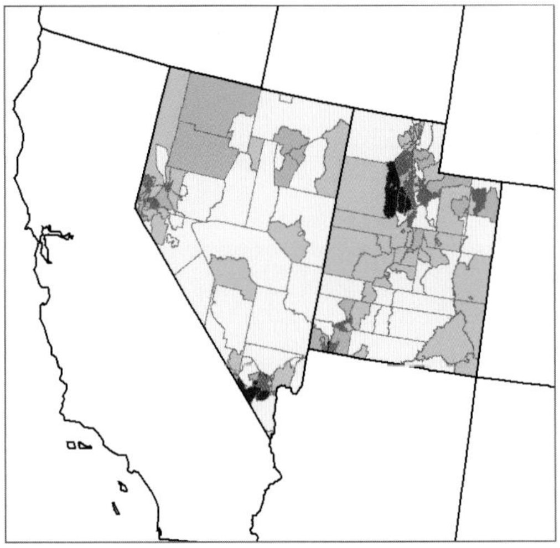

3 **Type 1:8,100,000 as the scale and press Enter.** The tracts for these states are now invisible because the scale is greater than the minimum scale.

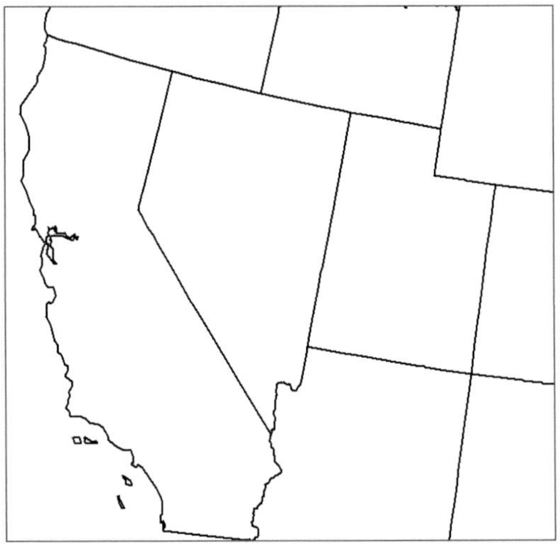

Tutorial 2-4

Create choropleth maps using custom attribute scales

Earlier in these exercises, you created a choropleth map from the Population By States layer using a classification method called natural breaks (Jenks) to divide the features in the map into five value classes. Although natural breaks is the default method, ArcMap allows you to choose other methods for classifying your data, including your own custom classification.

Create custom classes in a legend

1 Zoom to the full extent .

2 Turn off and minimize all layer groups except Population By State, and expand that layer in the TOC so that you can see its classes.

3 Right-click the Population By State layer and click Properties.

4 In the Layer Properties dialog box, click the Symbology tab.

5 In the Classification panel, click the Classes drop-down list, select 6, and click Classify.

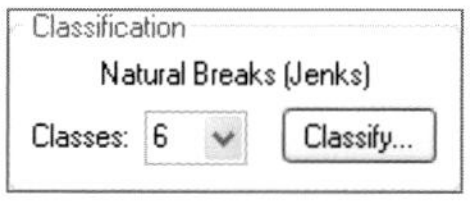

The Classification dialog box shows the current classifications, statistics, and break values for natural breaks in the data.

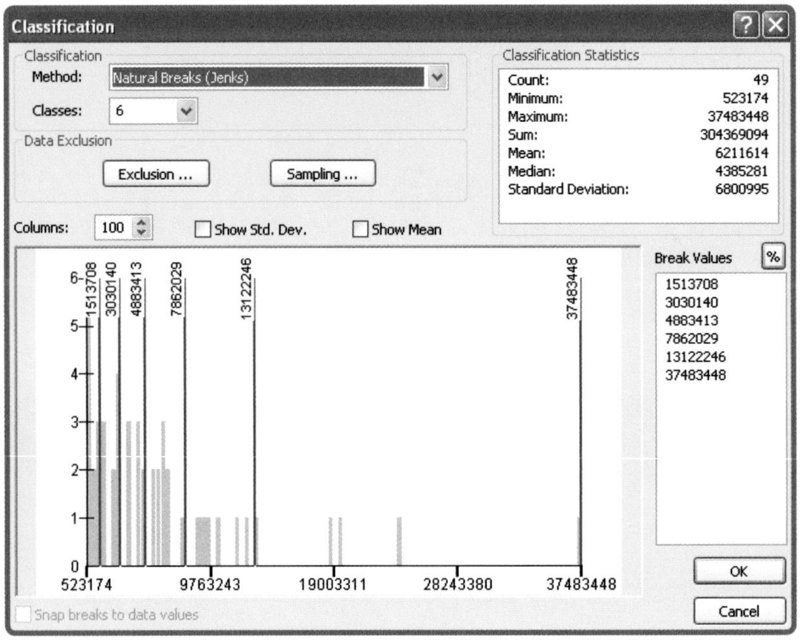

Manually change class values

Generally, it is easier to enter break values starting with the large values and working down. Here you will create new breaks using increasing interval widths that double with each successive class.

1 **Click the drop-down list for the Classification Method and click Manual.**

2 **In the Break Values panel, click the fifth value, 13122246, to highlight it.** Notice that the blue graph line corresponding to that value turns red.

3 **Type 32,000,000 and press Enter.**

4 Continue clicking the break values above the last one changed and enter the following: **16,000,000**; **8,000,000**; **4,000,000**; and **2,000,000**. Let the last (maximum) value remain 37,483,448.

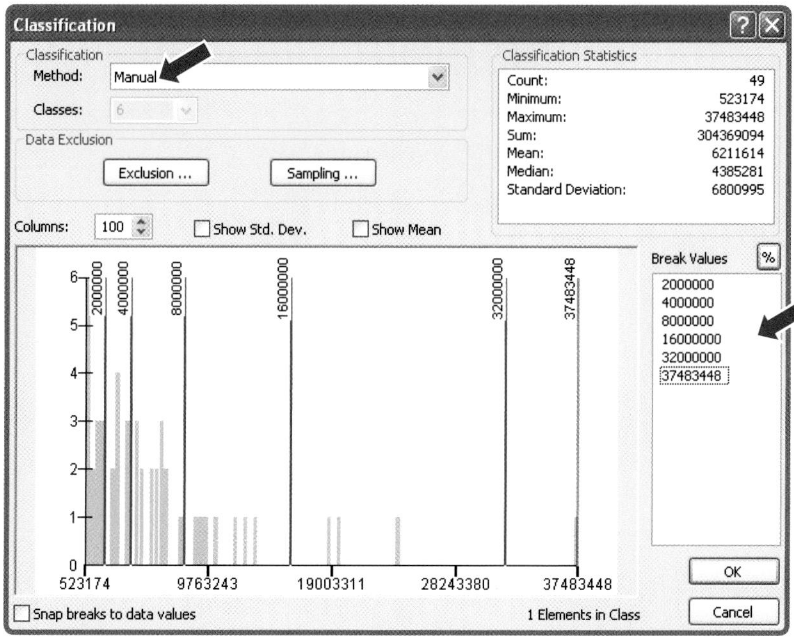

5 Click OK.

6 Click the gray Label heading to the right of the gray Range heading, then click Format Labels.

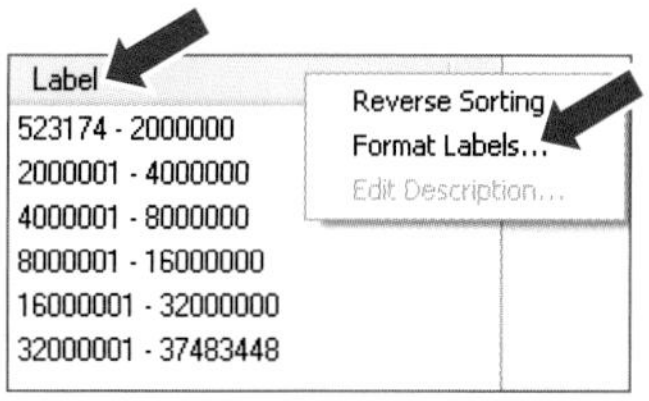

7 In the Number Format dialog box, select Show thousands separators and click OK.

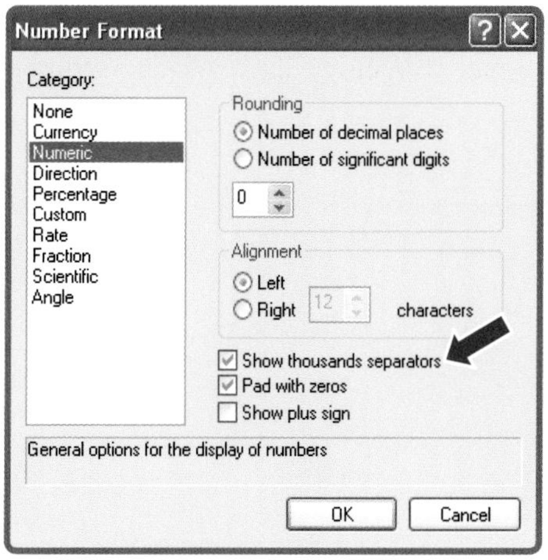

8 In the Label field of the Symbology tab, change the first value to **2,000,000 or less** and the last label to **32,000,001 or greater**.

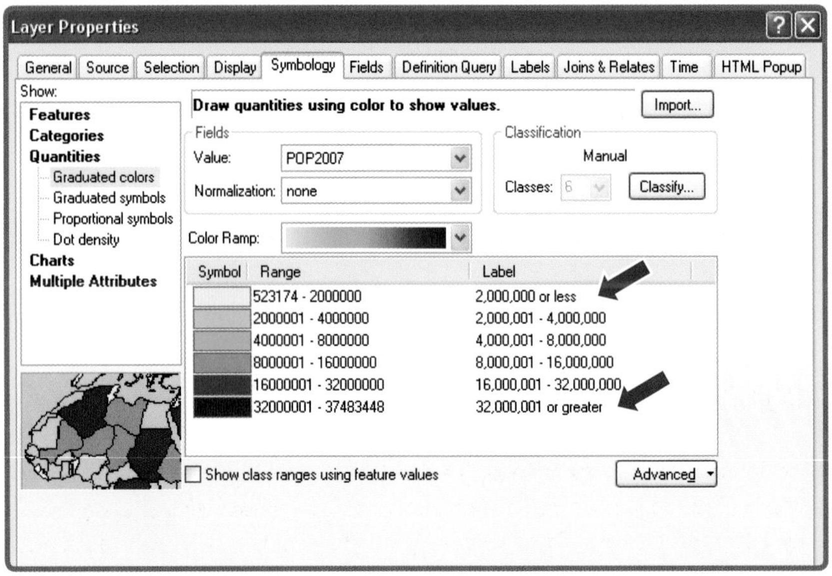

9 Click OK. The Population By State layer changes to reflect the new break values and labels. Besides being easier to read and interpret, the classification is appropriate for a long-tailed state population distribution, with its increasing-width intervals.

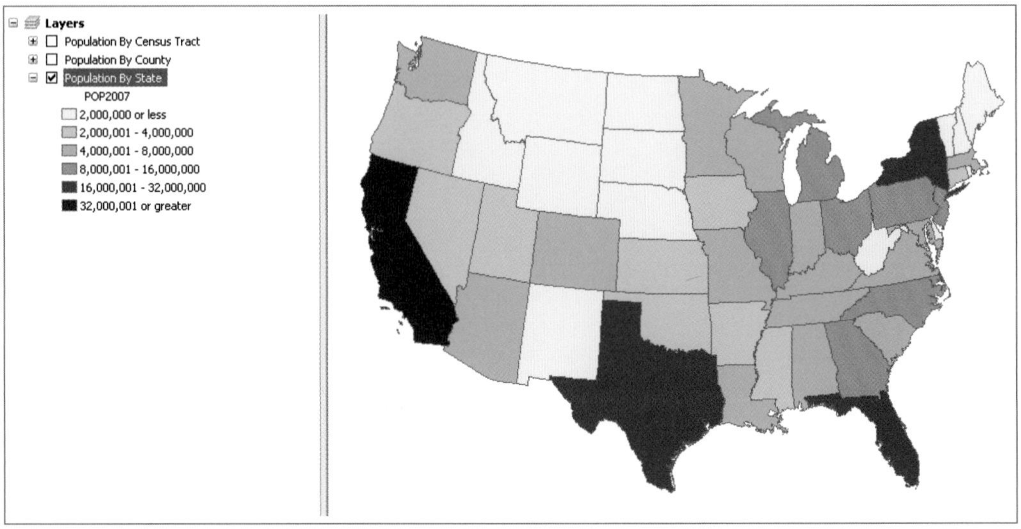

10 Right-click the Population By State layer in the TOC and click Save As Layer File.

11 Browse to the \ESRIPress\GIST1\MyExercises\Chapter2\ folder, type **PopulationByState** in the name field, and click Save.

YOUR TURN

Clear the minimal scale range for USCounties so they are visible when zoomed to full extent. Change the classification break values for the USCounties layer based on the population (POP2007) field. Use the same method as above to manually change the values using the following: **9,000 or less**; **9,001–18,000**; **18,001–36,000**; **36,001–72,000**; and **72,001 or greater**. Be sure to change the labels in the legend.

Change the classification break values for UTTracts and NVTracts to be the same: **3,000 or less**; **3,001–6,000**; **6,001–9,000**; **9,001–12,000**; and **12,001 or greater**. Be sure to change the labels in the legends.

Manually change class colors

While ArcMap provides color ramps with preselected colors, you can change colors for classes manually. Generally, it is best to have more classes with light colors and a few with dark colors (the human eye can differentiate light colors more easily than dark ones). So here you will create a custom monochromatic color ramp that starts with white and ends with a dark blue.

1 Turn off the Population By State layer and turn on the Population By County layer. Right-click the USCounties layer and click Properties.

2 In the Layer Properties dialog box, click the Symbology tab.

3 Click the Color Ramp drop-down list and scroll to and click the last color ramp.

4 Notice that the color variation of each break value is not very distinguishable.

5 Right-click the Color Ramp and Properties.

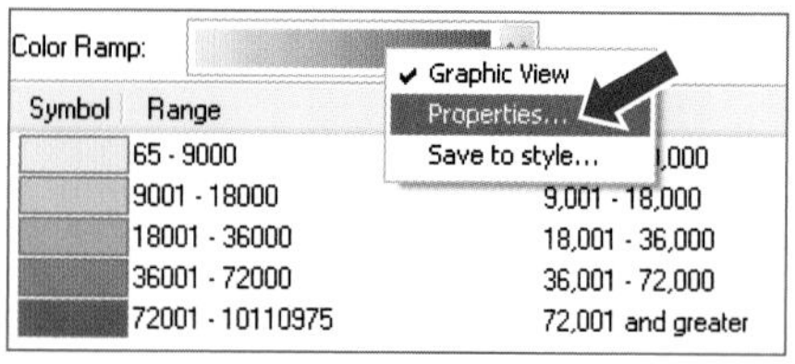

6 Click the color box beside Color 1 and click the Arctic White paint chip.

7 Click the color box beside Color 2, click the Dark Navy paint chip, and click OK twice.

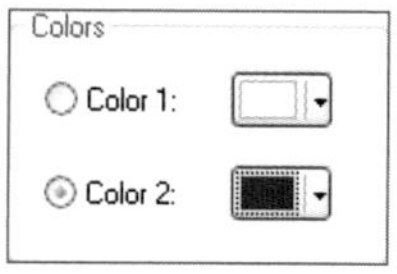

The Population By County map changes to reflect the new color ramp.

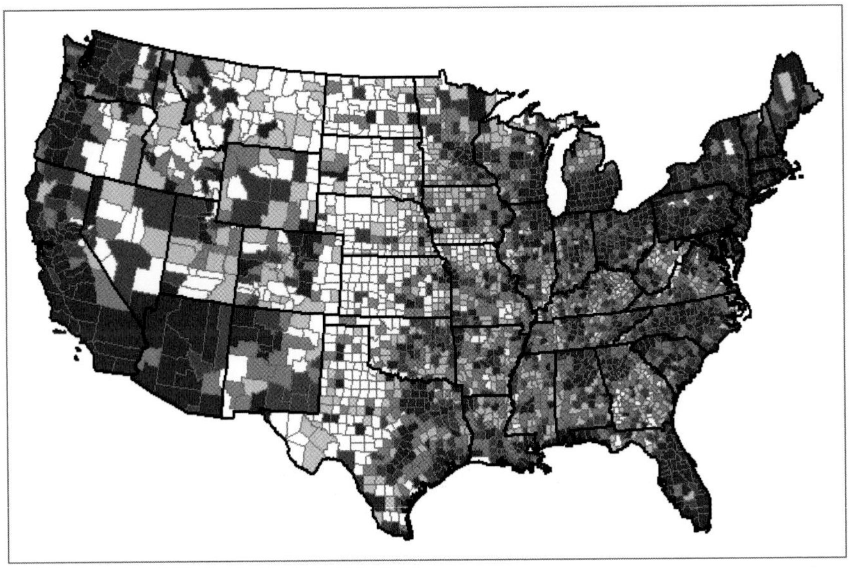

Note: You can also double-click each color symbol in the Symbology tab or TOC to change the classification colors individually.

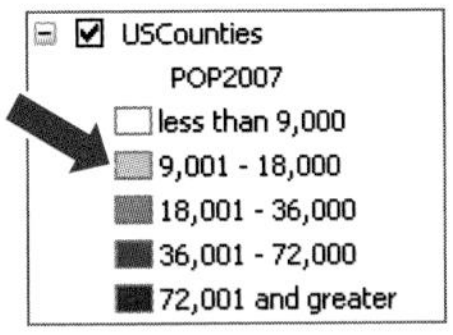

YOUR TURN

Change the class colors for UTTracts, NVTracts, and Population By State. When finished, save your map document.

Tutorial 2-5

Create point maps

Point maps show exact locations of data or events using individual point markers for each event. In the next exercise, you will create a point map showing U.S. cities as points with size-graduated point markers that represent population.

Create a point map of U.S. cities by population

1. Turn off and collapse all layers in your map and create a new group layer called **Population By City**.

2. To the new group layer, add the data layers \ESRIPress\GIST1\Data\UnitedStates.gdb\USStates and \ESRIPress\GIST1\Data\UnitedStates.gdb\USCities.

3. Double-click USStates in the Population By City group layer to open its Layer Properties window.

4. Click the Symbology tab, change the symbol to hollow (no color) fill with a medium gray outline of size 1, and click OK twice.

5. Double-click USCities in the Population By City group layer to open its Layer Properties window.

6. Click the Symbology tab and change the layer's symbology from Single Symbol to Quantities, Graduated Symbols.

7 In the Fields panel, change the Value to POP2000, the template symbol to a Mars red CIRCLE 1, symbol size to 2–18, and assign the break points and legend labels as follows:

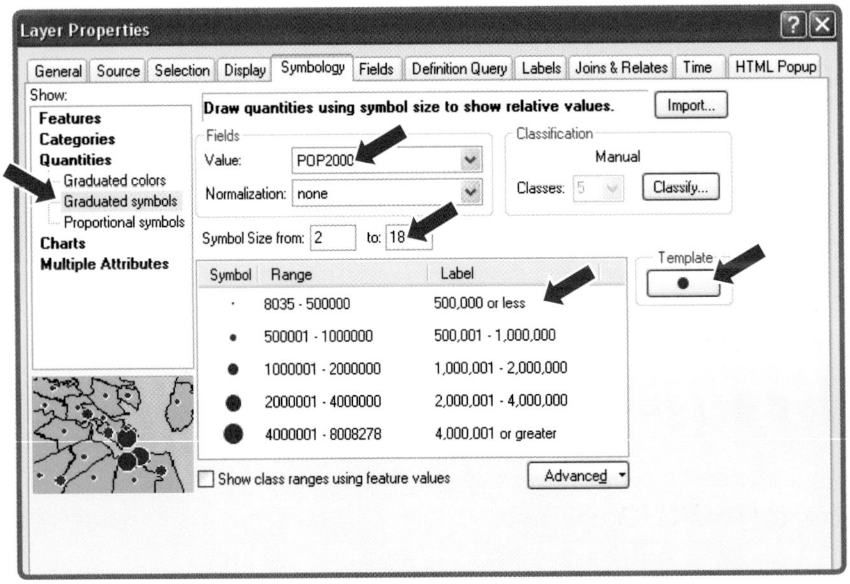

8 Click OK.

9 Zoom to the full extent. The resultant point map shows U.S. cities classified by population.

10 Click File, Save.

Tutorial 2-6

Create a point map based on a definition query

Suppose you have a layer containing all the cities in Pennsylvania, but you only want to display the cities with populations between 10,000 and 49,000. To accomplish this, you create a definition query to filter out all the cities with population values outside the desired range.

Create a new map document

1 Click File, New.

2 Click Blank Map from the New Document dialog box, and click OK.

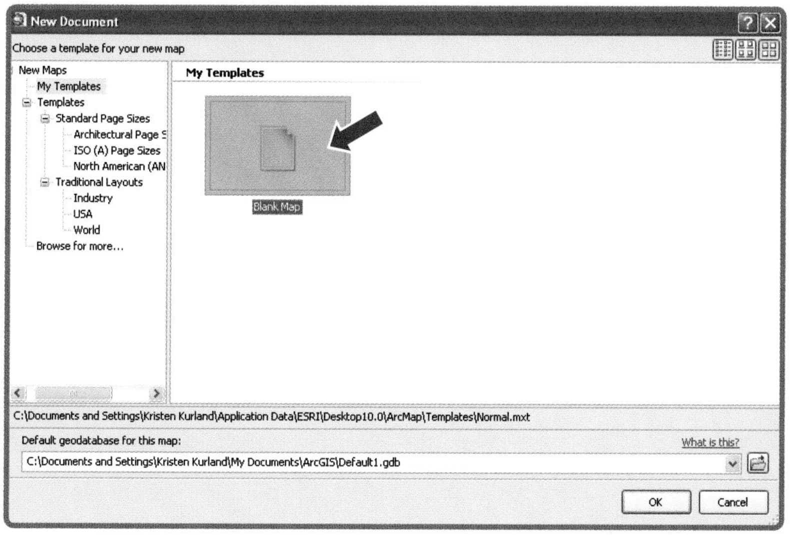

Add data to the map

1 Click the Add Data button.

2 Navigate to the folder where you have the \ESRIPress\GIST1\Data\ installed, click UnitedStates.gdb, and add the following layers: PACounties and PACities. The result is a map showing county polygon features for Pennsylvania and detailed cities. ArcMap picks an arbitrary color fill and point marker for the polygons and points, respectively.

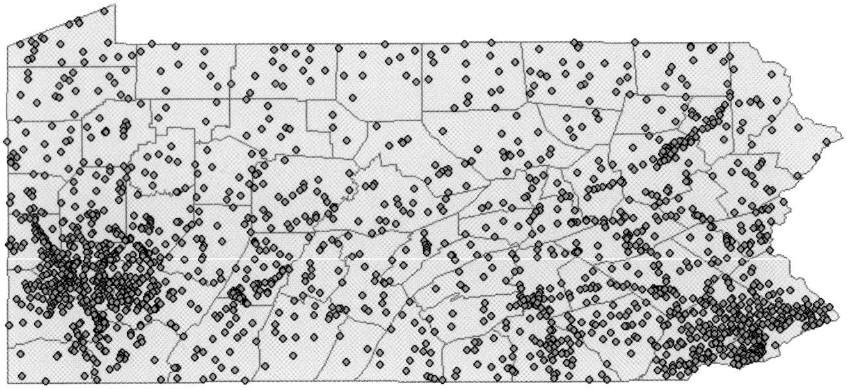

Symbolize polygons

To draw attention away from a feature, colors should be very light or, in the case of these polygons, have no color at all.

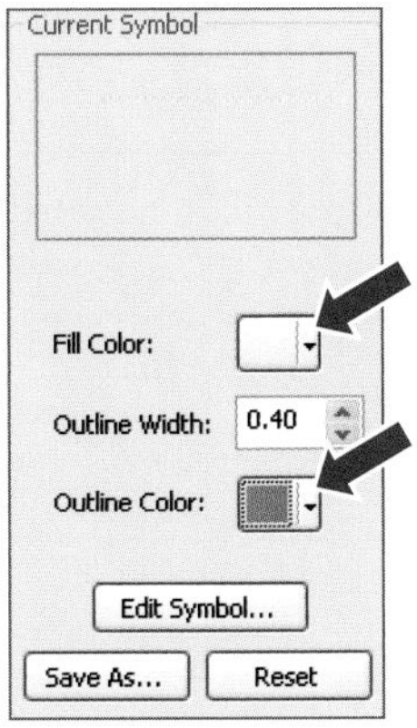

1 Right-click the PACounties layer and click Properties.

2 Click the General tab and change the name of the layer to **Pennsylvania Counties**.

3 Click the Symbology tab, click the symbol, choose No Color as the Fill Color, and choose a medium gray as the Outline Color.

4 Click OK twice.

Create a definition query

You might want to show only medium-sized cities in the state. To do so, you can query populations based on an attribute in the table called Feature.

1 Right-click the PACities layer and click Properties.

2 Click the Definition Query tab and Query Builder button.

3 In the Query Builder window, double-click "FEATURE".

4 Click = as the logical operator.

5 Click Get Unique Values. The resulting list has all unique values in the FEATURE attribute. Note that the attribute stores classification ranges for a numerical scale.

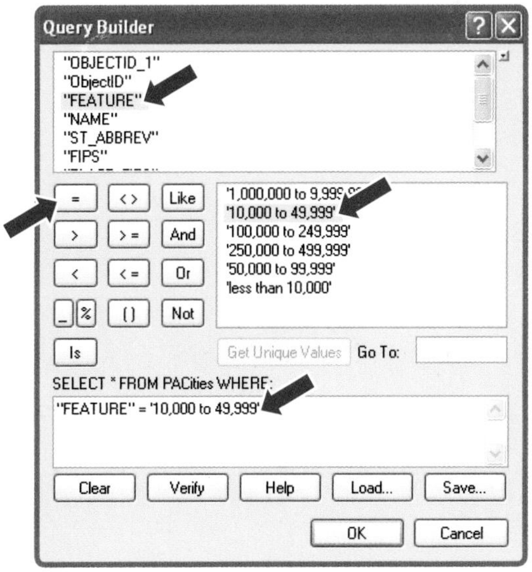

6 In the Unique Values list, double-click '10,000 to 49,999'. The completed query, "FEATURE" = '10,000 to 49,999', will yield a layer with only the cities in Pennsylvania with populations between 10,000 and 49,999. If the query has an error, edit it in the lower panel of the Query Builder, or click Clear and repeat steps 3 through 6.

7 Click OK twice to execute your query and close the Layer Properties dialog box.

Change a layer's name and symbol

1 Right-click the PACities layer and click Properties.

2 Click the General tab and change the name of the layer to **Population 10,000 to 49,999**.

3 Click the Symbology tab and the Symbol button.

4 Click the Circle 2 icon, change the color to Ultra Blue, and change the size to 6.

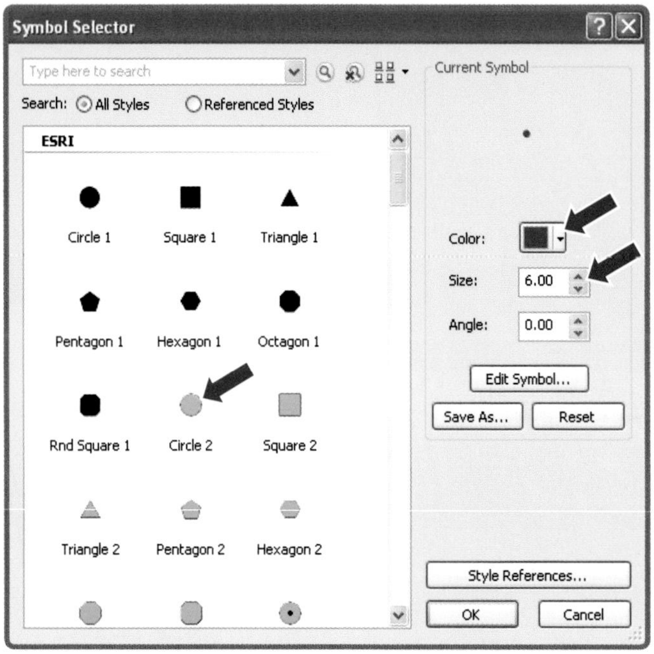

5 Click OK twice. The resulting map shows medium-sized cities in the state and ground polygons for counties.

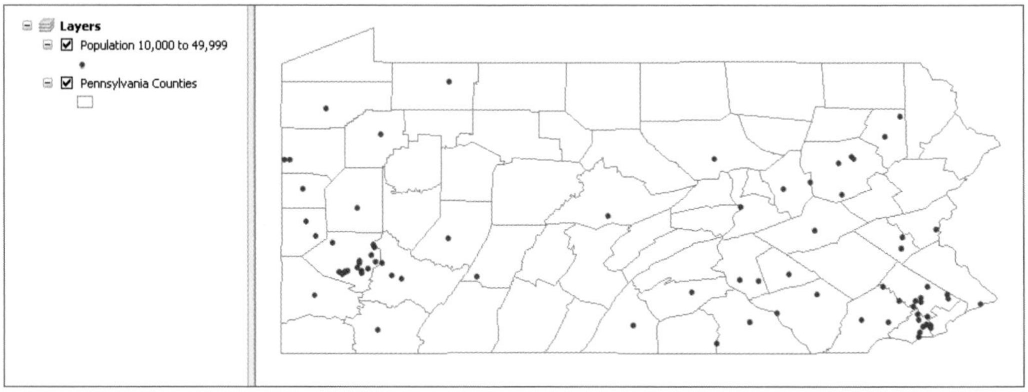

YOUR TURN

Note: You are required to complete this Your Turn exercise in order to finish the remaining exercises in this chapter.

Add the PACities layer again and create a definition query that displays Pennsylvania's county seats. **Hint:** The definition query will be "STATUS" = 'County Seat'. Change the layer name to **County Seats**, the symbol to Square 2, color to Leaf Green, and size to 6.

Add State Capital using Symbol Search

1 Click the Add Data button.

2 Navigate to the folder where you have the \ESRIPress\GIST1\Data\ installed, click UnitedStates.gdb, and add the PACities layer again.

3 Right-click the added PACities layer and click Properties.

4 Click the Definition Query tab and create the query "STATUS" = 'State Capital County Seat'. Click OK.

5 Click the Symbology tab and Symbol button.

6 In the Symbol Selector window, type **Capital** and press Enter in the search section. The resulting search shows all symbols that have the word "Capital" in their names.

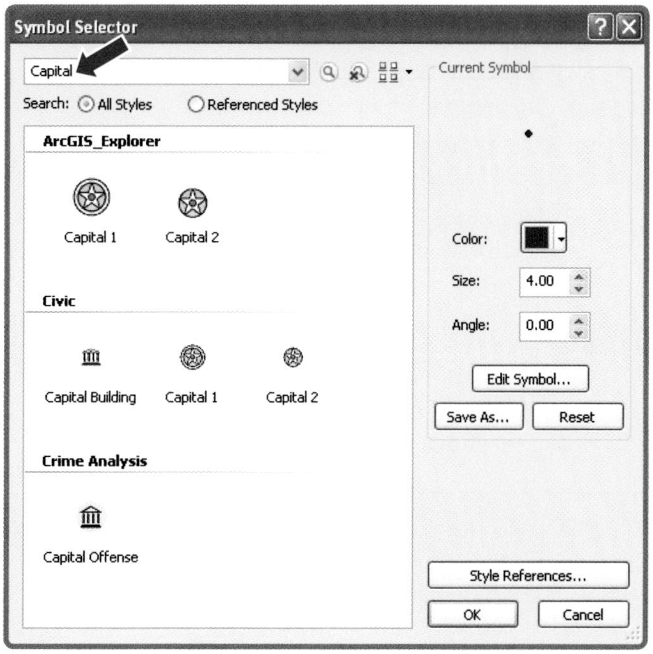

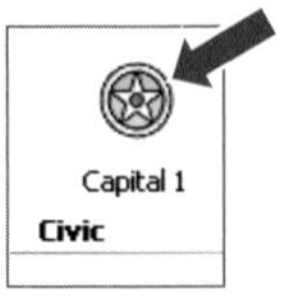

7 **Click the Capital 1 symbol from the Civic symbols. Click OK.**

8 **Click the General tab and change the layer name to State Capital.**

9 **Click OK.** The resultant map shows medium-sized cities, county seats, and the state capital, Harrisburg.

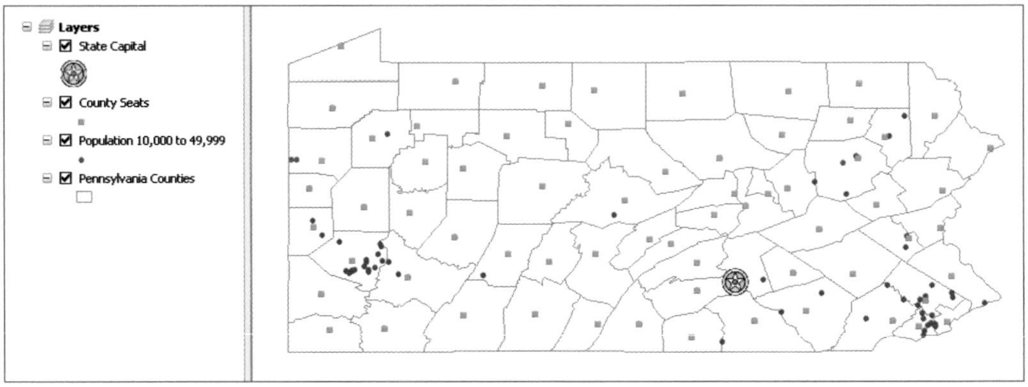

YOUR TURN

Add the PACities layer once more and create a definition query that displays Pennsylvania's three largest cities: Erie, Philadelphia, and Pittsburgh. Rename the layer "Major Cities" and show the three cities with a symbol that makes them stand out on the map. **Hint:** One solution is to use the query "NAME" = 'Erie' OR "NAME" = 'Philadelphia' OR "NAME" = 'Pittsburgh'. See if you can create halo labels for this layer and the State Capital layer.

Tutorial 2-7

Create hyperlinks

The Hyperlink tool allows access to documents or Web pages by clicking features. There are three types of hyperlinks: documents, URLs, and macros.

Create a dynamic hyperlink

1 On the Tools toolbar, click the Identify button.

2 Click the capital point symbol for Harrisburg.

3 In the Identify results window, right-click Harrisburg in the top panel, and click Add Hyperlink from the context menu.

4 Click the Link to a URL radio button and type `www.harrisburgpa.gov`.

5 Click OK, then close the Identify window.

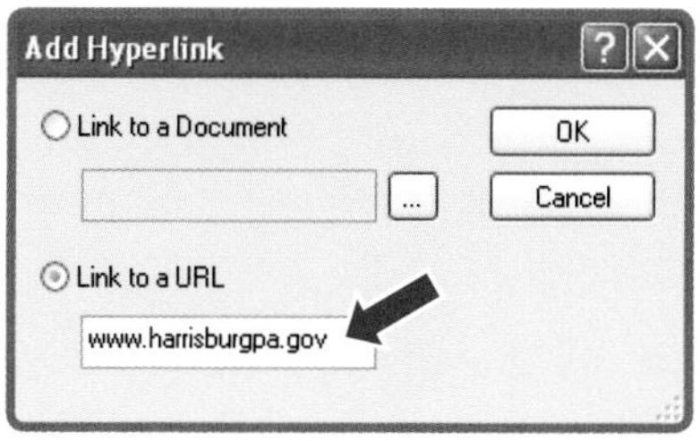

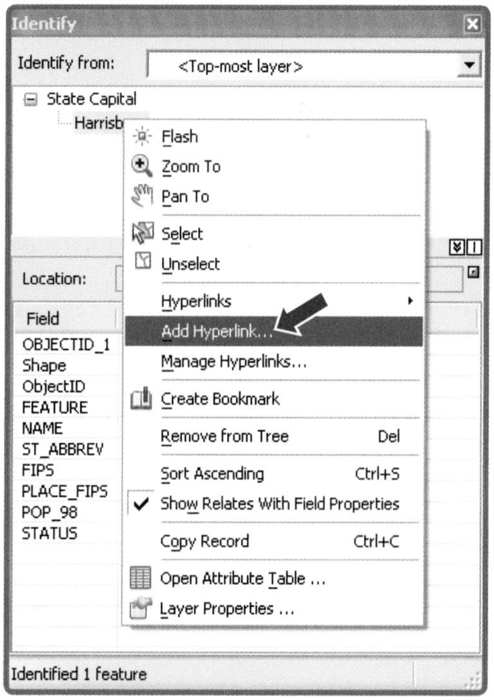

Launch a hyperlink

1 **From the Tools toolbar, click the Hyperlink button.** Features that have hyperlinks get a small blue circle drawn on them. In this case, the Harrisburg capital point marker is the only such feature.

2 **Move the cursor to the city of Harrisburg point feature.** When you are over a feature that has a hyperlink, the cursor turns from a yellow to a black lightning bolt and you see a pop-up tip with the name of the target.

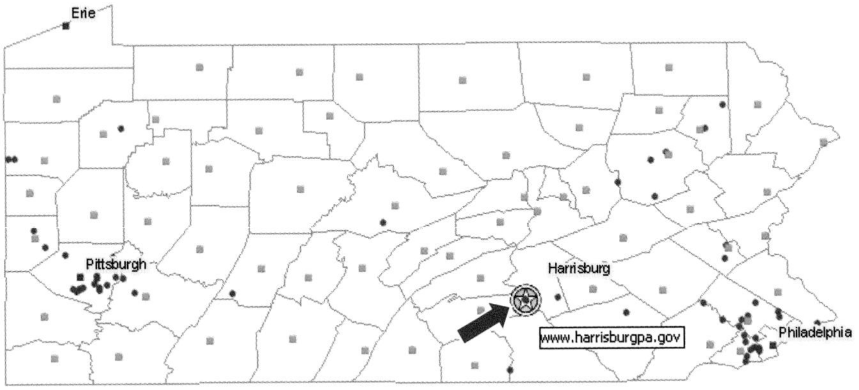

3 **Click the feature to go to the Harrisburg Web site.** Place the tip of the lightning bolt on the hyperlink's small blue circle and click. Your default Web browser will open to Harrisburg's Web page.

YOUR TURN

Add hyperlinks to the Erie, Pittsburgh, and Philadelphia point markers in the layer that just has those three cities. Use the following Web sites: `www.erie.pa.us` for Erie, `www.phila.gov/` for Philadelphia, and `www.city.pittsburgh.pa.us` for Pittsburgh. If these Web sites do not work, search the Internet for another site that links to each city.

Use the Hyperlink Pop-up tool

Hyperlink pop-ups are nice shortcuts for viewing attribute data for features on a map.

1 Right-click the County Seats layer and click Properties.

2 Click the HTML Popup tab and make selections as follows:

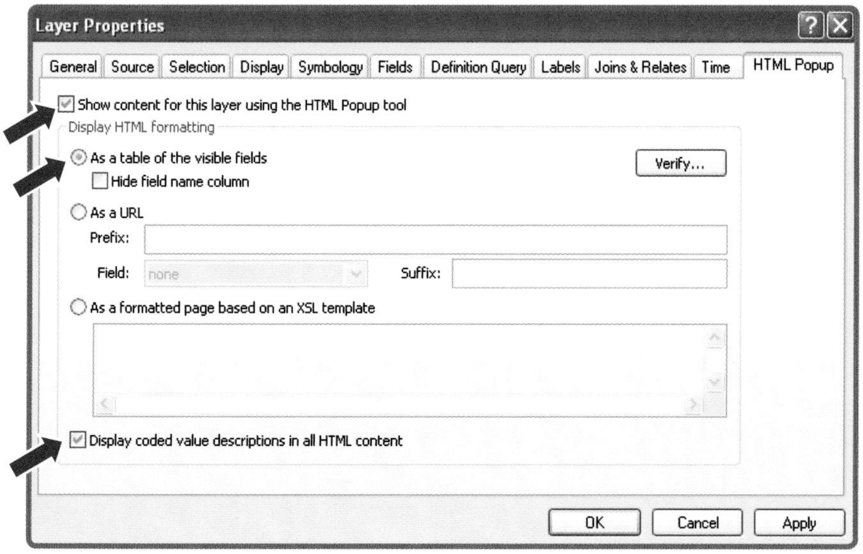

3 Click the Fields tab, turn on the fields as shown below, create an alias called **CITY NAME** for the NAME field, and click OK.

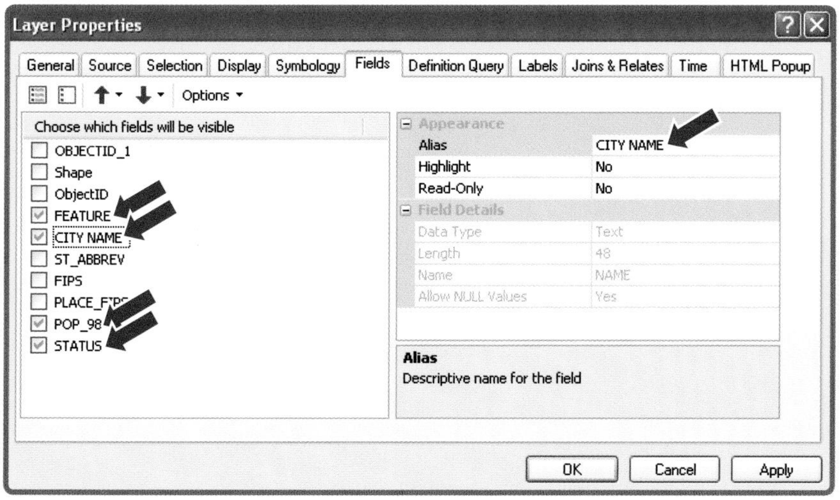

4 On the Tools toolbar, click the Hyperlink Popup button.

5 **Click any County Seats point feature in the map display.** The attribute information that you set in the field properties is displayed on the map in a pop-up window.

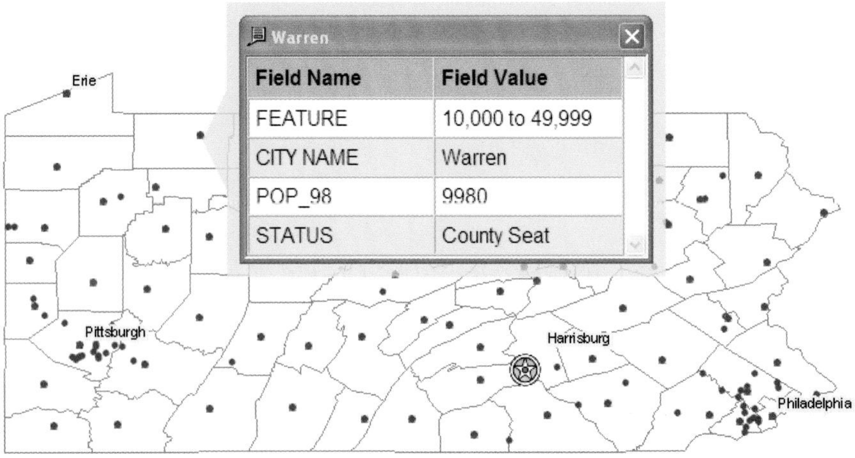

6 **Click additional point features.**

7 **Close the pop-up windows when finished.**

Tutorial 2-8

Create MapTips

When you hover your cursor over a feature on a map, it is possible to have just an attribute or an expression of that feature automatically displayed as a MapTip.

Create and display a MapTip

1 Right-click the County Seats layer in the TOC, click Properties, and click the Display tab.

2 Click NAME as the field in the Display Expression section.

3 Click Show MapTips using the display expression, and click OK.

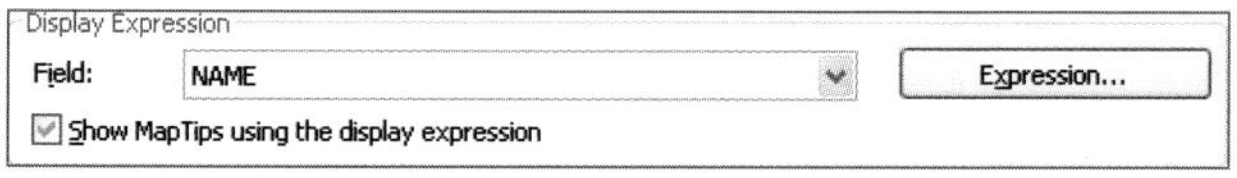

4 From the Tools toolbar, click the Select Elements button.

5 Hover over any city in the County Seats layer to see its name.

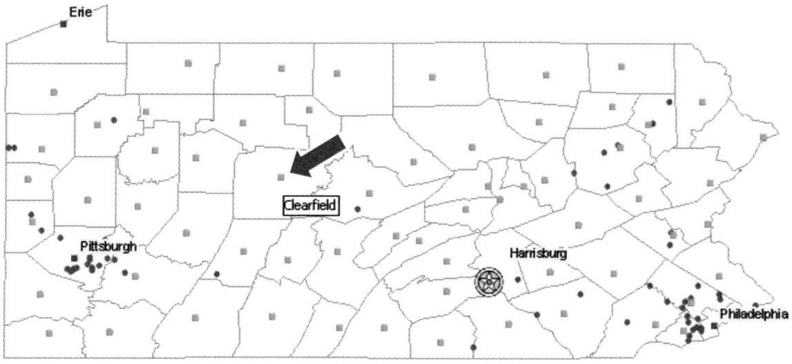

6 Save your map document.

YOUR TURN

Add MapTips to the Population 10,000 to 49,000 layer to show the city NAME. When finished, save your map document and close ArcMap.

Assignment 2-1

Create a map showing schools in the city of Pittsburgh by enrollment

Suppose that the City of Pittsburgh school board wants to do an extensive evaluation of local schools. Officials have collected data about all schools, public and private. The initial project identifies schools as either public or private and shows their enrollment. Your task is to make a map for the school board comparing the enrollment of public and private schools. You will use point features of different sizes to show this.

Start with the following:

- \ESRIPress\GIST1\Data\Pittsburgh\City.gdb\Neighborhoods—polygon layer of Pittsburgh neighborhoods
- \ESRIPress\GIST1\Data\Pittsburgh\City.gdb\Schools—point layer of all schools

Use the following fields of the Schools table:

DISTRICT—school type ("City of Pittsburgh" is a public school; "Pittsburgh Diocese" and "Private School" are private schools)

ENROLL—number of students enrolled

STATUS—open or closed

Create a point feature map with hyperlink

Create a map document called **\ESRIPress\GIST1\MyAssignments\Chapter2\Assignment2-1 YourName.mxd** showing separately the enrollment of students in public and private schools that are open. Include Pittsburgh neighborhood polygons for reference. Hyperlink the Web site `http://www.pghboe.net` (or a similar Web site for Pittsburgh schools) to the Allderdice school in the Squirrel Hill South neighborhood.

Hints

- Add two copies of Schools to your map document. Use one copy for public schools and the other for private schools.
- Use the same increasing-width interval scale for both public and private schools.
- Use MapTips for schools and label neighborhoods. Use a small, dark gray font for the labels. Under the Labels tab of the Layer Properties window, click Placement Properties, Conflict Detection, and type **1** in the Buffer field to improve appearance.

WHAT TO TURN IN

If your work is to be graded, turn in the following file:

ArcMap document: \ESRIPress\GIST1\MyAssignments\Chapter2\Assignment2-1 YourName.mxd

Assignment 2-2

Create a map showing K–12 population versus school enrollment

In this assignment, you will create a choropleth map showing the population by census tract for the entire state of Pennsylvania and also a map zoomed in to the city of Pittsburgh for the K–12 school-age population. Layers will turn on or off depending on the zoom level. You will also show schools by enrollment.

Start with the following:

- \ESRIPress\GIST1\Data\UnitedStates.gdb\PATracts—polygon layer of Pennsylvania census tracts, 2000
- \ESRIPress\GIST1\Data\UnitedStates.gdb\PACounties—polygon layer of Pennsylvania counties
- \ESRIPress\GIST1\Data\Pittsburgh\City.gdb\BlockGroups—polygon layer of Pittsburgh census block groups, 2000, that will be shown when zoomed in to the Pittsburgh area
- \ESRIPress\GIST1\Data\Pittsburgh\City.gdb\Neighborhoods—polygon layer of neighborhoods
- \ESRIPress\GIST1\Data\Pittsburgh\City.gdb\Schools—point layer of Pittsburgh schools

The value "City of Pittsburgh" for DISTRICT identifies public schools.

Create choropleth maps with scale thresholds

Create a new map document called **\ESRIPress\GIST1\MyAssignments\Chapter2\Assignment2-2 YourName.mxd** that shows the Pennsylvania census tracts for K–12 school-age population (ages 5–17) and county outlines for the entire state. For Pittsburgh, show the K–12 population using census block groups and neighborhood outlines. Include the point layer for Pittsburgh public schools that are open but with low enrollment (over 0 and under 200 students). Use MapTips for schools. Label counties and neighborhoods.

When zoomed to the entire state, do not show details of the city of Pittsburgh, but turn on these layers when zoomed into that area. Have the Pennsylvania details turned off when zoomed in to the Pittsburgh details. Create a bookmark to help you easily zoom in to the Pittsburgh details.

Hints

- Create two layer groups: one for the state of Pennsylvania and one for Pittsburgh details so you can turn them on or off as necessary.
- Add a halo to labels to make them easier to read. In the Labels tab of the Layer Properties window, click Symbol, Edit Symbol, the Mask tab, and the Halo radio button. Then type **1.5** for the size and click Symbol to use a light gray halo. Use a size 7 or 8 text symbol.

STUDY QUESTIONS

Create a Microsoft Word file called **\ESRIPress\GIST1\MyAssignments\Chapter2\Assignment2-2YourName.doc** with answers to the following questions:

1. The seven public schools meeting the criteria (over 0 and under 200 enrollment) are in what neighborhoods?
2. Name a school that may close. Explain why you picked this school.

WHAT TO TURN IN

If your work is to be graded, turn in the following files:

ArcMap document: \ESRIPress\GIST1\MyAssignments\Chapter2\Assignment2-2YourName.mxd

Word document: \ESRIPress\GIST1\MyAssignments\Chapter2\Assignment2-2YourName.doc

If instructed to do so, instead of the above individual files, turn in a compressed file, **Assignment2-2YourName.zip**, with all files included. Do not include path information in the compressed file.

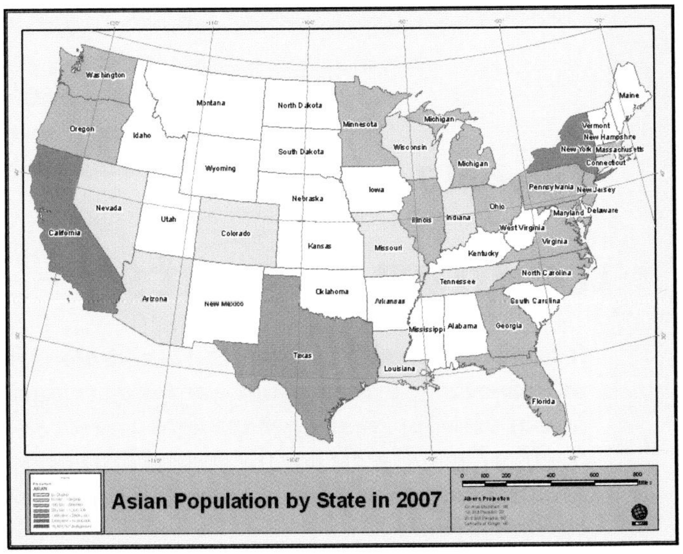

6 Use the Zoom In button and Pan button to make the map larger and centered.

7 Select 100% on the Layout toolbar and use scroll bars to view the legend, scale bar, and text.

8 Click the Zoom Whole Page button on the Layout toolbar.

9 Click File and Save As, and save your map as **\ESRIPress\GIST1\MyExercises\Chapter3\Tutorial3-2a.mxd**.

YOUR TURN

Save Tutorial3-2a.mxd as **\ESRIPress\GIST1\MyExercises\Chapter3\Tutorial3-2b.mxd**. Click the Change Layout button and use a layout template of your choice. Change the title of the layout to **Asian Population by State in 2007**.

Save a layer file

The set of layouts that you will produce below are for comparing populations by race or ethnicity for states. Next, as a preliminary step, you will create a layer file that saves the symbology of a map layer for reuse. To facilitate comparisons of populations by race on separate maps, it is desirable to use the same numeric scale for all maps. So as part of the work, you will save a layer file that allows easy reuse of a numeric scale.

1 Click File and Open, browse to your \ESRIPress\GIST1\Maps\ folder and double-click Tutorial3-2.mxd (the same map document you used in the previous exercise).

2 Click File and Save As, and save your map document as **\ESRIPress\GIST1\MyExercises\Chapter3\Tutorial3-Asians.mxd**.

3 Right-click Population in the TOC, click Save As layer File, browse to \ESRIPress\GIST1\MyExercises\Chapter3\, type **StatesPopulation.lyr** for Name, and click Save.
Below is the custom layout you will build. The scale in the legend will come from your saved layer file rather than you having to build it for each map that you produce. The blue horizontal and vertical lines are guides that you will create for precise alignment and placement of elements.

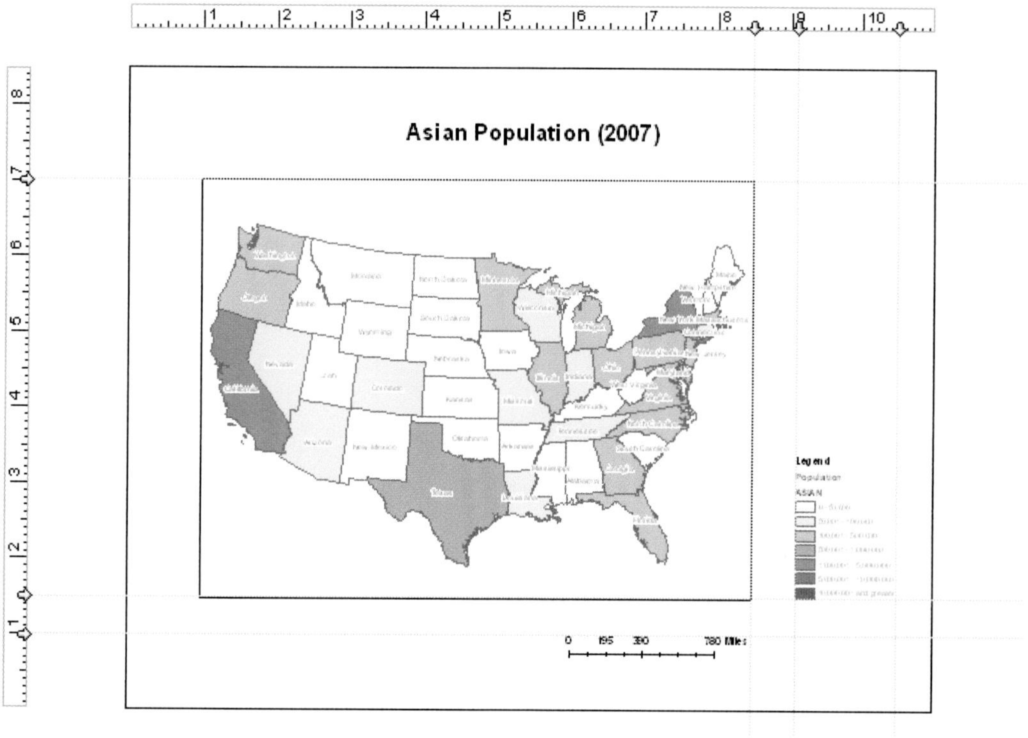

Verify Layout View options

You need to verify option settings for working with layouts. In particular, you need settings to create and show guidelines and have layout elements snap to them.

1 On the main menu, click View, Layout View.

2 On the main menu, click Customize, ArcMap Options, and the Layout View tab.

3 Verify that your settings match those at right.

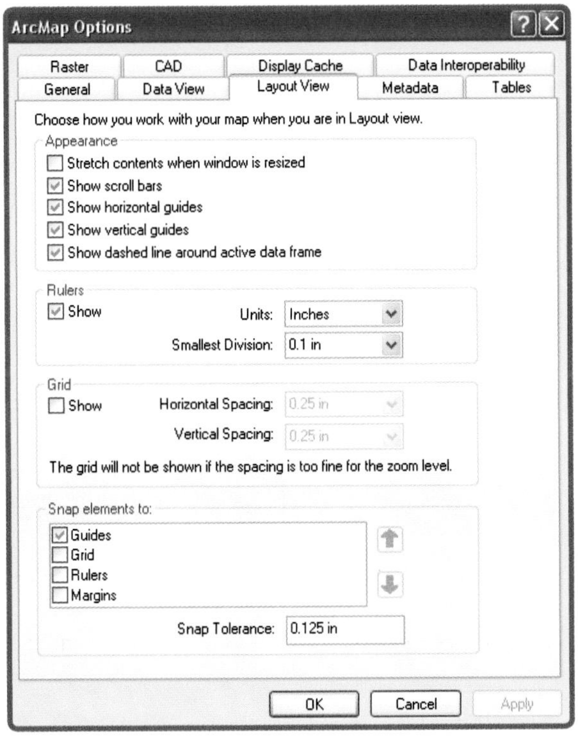

4 When finished, click OK.

Set up layout page orientation and size

Next, set up the layout page assuming that you will use an 8.5-by-11-inch document.

1 Right-click anywhere inside the layout and click Page and Print Setup.

2 If you have access to a printer, select it and desired properties, and then close any windows for setting printer properties.

3 In ArcMap's Page and Print Setup window, select Letter (8.5 × 11 inch) for paper size and Landscape for the Orientation in both the Paper and Page panels.

4 Click OK.

5 Right-click anywhere inside the layout and click the Zoom Whole Page button.

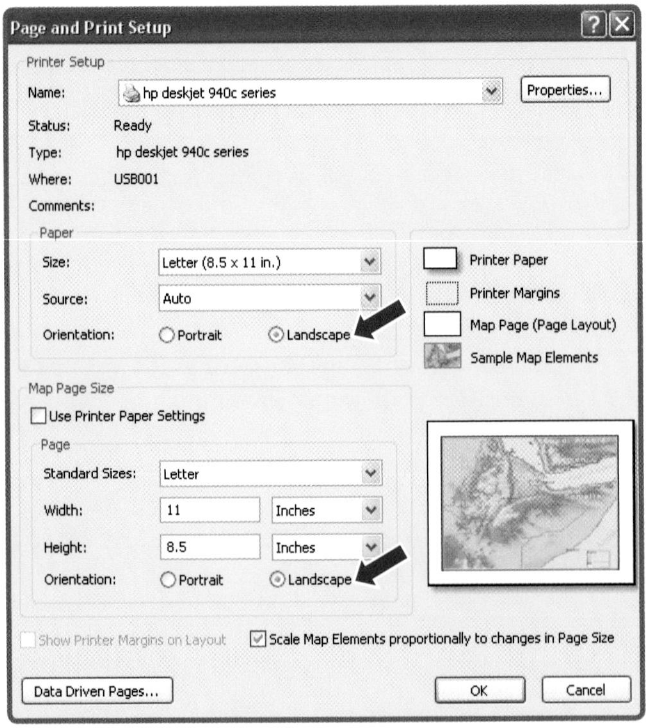

Create and use guidelines in the layout view

In the next steps, you will use vertical and horizontal rulers to create guidelines.

1 **Click at 8.5 inches on the top horizontal ruler to create a vertical blue guide at that location.** If you place your guide at the wrong location, right-click its arrow on the ruler, click Clear Guide, and start over.

2 **Do the same at 7 inches on the left vertical ruler.**

3 **Click the map to select it (dashed outlines and grab handles appear), right-click the map and click Properties, then click the Size and Position tab.**

4 **Click the Preserve Aspect Ratio check box, type 7.5 in the Size Width field, press the Tab key, and click OK.**

5 **Drag the map so that its upper right corner is at the intersection of the two guides, and release.** The map snaps precisely to the intersection of the guides when you release. The objective of the next step is to fill the map element rectangle with the map as large as possible.

6 **Use the Zoom In button on the Tools toolbar to drag a rectangle around just the physical map itself to increase the size of the map within its map element rectangle.** If you need to start over, click the Full Extent button .

Insert a title

1. On the main menu, click Insert, Title.
2. Type **Asian Population (2007)** in the Insert Title text box, and then click OK.
3. Double-click the title, click Change Symbol, select 22 for size, and click B (Bold) for Style.
4. Click OK, then click OK again.
5. Center the title over the map.

Insert a legend

1. Click the horizontal ruler at 10.5 inches to create a new vertical guideline, and click the vertical ruler at 1.5 inches to create a new horizontal guideline.
2. Click Insert and Legend, click Next four times, and click Finish. Resizing and placing the legend takes the next three steps.
3. Click, hold, and drag the legend so that its right side snaps to the 10.5-inch vertical guide and bottom to the 1.5-inch horizontal guide.
4. Click the horizontal ruler at 9 inches to create a new vertical ruler.
5. Click the top left grab handle of the legend and drag to the right and down to make the legend smaller. Snap it to the 9-inch vertical guideline while staying locked to the 10.5-inch vertical and 1.5-inch horizontal guidelines.

Insert a scale bar

1. Click the vertical ruler at 1 inch to create a new horizontal guide.
2. Click Insert, Scale Bar.
3. Click Scale Line 2, Properties.
4. Select Miles for the Division Units, click OK, then click OK again.

5 Drag the resulting scale bar so that its top is at the 1-inch horizontal guideline and its right side is at the 8.5-inch vertical guideline.

6 Drag the left side of the scale bar to the right until its width is 1,000 miles. This takes trial and error, with you dragging and releasing to see the resultant width in miles.

Insert text

1 Click the vertical ruler at 0.5 inches.

2 Click Insert, Text. ArcMap places a small text box in the center of your map (it is difficult to see at this scale).

3 Double-click the text box, type **Source: U.S. Census Bureau**, and click OK.

4 Drag the text box so that its bottom left corner is at the intersection of the 8.5-inch vertical and 0.5-inch horizontal guides.

5 Save your map document.

Tutorial 3-3

Reuse a custom map layout

Using your custom map to produce additional maps will save time, but just as important is the consistency of the resulting maps. Reuse guarantees that sizes and placements of objects match perfectly for a collection of maps.

Use a layer file

1 Click File and Open, browse to your \ESRIPress\GIST1\Maps\ folder, and double-click Tutorial3-3.mxd. You will replace the Asian attribute with the Black attribute of the Population layer.

2 Click View, Data View.

3 Right-click the Population layer in the TOC. Click Properties, the Symbology tab, and the Import button.

4 In the Import Symbology dialog box, click the Browse button. Browse to the \ESRIPress\GIST1\MyExercises\FinishedExercises\Chapter3\ folder, double-click StatesPopulation.lyr, and click OK.

5 In the Import Symbology Matching dialog box, click the Value Field drop-down arrow, click Black, and click OK.

6 In the Layer Properties window, change the color ramp to a monochromatic blue ramp, right-click the color ramp (to the right of the label, Color Ramp), click Properties, click the drop-down arrow for Color 1, click the white color chip, click the button for Color 2, click its drop-down arrow, click a dark blue, click OK, then click OK again.

7 Click View, Layout View.

8 Change the map title to **Black Population (2007)**.

9 Click File and Save As, browse to the \ESRIPress\GIST1\MyExercises\Chapter3\ folder, name the file **Tutorial3-Blacks.mxd**, and click Save.

Export a layout as an image file

Let's have you export this layout to a high-quality image file, which you could use in a Word document or PowerPoint presentation.

1 **Click File and Export Map, browse to the \ESRIPress\GIST1\MyExercises\Chapter3\ folder. Make sure that JPEG (*.jpg) is selected as the file type, choose 300 dpi for resolution under options, and click Save.**

2 **Using My Computer, browse to \ESRIPress\GIST1\MyExercises\Chapter3\, right-click Tutorial3-Blacks.jpg, click Open With, and open the image in a viewer.** While a simple layout, it is quite professional and attractive in appearance.

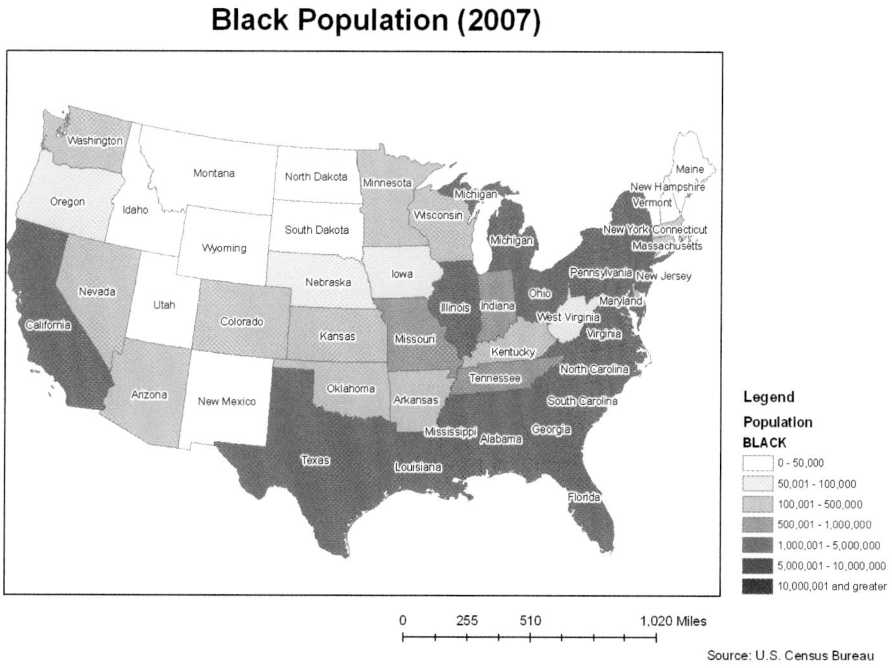

3 **When finished examining the image, close the viewer and the My Computer window.**

YOUR TURN

Make progress on completing the map collection of population by race by making a map document for one of the following populations from your template: Whites, Hispanics, or Native Americans. Name the map **Tutorial3-Whites.mxd**, **Tutorial3-Hispanics.mxd**, or **Tutorial3-NativeAmericans.mxd**, respectively, and save it in the \ESRIPress\GIST1\MyExercises\Chapter3\ folder. Use a monochromatic color ramp of your choice.

Tutorial 3-4

Create a custom map template with two maps

To facilitate comparisons, you can place two or more maps on the same layout. Your population maps by racial/ethnic groups are ideal for this purpose because they share the same numeric scale, making comparisons easy.

Create a map document with two frames

1 Click File and Open, browse to your \ESRIPress\GIST1\Maps\ folder, and double-click Tutorial3-4.mxd.

2 On the Tools toolbar, click the Full Extent button .

3 Click File and Save As, browse to the \ESRIPress\GIST1\MyExercises\Chapter3\ folder, name the map **Tutorial3-AsiansBlacks.mxd**, and click Save.

4 In the TOC, right-click the Layers icon **Layers**, click Properties and the General tab, and change the name from Layers to **Asians** (but do not click OK).

5 Click the Coordinate System tab, click Add to Favorites, and click OK. You will need to apply the coordinate system of the Asians data frame to the new data frame that you will create next. Creating a favorite makes this easy.

6 On the main menu, click Insert, Data frame.

7 Right-click the New Data Frame, click Properties, the General tab, and change the name from New Data Frame to **Blacks** (but do not click OK).

8 Click the Coordinate System tab, expand Favorites in the bottom left panel, click USA-Contiguous_Albers_Equal_Area_Conic under Favorites, and click OK.

9 In the Asians data frame, right-click Population, click Copy, right-click the Blacks data frame, and click Paste Layer(s).

YOUR TURN

Replace the color gradient of Population using the Black attribute as you did above in tutorial 3-3. Your finished map document will appear as follows (note that to switch from one data frame to another, you right-click the data frame and click Activate). The active data frame's label in the TOC is in bold type).

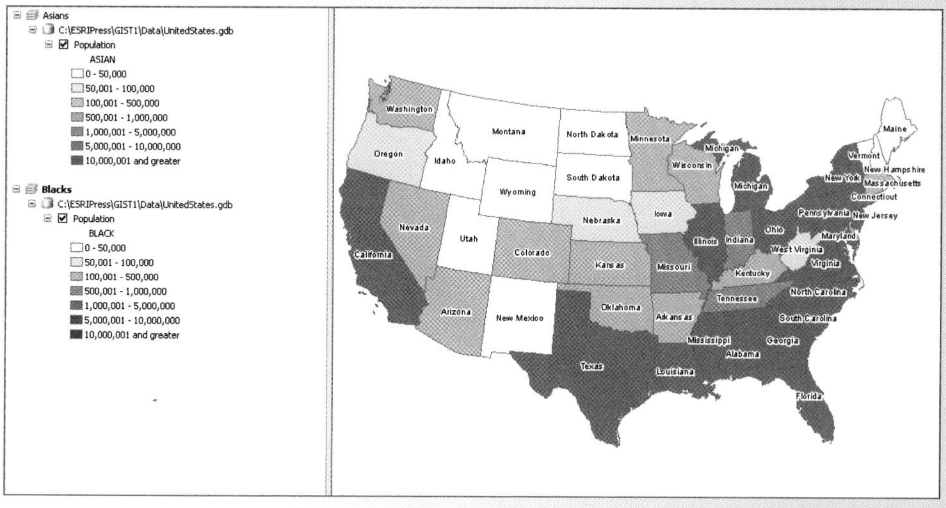

Set up Layout View

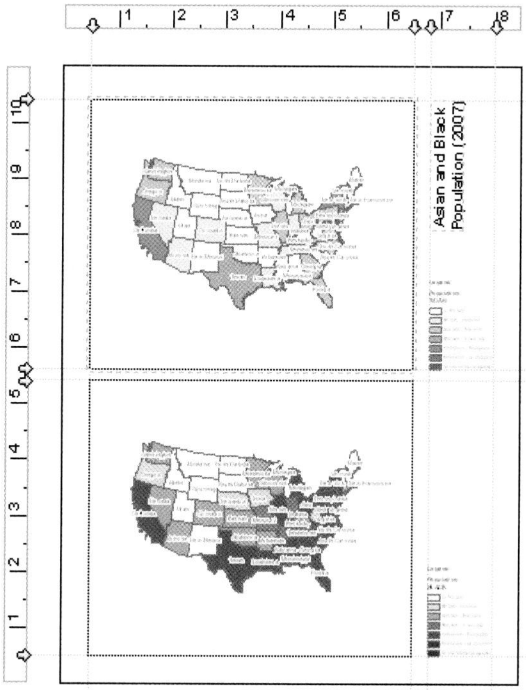

At right is the layout that you will create in the steps of this exercise. It makes comparisons between two population distributions easy.

1. Click View, Layout View.
2. In layout view, right-click in the layout, click Page and Print Setup, make sure Size is Letter, make sure that both Portrait radio buttons are selected in the Page and Paper frames, then click OK.
3. Click the horizontal ruler at the 0.5-, 6.5-, approximately 6.8-, and 8.0-inch marks.

4 Click the vertical ruler at the 0.5-, approximately 5.4-, approximately 5.6-, and 10.5-inch marks.

5 Click and drag the Asian data frame so that its upper left corner snaps to the intersection of 10.5-inch horizontal guide and 0.5-inch vertical guide.

6 Click and drag the lower right grab handle of the data frame to snap it at the 5.6-inch horizontal guide and 6.5-inch vertical guide.

7 Click and drag the Blacks data frame so that its upper left corner snaps at the intersection of the horizontal 5.4-inch guide and vertical 0.5-inch guide.

8 Drag the Blacks data frame so that its lower right grab handle snaps at the 0.5-inch horizontal guide and 6.5-inch vertical guide.

9 On the Tools toolbar, click the Full Extent button in both data frames.

Add elements to layout

1 Click the Asian map element to activate its frame.

2 Click Insert, Legend.

3 Click Next four times, and then click Finish.

4 Drag the legend so that it snaps on the lower right to the 8-inch vertical guide and 5.6-inch horizontal guide intersection, then resize it to fit between the 6.8-inch and 8-inch vertical guides and on the 5.6-inch horizontal guide.

5 Click the Blacks data frame and repeat steps 2–4 so that the second legend's lower right is at the intersection of the 8-inch vertical and 0.5-inch horizontal guides.

6 Click Insert, Text, click anywhere outside the text box, right-click the text box, and click Properties and the Text tab.

7 In the Text panel, type **Asians and Blacks**, press Enter to jump to a new line, type **Populations (2007)**, type **90** for Angle, click Change Symbol, change the Size to 20, choose B (bold) for style, and click OK twice.

8 Position the top left of the text box at the intersection of the 6.8-inch vertical and 10.5-inch horizontal guides.

9 Save your map document.

YOUR TURN

Export your layout to the \ESRIPress\GIST1\MyExercises\Chapter3\ folder as a JPEG image and view it in an image viewer. It is quite a nice layout and image. There are some remarkable similarities in the distributions of the two races.

Tutorial 3-5

Add a report to a layout

ArcMap has a built-in capability to make tabular reports. You can add reports to layouts to provide detailed information.

Open a map document

1 Click File and Open, browse to the \ESRIPress\GIST1\Maps\ folder, and double-click Tutorial3-5.mxd.

2 Click File and Save As, browse to \ESRIPress\GIST1\MyExercises\Chapter3\, type **Tutorial3-AsiansReport.mxd** for the File Name, then click Save.

3 Click View, Data View.

Make a selection of records

You will generate a report for the selected records only.

1 Right-click the Population layer in the table of contents and click Open Attribute Table.

2 Scroll to the right in the Attributes of Population to find the ASIAN column, right-click the ASIAN column heading, and click Sort Descending.

3 Scroll left in the table until you see the STATE_NAME column.

4 If necessary, make the table large enough so you can see the first 10 state records.

5 Click the row selector for the top row, then hold and drag down to select the top 10 rows in the table (California through Pennsylvania).

6 Close the Attributes of Population window.

Create a report

ArcMap has a wizard for creating reports that you will use next.

1. Click View, Layout View.
2. Click View, Reports, Create Report.
3. In the Report Wizard dialog box, click Dataset Options, the Selected Set option button, OK.
4. In the Available Fields box, double-click STATE_NAME, ASIAN, and POP2000. Click Next twice.
5. For sorting, select ASIAN for the field and Descending for the sort. Click Next.
6. Click the Outline option button for Layout and click Next.
7. Click the Monaco report format and click Finish.

Modify a report

The report wizard provides a good start, but you can improve this report by making it more compact, formatting the long numeric values to have comma separators, and improving labels.

1. In the Report Viewer window, click Edit on the menu bar.
2. Drag the bottom of the detail panel down as seen at right to expose all three data fields.

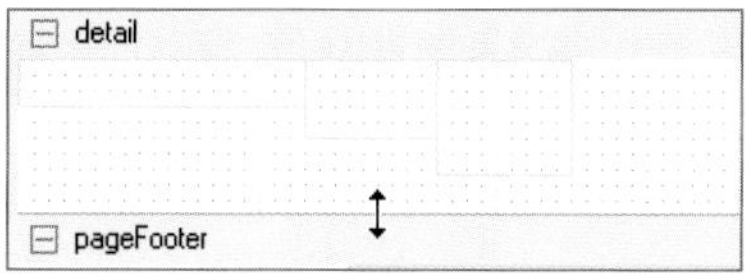

3. Drag up the bottom of each data field in turn so that it just touches the third row of dots. Move and increase the widths of the ASIAN and POP2000 fields as seen at right (observe the horizontal ruler for positioning). Drag the bottom of the detail panel up to just below the fields. This step eliminates extra space between data rows.

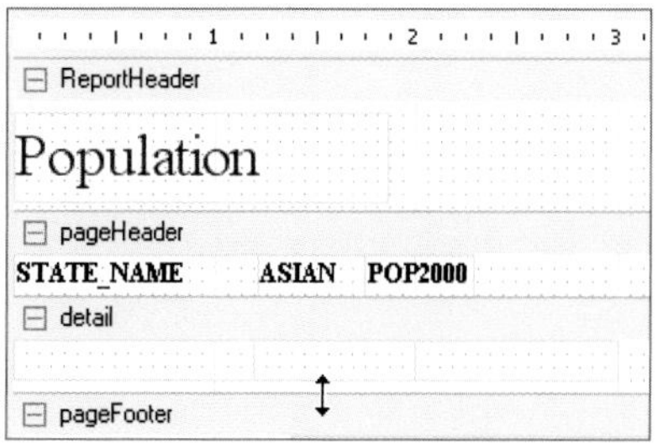

4 Click The Run Report button to see the results thus far.

Population

STATE_NAME	ASIAN	POP2000
California	3697513	33871648
New York	1044976	18976457
Texas	562319	20851820
New Jersey	480276	8414350
Illinois	423603	12419293
Washington	322335	5894121
Florida	266256	15982378
Virginia	261025	7078515
Massachusetts	238124	6349097
Pennsylvania	219813	12281054

5 Click Edit, click the ASIAN detail data field to select it, click in the Output Format property field in the Element Properties panel, then click its Builder button. Click Number in the Category panel, change Decimal Places to **0**, make sure that Use 1000 Separator is selected, and click OK.

6 Repeat step 5 for the POP2000 field.

7 Click the STATE_NAME label field in the pageHeader area, and change the Text property from STATE_NAME to **State**.

8 Do the same for ASIAN and POP2000, changing the text to **Asian** and **Total**.

9 Move the Asian and Total labels so that they are above their detail fields.

10 Click The Run Report button to see the results.

Population

State	Asian	Total
California	3,697,513	33,871,648
New York	1,044,976	18,976,457
Texas	562,319	20,851,820
New Jersey	480,276	8,414,350
Illinois	423,603	12,419,293
Washington	322,335	5,894,121
Florida	266,256	15,982,378
Virginia	261,025	7,078,515
Massachusetts	238,124	6,349,097
Pennsylvania	219,813	12,281,054

Add a report to a layout

1 Click the Save Report to Output File button in the Report Viewer window, browse to the \ESRIPress\GIST1\MyExercises\Chapter3\ folder, type **AsianTop10** for the File name, click Save, then OK. This saves a version of the report that you can open and edit in ArcMap again in the future if needed. The next version you will save is convenient for inserting a compact version of the report into your layout. If instead you use the report wizard to insert the report into your layout, it inserts an entire page with many blank lines at the bottom. From Excel, you will be able to copy and paste only desired content.

2 Click the Export Report to File button, select Microsoft Excel (XLS) as the Export Format, click the Builder button for the file name field, browse to the \ESRIPress\GIST1\MyExercises\Chapter3\ folder, type **AsianTop10** for the File name, and click Save and OK.

3 Open My Computer, browse to \ESRIPress\GIST1\MyExercises\Chapter3\, and double-click AsianTop10.xls. The file opens in Microsoft Excel. If you are not familiar with Excel or have difficulty, you can find a finished copy of the file as \ESRIPress\GIST1\MyExercises\FinishedExercises\Chapter3\AsianTop10.xls.

4 In Excel, if necessary, make the column wider for data fields (double-click the right-side of column headers of columns that have data), and make any other edits that you would like such as deleting extra blank columns (right-click column headers and click Delete).

5 Click cell A2 for the State label, hold your mouse button down, and select the entire table.

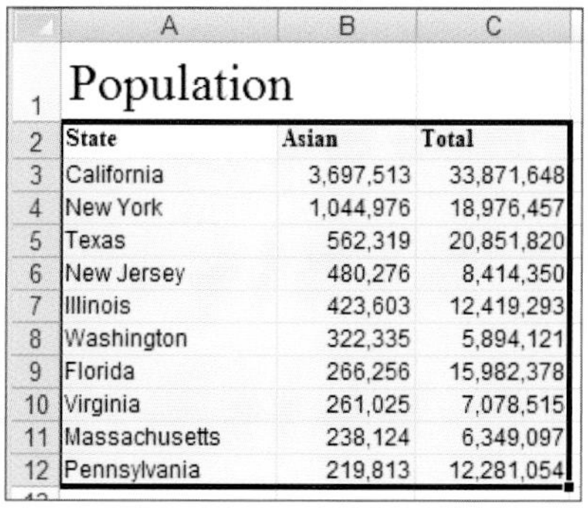

	A	B	C
1	Population		
2	State	Asian	Total
3	California	3,697,513	33,871,648
4	New York	1,044,976	18,976,457
5	Texas	562,319	20,851,820
6	New Jersey	480,276	8,414,350
7	Illinois	423,603	12,419,293
8	Washington	322,335	5,894,121
9	Florida	266,256	15,982,378
10	Virginia	261,025	7,078,515
11	Massachusetts	238,124	6,349,097
12	Pennsylvania	219,813	12,281,054

6 Press Ctrl+C to copy the selection, click your ArcMap window to activate it, and press Ctrl+V to paste the selection into your layout.

7 Close Excel without saving your work.

Modify a layout for displaying report

Below is the finished layout with report.

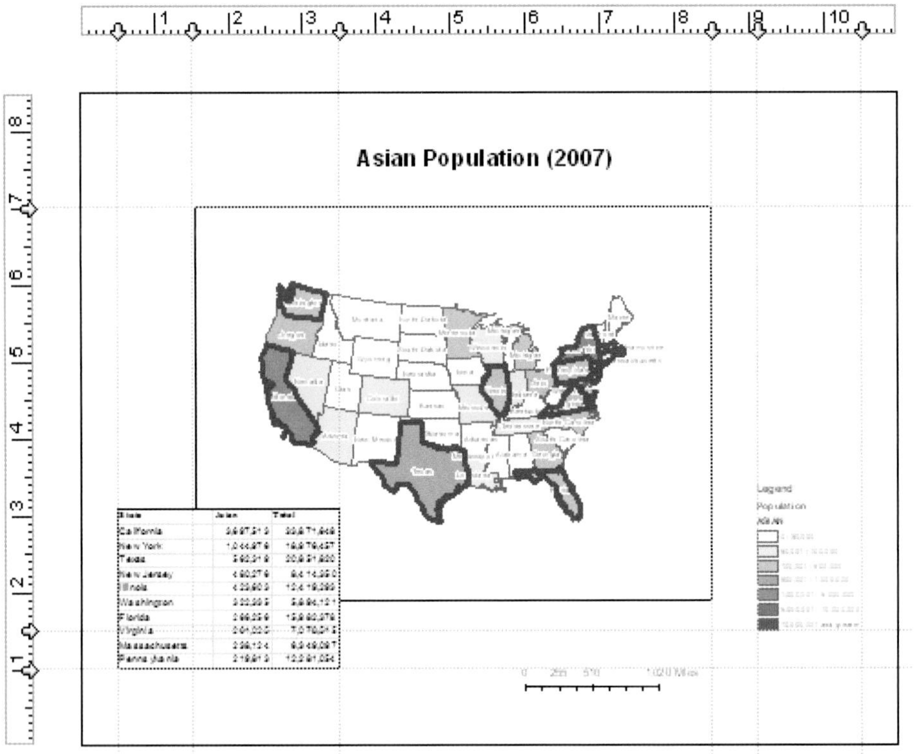

1 Add vertical guides at the 0.5-, 1.5-, and 3.5-inch horizontal ruler positions in your layout.

2 Move and resize the map so that its upper left point is at the intersection of the 1.5-inch vertical and 7-inch horizontal guides, and the upper right corner is at the 8.5-inch vertical and 7-inch horizontal guides.

3 Right-click the map and click the Full Extent button.

4 Drag the report so that its lower left corner snaps at the intersection of the 1-inch horizontal and 0.5-inch vertical guides.

5 Click the upper right grab handle of the report and drag down to snap it to the 3.5-inch vertical guide.

6 Right-click the report, click Properties, click the Frame tab, click in the Border area, click 1.0 point, click in the Background area, click White, and click OK.

7 Save your map document.

Tutorial 3-6

Add a graph to a layout

It took quite a bit of work to get the report added to your layout. Adding a graph of the same data is quite easy by comparison.

Create a graph and add it to a layout

1 Click File and Open, browse to the \ESRIPress\GIST1\Maps\ folder, and double-click Tutorial3-6.mxd. Make a selection of records.

2 Click View, Graphs, and Create. Then make selections as follows:

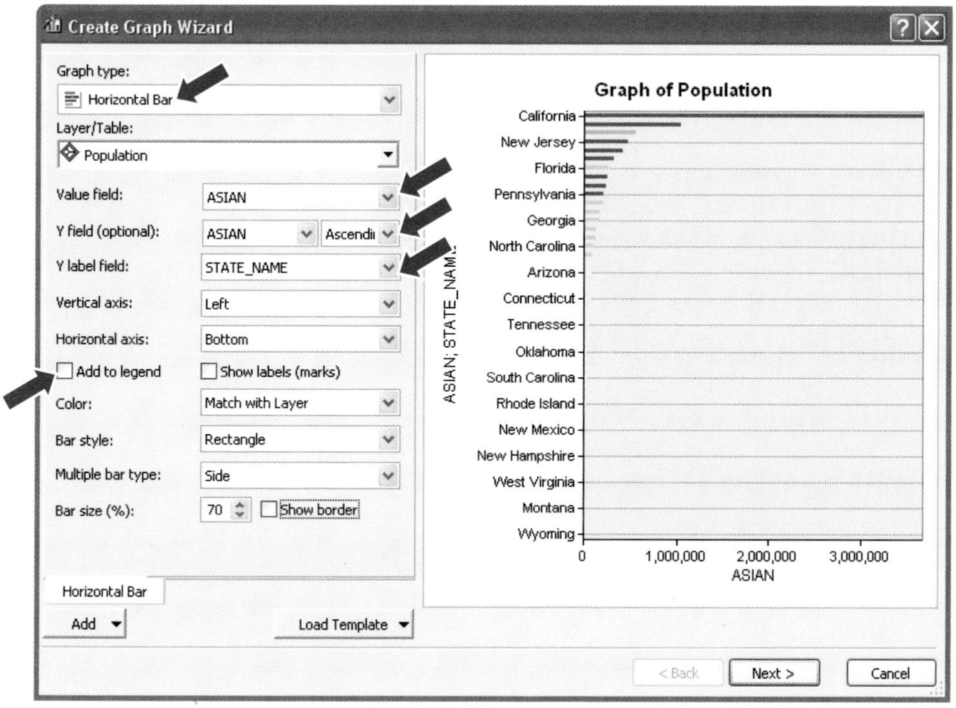

3 Click Next and make selections as follows:

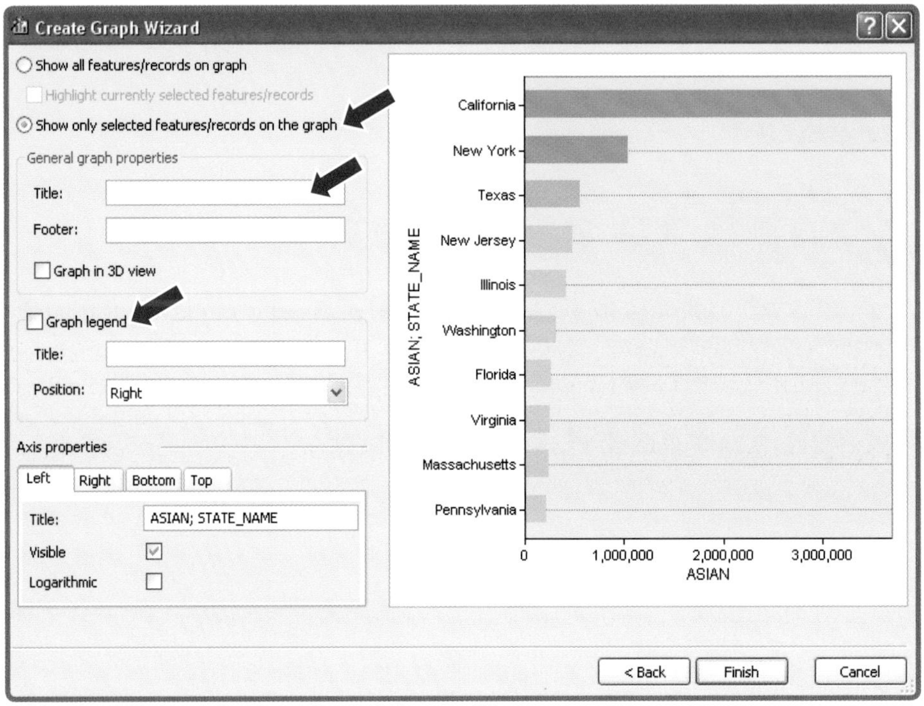

4 Click Finish.

5 Right-click the graph, click Save, browse to \ESRIPress\GIST1\MyExercises\Chapter3\, type **AsianTop10Graph.grf**, and click Save.

6 Right-click the graph, click Add to Layout, and close the graph window.

7 Snap the lower left corner of the graph to the intersection of the 0.5-inch vertical and 1-inch horizontal and guides.

8 Click File and Save As, browse to \ESRIPress\GIST1\MyExercises\Chapter3\, type **Tutorial3-AsiansTop10Chart.mxd** for the File Name, then click Save.

YOUR TURN

Export the map document as **AsianTop10.jpg** to \ESRIPress\GIST1\MyExercises\Chapter3\ and view it.

Tutorial 3-7

Create multiple output pages

Sometimes it is desirable to produce many maps from a single layout, with each map for a different extent—for example, each municipality within a county, or each state or province of a country. ArcMap's Data Driven Pages serve this purpose. You have to define each extent in an index layer that has the collection of extents as polygons, such as counties or states/provinces. The output is a collection of images or even a PDF document.

Apply Data Driven Pages

1 **Click File and Open, browse to your \ESRIPress\GIST1\Maps\ folder if necessary, and double-click Tutorial3-7.mxd.** The map document opens zoomed in to several municipalities of Allegheny County, Pennsylvania. The map has school locations and public parks. Suppose that you need a PDF document with a map for each municipality (although the PDF you will produce will only have a few of the municipalities to keep the file small).

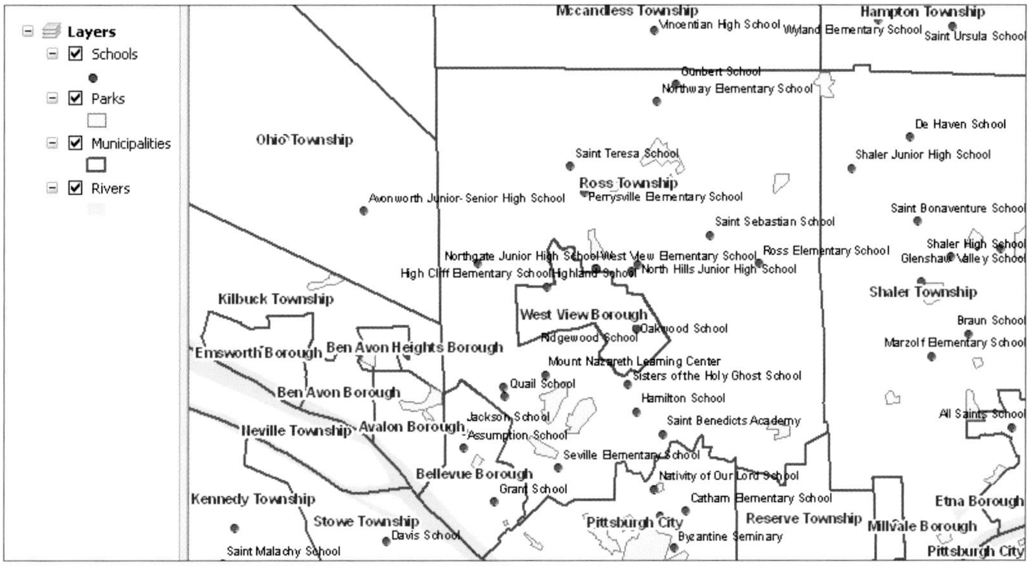

2 **Click File and Save As, browse to \ESRIPress\GIST1\MyExercises\Chapter3\, and click Save.**

3 On the main menu, click Customize, Toolbars, Data Driven Pages.

4 On the Data Driven Pages toolbar, click the Data Driven Page Setup button .

5 In the resulting window, make sure that Enable Data Driven Pages is clicked, and select Municipalities for Layer (second drop-down list in the left panel). Set both Name and Sort Fields to NAME. Click OK. The map's extent switches to the first municipality alphabetically, Aleppo Township, which has no schools or parks.

6 On ArcMap's main menu, click View, Layout View.

7 On the main menu click Insert, Dynamic Text, and Data Driven Page Name. ArcMap places a small text box in the center of your map with the municipality name, Aleppo Township.

8 Drag the inserted text box to the top, center of the page; double-click it; click Change Symbol; set Size to 22 and Style to B for bold; click OK; and click OK again.

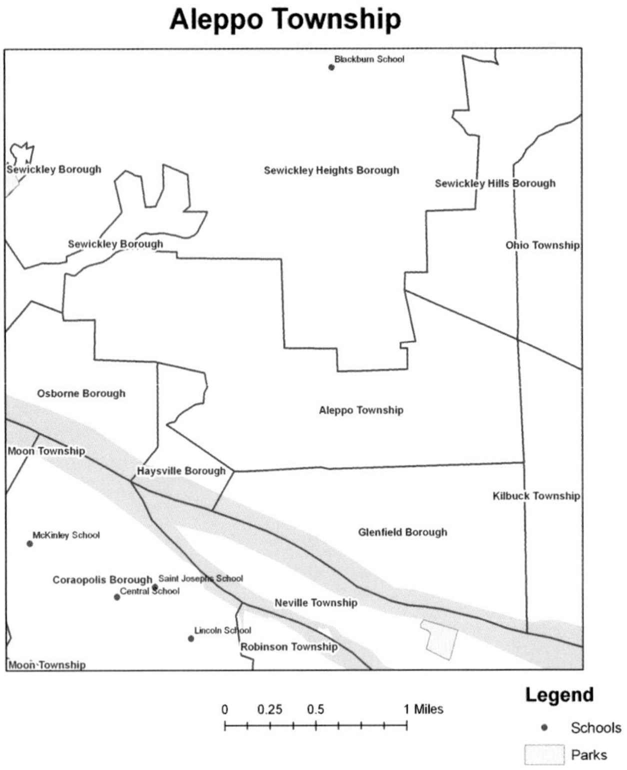

3-1
3-2
3-3
3-4
3-5
3-6
3-7
3-8
A3-1
A3-2
A3-3

YOUR TURN

Resize and relocate the map to your liking on the layout and insert a legend and scale bar.

Output Data Driven Pages

1 Click File, Export Map.

2 Browse to \ESRIPress\GIST1\MyExercises\Chapter3\, change the Save as type to PDF (*.pdf), change the File Name to **MunicipalitySchoolsParks.pdf**, and click the Options Button ▷ Options to expose options.

3 Under the General tab, type **300** for dpi, click the Pages tab and its Page Range option button, type **1–3** in the associated field, click the General tab again, and click Save.

4 Open a My Computer window, browse to \ESRIPress\GIST1\MyExercises\Chapter3\, double-click MunicipalitySchoolsParks.pdf to open the file in Adobe Acrobat or an equivalent program, and view the resulting document. The PDF document has maps for Aleppo, Aspinwall, and Avalon.

5 Close the document you just opened, close the Data Driven Pages toolbar, and save your map document.

Tutorial 3-8

Build a map animation

Police want to identify new, persisting, and fading spatial clusters of crime locations. These clusters make good areas for police to patrol for enforcement and prevention. Animations of crime data for this purpose are effective if they display two tracks of points: new crime points along with recent, but older crime points for context. Then the observer can detect the emergence of new clusters (where there were none before), the persistence of existing clusters getting new crime points, and diminishing clusters as no new points are added and the cluster fades away. So the approach you will take calls for two animation layers—one for each day's events and a second with the past two week's events. Computer-aided dispatch drug calls and shots-fired calls are important crime events for this kind of analysis.

Open the map document for animation

First you will build an animation just showing the sequence of one set of points in a track, one day at a time. About all this accomplishes is to get you started with animation and convince you that the drug and shots-fired calls jumped around in a portion of the Middle Hill neighborhood of Pittsburgh. Then you will add a second set of points that provide context, allowing you to better see hot-spot patterns.

1 **Click File and Open, browse to your \ESRIPress\GIST1\Maps\ folder if necessary, and double-click Tutorial3-8.mxd.** The map document opens to the Middle Hill neighborhood. The computer-aided dispatch (CAD) data, with dates ranging from 7/1/2009 through 8/31/2009, represents calls from citizens reporting illegal drug dealing and shots fired. The animation that you will build will show the daily sequence of CAD call locations, starting with 7/1/2009.

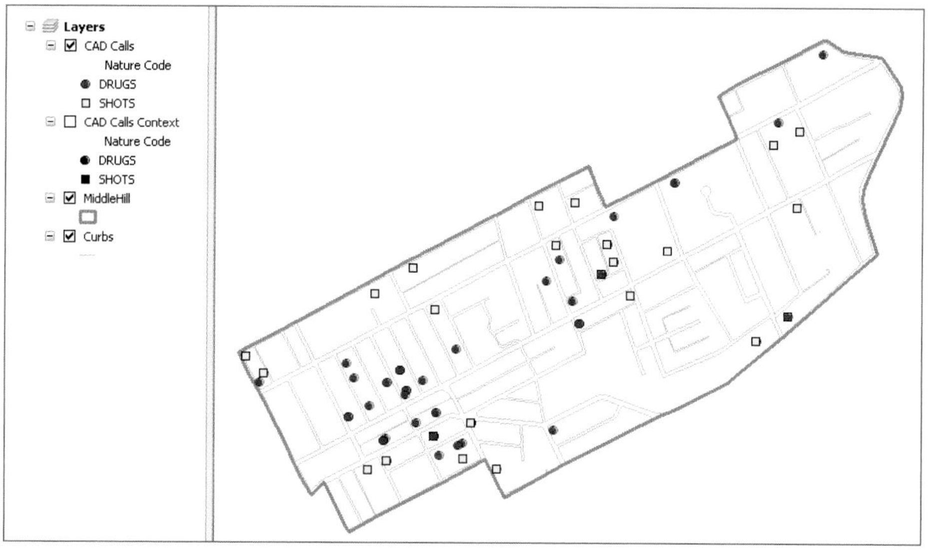

2 Click File and Save As, browse to \ESRIPress\GIST1\MyExercises\Chapter3\, and click Save.

Set time properties of a layer

1 Right-click CAD Calls in the TOC, click Properties and the Time tab, and type or make selections as follows. CALLDATE has dates such as 7/1/2009. The Time Step Interval is the unit of time for measurement, here 1 day.

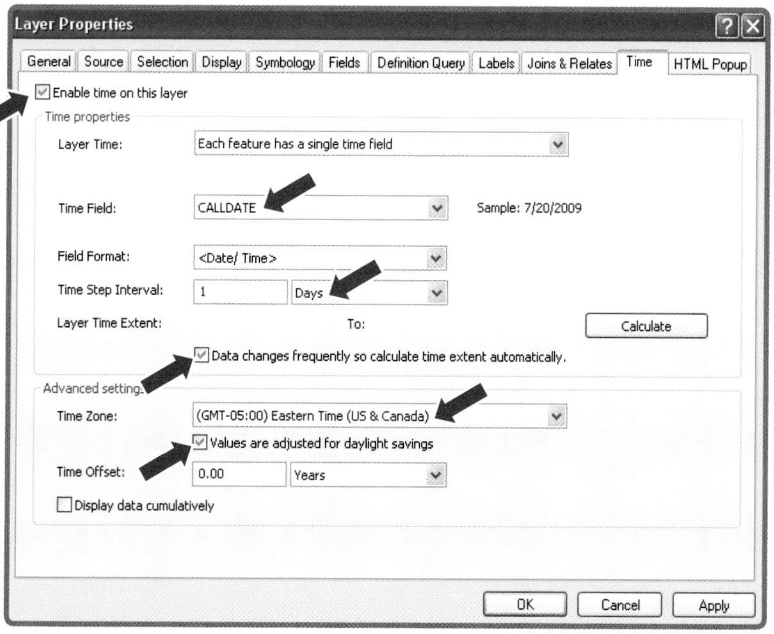

2 Click Calculate, change the Time Step Interval from 4 to 1, and click OK.

Use the Time Slider window for viewing

With time properties set, you are ready to use the Time Slider interface to play a simple video of daily crime points.

1 On the Tools toolbar, click the Open Time Slider Window button.

2 In the Slider window, click the Options button and the Playback tab.

3 Drag the speed selector to roughly half way between slower and faster.

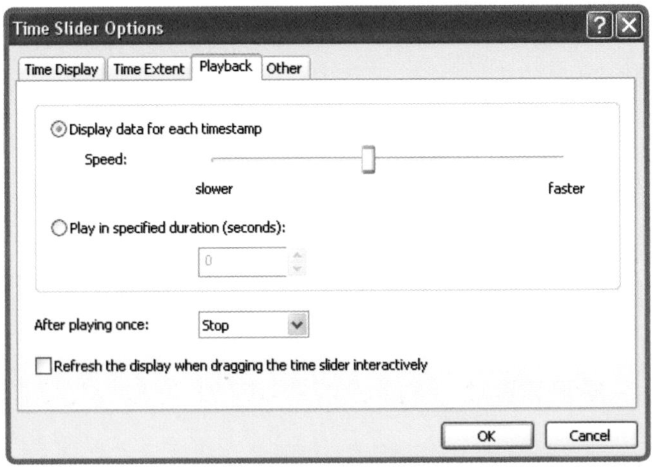

4 Click OK.

5 Click the Enable time on map button.

6 **Click the play button on the slider.** The video plays, one day at a time, shown below at July 30, 2009. Play the video a few more times to see if you can spot any patterns. This is hardly possible until you add the crime time context to the video.

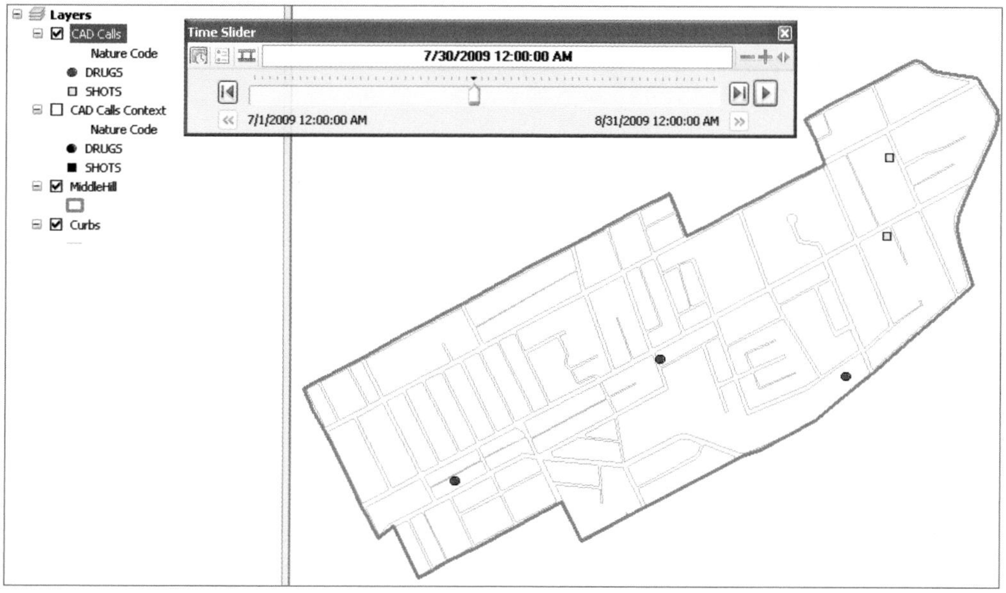

7 **Close the Slider window and save your map document.**

Create a new date column for animating a window of crime points

On any day of the animation, you need to show the current day's crimes with point markers in bright colors and the crime context consisting of two weeks' crime points ending on the same day with black point markers. Displaying a moving window of crime points for two week's data requires starting and ending dates for the window. For this purpose you will create a new date column with 14 days added to CALLDATE to yield the end date.

1 **Right-click CAD Calls Context in the TOC and click Open Attribute Table.**

2 **Click the Table Options button, select Add Field, type EndDate for Name, select Date for Type, and click OK.**

3 **Right-click the column heading for EndDate and click Field Calculator.**

4 **Double-click CALLDATE in the Fields panel, click the + button, and type a blank space and 14 to yield [CALLDATE] + 14 for EndDate's expression.**

5 Click OK.

6 Close the attribute table.

Set advanced time properties of a layer

1 Right-click CAD Calls Context in the TOC, click Properties and the Time tab, and type or make selections as follows:

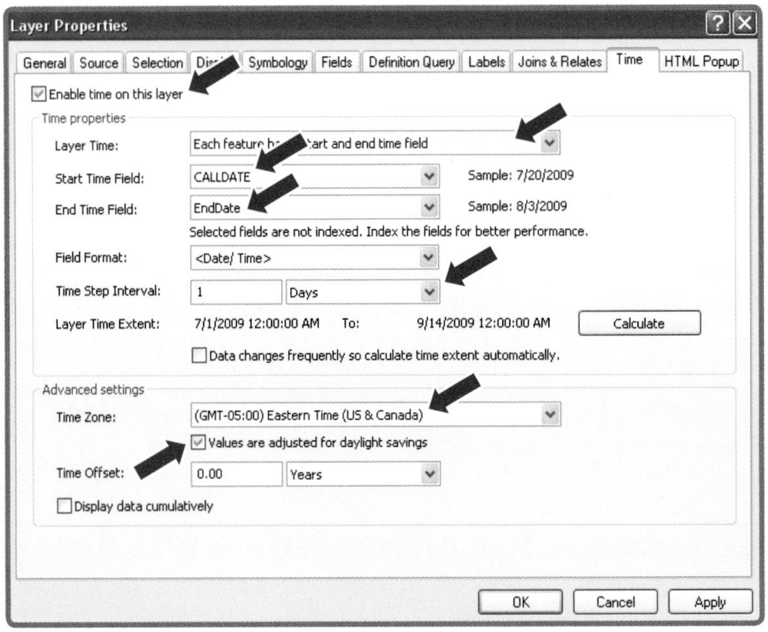

2 Click Calculate, change the Time Step Interval from 4 to 1, and click OK. When you play the animation next, both layers with time properties will animate. CAD Calls will show the current day's crime locations while CAD Calls Context will show all crimes in the interval, including two weeks ending on the current day.

Use the Time Slider window for advanced viewing

1 On the Tools toolbar, click the Open Time Slider Window button.

2 Turn on the CAD Calls Context layer. Click the play button on the slider. Following is the animation at July 30, 2009. You see all of the crime locations for the last two weeks with the last day's crimes in red or yellow and older crimes in black. At this time there is much persistence in crime clusters.

3 Save your map document and close ArcMap.

Assignment 3-1

Create a layout comparing elderly and young populations in Orange County, California

Sometimes you will want to compare two or more maps in layout view to visualize and analyze information about multiple attributes. In this exercise, you will create a map layout with two maps with population percentages for the elderly and youths in Orange County, California.

Start with the following:

- \ESRIPress\GIST1\Data\UnitedStates.gdb\CAOrangeCountyTracts—census tract polygon boundaries for Orange County, California, Census 2000

Create a comparison map of census data

Create a new map document called **\ESRIPress\GIST1\MyAssignments\Chapter3\Assignment3-1 YourName.mxd** that includes an 8.5-by-11-inch portrait layout with two data frames: one with the percentage of 2000 population who are 5–17 years old and the second with the percentage who are 65 or older. Use the same numerical scale for both maps. Include a graphic scale bar in miles.

Hints

- Use the Symbology tab in Layer Properties to show a population as a percentage of the total population for the year 2000 (use POP2000 field to normalize the data).
- Use your judgment as to the color, sizes, titles, and other map elements to add or modify.

Export the map as a JPEG file called **\ESRIPress\MyAssignments\Chapter3\Assignment3-1 YourName.jpg**.

WHAT TO TURN IN

If your work is to be graded, turn in the following files:

ArcMap document: \ESRIPress\GIST1\MyAssignments\Chapter3\Assignment3-1YourName.mxd

Image file: \ESRIPress\GIST1\MyAssignments\Chapter3\Assignment3-1YourName.jpg

If instructed to do so, instead of the above individual files, turn in a compressed file, **Assignment3-1YourName.zip**, with all files included. Do not include path information in the compressed file.

Assignment 3-2

Create a walking map of historic districts in downtown Pittsburgh

Walking tours are great attractions for tourists. You will create an overall view of the historic sites in the Central Business District of Pittsburgh as well as a zoomed-in map for one area. Visit `http://www.city.pittsburgh.pa.us/wt/html/walking_tours_main.html` for examples. For photos of historic buildings that you will need for this assignment, click the City Legacies link on the home page, click any of the photos on that page, and then any building name link to the right of the photo on the resulting page.

Start with the following:

- \ESRIPress\GIST1\Data\Pittsburgh\CentralBusinessDistrict.gdb\CBDOutline—polygon feature of Pittsburgh's Central Business District neighborhood outline
- \ESRIPress\GIST1\Data\Pittsburgh\CentralBusinessDistrict.gdb\CBDBLDG—polygon features of Pittsburgh's Central Business District buildings
- \ESRIPress\GIST1\Data\Pittsburgh\CentralBusinessDistrict.gdb\CBDStreets—line features of Central Business District streets
- \ESRIPress\GIST1\Data\Pittsburgh\CentralBusinessDistrict.gdb\Histsite—polygon features of historic areas in Pittsburgh's Central Business District
- \ESRIPress\GIST1\Data\Pittsburgh\CentralBusinessDistrict.gdb\Histpnts—point features of historic sites in Pittsburgh's Central Business District

Create a large-scale map

Create a new map called **\ESRIPress\GIST1\MyAssignments\Chapter3\Assignment3-2 YourName.mxd** with an 8.5-by-11-inch layout containing two data frame maps—one scaled at 1:14,000 showing all of the historic districts in the Central Business District, and one scaled at 1:2,400 showing one of the historic districts in detail. A suggestion is to show Market Square in detail and a photo of Burke Building, which is in Market Square. To set a scale, click the frame of a map in layout view and type the desired scale in the scale text box on the Standard toolbar.

Keep in mind basic mapping principles such as colors, ground features, and so forth, covered in previous chapters. Choose labels and other map elements that you think are appropriate for each map as well as the overall layout. Include the photograph of the building that you download from the City of Pittsburgh Web page and save to the \ESRIPress\GIST1\MyAssignments\Chapter3\ folder. Include the photo's source in your layout. Click Customize, Toolbars, Draw to add the Draw toolbar to ArcMap. Click the list arrow of the fourth tool from the left and select Line. Draw two lines from two ends of the historic area on your CBD map to the same two points on your detail map. That provides a graphic guide for interpretation of the detail map. Also, draw a line from the building on the detail map to its picture that you inserted into the layout.

Symbolize the Historic Sites polygons as a transparent layer (see "Hint" below) so you can see the buildings under the sites and the Central Business District as a thick outline.

Export your map as a PDF file called **\ESRIPress\MyAssignments\Chapter3\Assignment3-2 YourName.pdf.**

Hint

Drawing a layer transparently:

- Right-click the layer needing transparency.
- Click Properties and the Display tab.
- Type a transparency percentage, such as 50.
- Click OK.

WHAT TO TURN IN

If your work is to be graded, turn in the following files:

ArcMap document: \ESRIPress\GIST1\MyAssignments\Chapter3\Assignment3-2YourName.mxd

Exported map: \ESRIPress\GIST1\MyAssignments\Chapter3\Assignment3-2YourName.pdf

Downloaded image of a building: \ESRIPress\GIST1\MyAssignments\Chapter3\BuildingNameYourName.jpg

If instructed to do so, instead of the above individual files, turn in a compressed file, **Assignment3-2YourName.zip**, with all files included. Do not include path information in the compressed file.

Assignment 3-3

Create an animation for an auto theft crime time series

Auto thieves are often creatures of habit; they return to the same areas and repeat other patterns that led to successful thefts in the past. An animation of successive auto theft locations can help determine the space-time pattern of such a thief. Suppose that police suspect a serial auto thief who steals cars for basic transportation and then vandalizes abandoned stolen cars in a unique way with spray paint.

Start with the following:

- \ESRIPress\GIST1\Data\DataFiles\AutoTheftCrimeSeries.shp—point layer of the suspected crime series of auto thefts
- \ESRIPress\GIST1\Data\Pittsburgh\MidHill.gdb\MiddleHill—polygon layer of the Middle Hill neighborhood boundary
- \ESRIPress\GIST1\Data\Pittsburgh\MidHill.gdb\Streets—line layer of streets in the Middle Hill neighborhood
- \ESRIPress\GIST1\Data\Pittsburgh\MidHill.gdb\Curbs—line layer of curbs in the Middle Hill neighborhood

Create an animation of serial auto thefts

Create a new map document called **\ESRIPress\GIST1\MyAssignments\Chapter3\Assignment3-3YourName.mxd** that includes the above map layers as follows:

- Thick outline for the study area using MiddleHill
- Streets in background color
- Two copies of AutoTheftsCrimeSeries.shp, one called Auto Theft Crime Series with a size 10 Circle 2 point marker and bright color fill, and the second called Auto Theft Crime Series Context with the same point marker but black color fill and placed under the first copy in the TOC

Set the time properties for the two copies of AutoTheftCrimeSeries similar to those in tutorial 3-8. Have Auto Theft Crime Series display a single day's auto thefts in each frame and have Auto Theft Crime Series Context display the cumulative set of crimes. Set the playback speed to a medium level.

When you have the animation working to your satisfaction, label streets using the Streets layer, but use No Color so that only the labels appear and not the streets. The labels blink as the animation plays using the time slider, but will not do so when you take the next step. Use the Export to Video button on the Time Slider window to create a movie file, **\ESRIPress\GIST1\MyAssignments\Chapter3\AutoTheftCrimeSeriesYourName.avi**. Try playing the movie by double-clicking it in a My Computer window.

WHAT TO TURN IN

If your work is to be graded, turn in the following files:

ArcMap document: \ESRIPress\GIST1\MyAssignments\Chapter3\Assignment3-3YourName.mxd

Movie file: \ESRIPress\GIST1\MyAssignments\Chapter3\AutoTheftCrimeSeriesYourName.avi

If instructed to do so, instead of the above individual files, turn in a compressed file, **Assignment3-3YourName.zip**, with all files included. Do not include path information in the compressed file.

Part 2
Working with spatial data

4

File geodatabases

ArcGIS can directly use or import most GIS file formats in common use for geoprocessing and display. The recommended native file format for use in ArcGIS is the file geodatabase that stores map layers, data tables, and other GIS file types in a system folder that has the suffix extension .gdb in its name. In this chapter you will learn about working with file geodatabases.

Learning objectives

- *Build a file geodatabase*
- *Use ArcCatalog utilities*
- *Modify an attribute table*
- *Join tables*
- *Create centroid coordinates in a table*
- *Aggregate data*

Tutorial 4-1

Build a file geodatabase

A file geodatabase is quite simple and flexible, being merely a collection of files in a file folder. Nevertheless, you need a special utility program to build and maintain a file geodatabase. That program is ArcCatalog, which you will use next. Some of the functionality of ArcCatalog is also available in ArcMap in its Catalog window. The Catalog window allows you to do some utility work while in ArcMap without opening the separate application program ArcCatalog.

Open ArcCatalog

1 On the Windows taskbar, click Start, All Programs, ArcGIS, ArcCatalog 10.

2 Click the Connect to Folder button, expand the folder and file tree for ESRIPress, click the GIST1 folder icon to select it, and click OK.

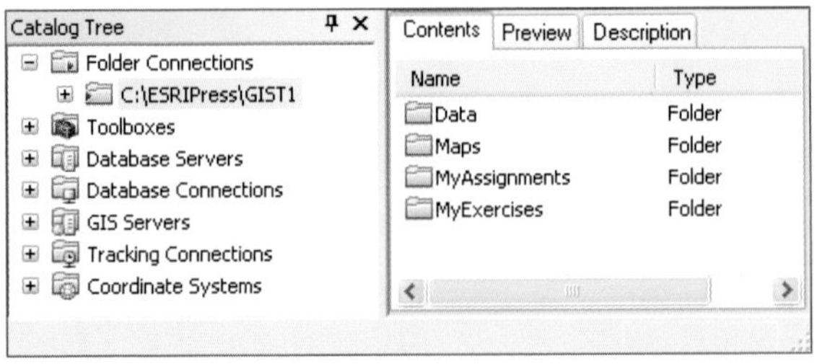

Create an empty file geodatabase

You must create a file geodatabase using ArcCatalog or Catalog. Windows Explorer or My Computer is not capable of building all the parts of a file geodatabase.

1 In the Catalog Tree panel, expand the \ESRIPress\GIST1\ and the MyExercises folder.

2 Click the Chapter4 folder to display its contents in ArcCatalog's right panel.

3 Right-click Chapter4 in the left panel and click New, File Geodatabase.

4 Change the name from New File Geodatabase.gdb to **MaricopaCountyFiles.gdb**.
ArcCatalog creates a file geodatabase that you can now populate with feature classes and stand-alone tables. Feature classes are map layers stored in a geodatabase. Next, you will import map layers in shapefile format into your new file geodatabase feature classes.

Import shapefiles

A shapefile is an older ESRI file format that many GIS suppliers still use to make GIS map layers widely available. ArcCatalog and Catalog allow you to import shapefiles and other map file formats into a file geodatabase.

1 In the Catalog's right panel, right-click the MaricopaCountyFiles file geodatabase, click Import, Feature Class (multiple). The multiple import option provides the convenience of importing several features at the same time.

2 In the Feature Class to Geodatabase (multiple) dialog box, click the browse button to the right of the Input Features field, browse to \ESRIPress\GIST1\Data\MaricopaCounty\, double-click to open that folder, hold the Shift key down, select both tgr04013ccd00.shp and tgr04013trt00.shp, and click Add. That action adds tgr04013ccd00.shp and tgr04013trt00.shp to the input panel.

3 Click OK. ArcCatalog imports the shapefiles into the file geodatabase.

Import a data table

Next, you will import a 2000 census data table at the tract level.

1 Right-click the MaricopaCountyFiles file geodatabase, then click Import, Table (single).

2 In the Table to Table dialog box, click the browse button to the right of Input Rows, browse to \ESRIPress\GIST1\Data\MaricopaCounty\, click CensusDat.dbf, and click Add.

3 Type **CensusDat** in the Output Table field.

4 Click OK.

Tutorial 4-2

Use ArcCatalog utilities

Now that you've created a file geodatabase, you can start using ArcCatalog's utilities. First are the preview utilities, which give you a good overview of a feature layer or table.

Preview layers

1 Click MaricopaCountyFiles.gdb to expose its contents in the right panel.

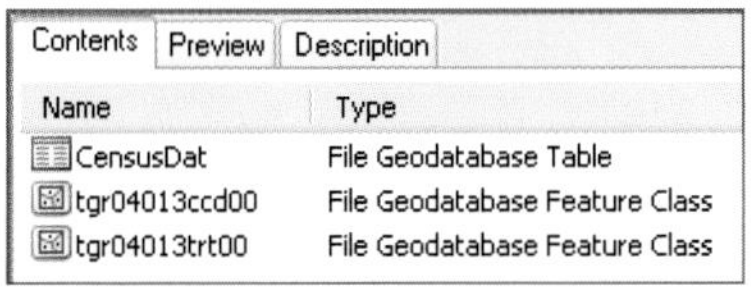

Name	Type
CensusDat	File Geodatabase Table
tgr04013ccd00	File Geodatabase Feature Class
tgr04013trt00	File Geodatabase Feature Class

2 In the right panel, click tgr04013ccd00 and click the Preview tab. ArcCatalog previews the tgr04013ccd00 map layer's geography.

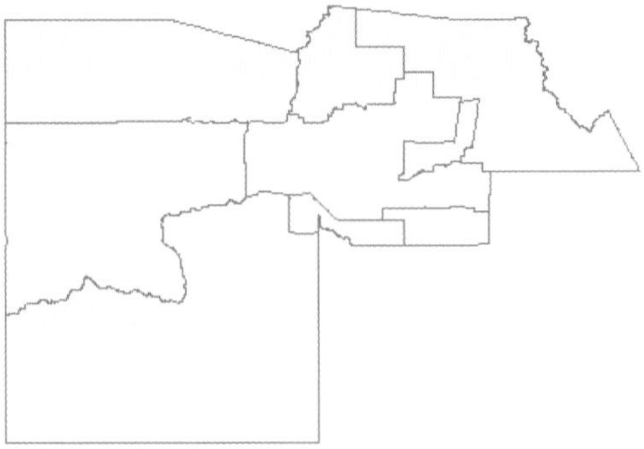

3 At the bottom of the Preview tab, select Table as the Preview. ArcCatalog previews the tgr04013ccd00 map layer's attribute table.

ID	COUNTY	MCD2000	NAME
1	04013	90459	Buckeye
2	04013	90561	Chandler
3	04013	90867	Deer Valley
4	04013	91377	Gila Bend
5	04013	92601	Phoenix
6	04013	93009	St. Johns
7	04013	93060	Salt River
8	04013	93468	Tonto
9	04013	93774	Wickenburg

4 **Click the Description tab.** ArcCatalog previews the tgr04013ccd00 map layer's metadata in a report format.

5 **Click the Contents tab.**

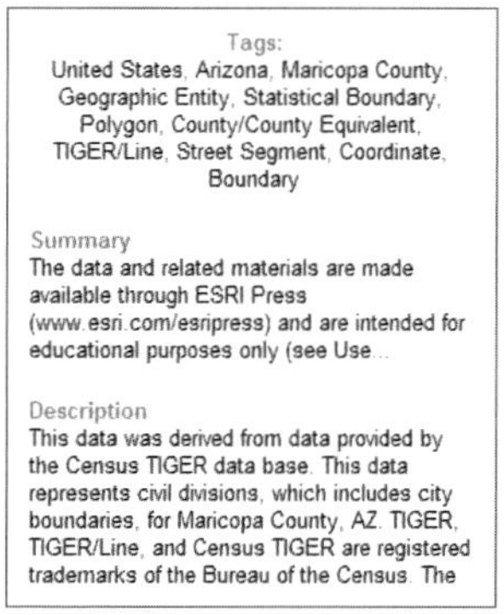
Tags:
United States, Arizona, Maricopa County, Geographic Entity, Statistical Boundary, Polygon, County/County Equivalent, TIGER/Line, Street Segment, Coordinate, Boundary

Summary
The data and related materials are made available through ESRI Press (www.esri.com/esripress) and are intended for educational purposes only (see Use...

Description
This data was derived from data provided by the Census TIGER data base. This data represents civil divisions, which includes city boundaries, for Maricopa County, AZ. TIGER, TIGER/Line, and Census TIGER are registered trademarks of the Bureau of the Census. The

YOUR TURN

Preview tgr04013trt00 and CensusDat.

Rename feature layers

Because a file geodatabase has a special file format, you must use ArcCatalog for many file-management purposes, including renaming and copying items.

1 **In the left panel under MaricopaCountyFiles, right-click tgr04013ccd00, click Rename, and type Cities.**

2 **Repeat step 1 to rename CensusDat CensusTractData, and tgr04013trt00 Tracts.**

Copy and delete feature layers

1 **In the left panel under MaricopaCountyFiles, right-click Cities, click Copy, right-click MaricopaCountyFiles.gdb, click Paste, and OK.** ArcCatalog creates the copy, Cities_1.

2 **Right-click Cities_1, and click Delete and Yes.**

YOUR TURN

Open a My Computer window, browse to \ESRIPress\GIST1\MyExercises\Chapter4\MaricopaCountyFiles.gdb, right-click the folder to get its properties and size, and take a look at the files inside of it comprising the cities, tracts, and census tract data. You should find that the folder size is 1.55 megabytes on the disk and that the files are incomprehensible. You need ArcCatalog or the Catalog in ArcMap to use and manipulate these files. Leave the My Computer window open for use in the following steps.

Compress a file geodatabase

1 In the left panel of ArcCatalog, right-click MaricopaCountyFiles.gdb, click Compress File Geodatabase, and click OK.

2 Use a My Computer window to check the size of the MaricopaCountyFile.gdb folder. In this case there was hardly any reduction in file size. While ArcMap can display compressed feature layers by uncompressing them on the fly, you will use the next step to uncompress the folder and get the layers back to original size.

3 In the left panel, right-click MaricopaCountyFiles.gdb, click Uncompress File Geodatabase, and click OK.

4 Close ArcCatalog.

YOUR TURN

Open ArcMap and create a new map document called **Tutorial4-1.mxd** stored in your \ESRIPress\GIST1\MyExercises\Chapter4\ folder. Add the two layers and table from MaricopaCountyFiles.gdb and symbolize the two layers to your liking

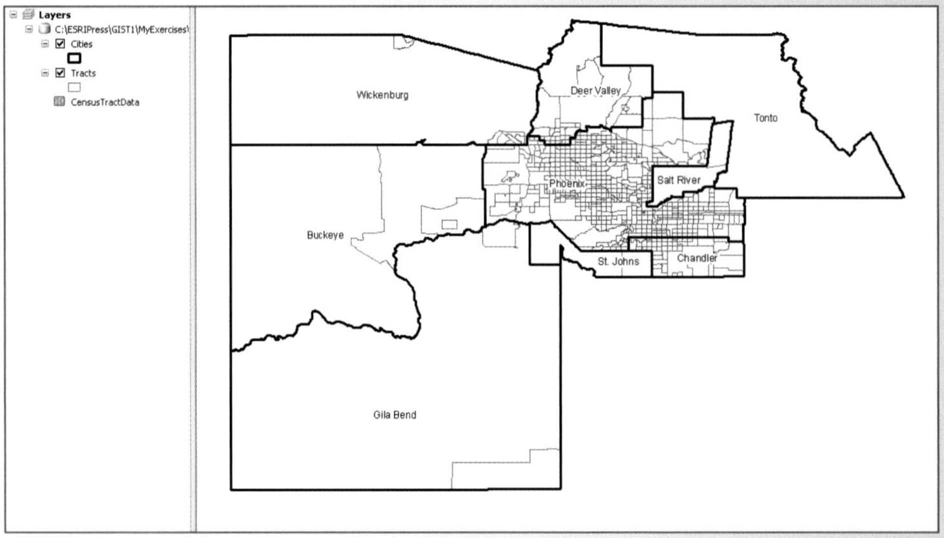

Tutorial 4-3

Modify an attribute table

Most of what gets processed or displayed in a GIS depends on attribute table values. There are many tasks, therefore, that you need to know how to perform on these tables, including modifying columns and values.

Delete unneeded columns

Many map layers have extra or unnecessary attributes that you can delete when tidying up.

1 **In the TOC, right-click Tracts and click Open Attribute Table.** The key identifier, or primary key, created by ArcGIS that you will retain is OBJECTID, which has sequence numbers for values. STFID is a candidate key, meaning that it, too, is a unique identifier—unique for every census tract in the United States. FIPSSTCO, however, is an extra identifier that you do not need.

2 **In the table, right-click the column heading for the FIPSSTCO column, click Delete Field, and click Yes.**

3 **Similarly delete the ID, TRT2000, and TractID fields.**

4 **Close the Tracts table.**

YOUR TURN

MCD2000 is a unique city identifier in the United States. Delete the following fields from the Cities layer: ID, County, SubMCD, and SubName. When finished, close the Cities table.

Modify a primary key

It is often necessary to join two tables to make a single table. For example, there are hundreds of census variables, so it is impractical to have all needed census variables for tracts stored in the tract polygon table. Instead, you select the variables you wish, download a corresponding table from the Census Bureau Web site, and join the table to the tract polygon table.

For tables to join, they must share unique identifiers or keys. The STFID column of the Tracts table and the GEO_ID column of the CensusTractData table are the corresponding unique identifiers for these tables. These attributes would match, except that GEO_ID has the extra characters "14000US" at the beginning of each value. Next, you will use a string function, Mid([GEO_ID], 8,11), that extracts an 11-character string from GEO_ID starting at position 8 and creates a new column in Attributes of CensusTractData to match STFID of Attributes of Tracts.

1 **Right-click Tracts in the TOC and click Open Attribute Table.** Note that STFID in this table has values such as 04013010100.

2 **Close the Tracts table, right-click CensusTractData in the TOC, and click Open.** GEO_ID in this table has values such as 14000US04013010100, with the extra seven beginning characters.

3 **In the CensusTractData table, click the drop-down arrow of the Table Options button ▾ and click Add Field.**

4 **In the Add Field dialog box, type STFID in the Name field, change the Type to Text and Length to 11, and click OK.**

5 **Scroll to the right in the CensusTractData table, right-click the STFID column heading, and click Field Calculator.**

6 **In the Field Calculator dialog box, change the Type from Number to String; double-click the Mid() function; and in the STFID= box, edit the Mid() function to Mid([GEO_ID], 8,11) and click OK.** That calculates values for STFID in this table such as 04013010100.

Calculate a new column

1 In the CensusTractData table, click the drop-down arrow of the Table Options button and click Add Field.

2 In the Add Field window, type **RNatWht** in the Name field, change the Type to Float, and click OK. The new column will contain the ratio of Native American per capita income to White per capita income. Wherever this ratio is greater than one, Native Americans earn more than Whites. Next, you must select only records where PCIncWht is greater than zero, because PCIncWht is the divisor for this ratio and will be used to calculate values for RNatWht. Anything divided by zero is undefined, so this case must be avoided.

3 In the CensusTractData table, click the drop-down arrow of the Table Options button and click Select by Attributes.

4 In the Select by Attributes dialog box, scroll down the list of fields, double-click PCINCWHT to add it to the lower Select panel, click the > symbol button, click Get Unique Values, and double-click 0 in the Unique Values list.

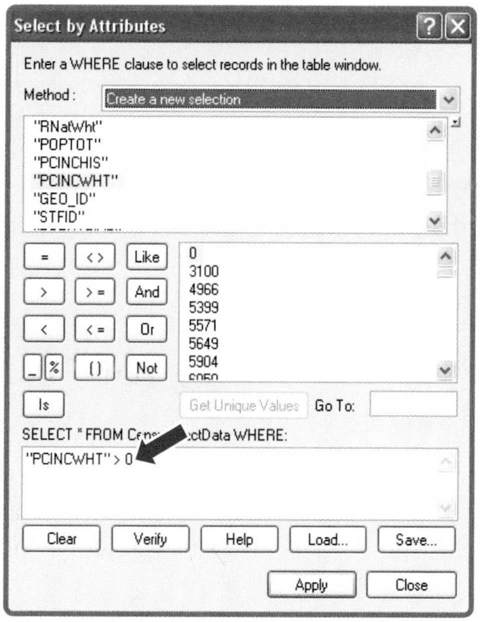

5 Click Apply, Close.

6 Right-click the RNatWht column heading and click Field Calculator.

7 In the Field Calculator window, click the Clear button, double-click PCINCNAT in the Fields panel, click the / button, double-click PCINCWHT in the Fields panel, and click OK.

8 In the CensusTractData table, click the drop-down arrow of the Table Options button and click Clear Selection.

9 Close the Attributes of Tracts table.

YOUR TURN

Repeat the previous steps to calculate a new column in the CensusTractData table called RHisWht, which is the ratio of PCINCHIS divided by PCINCWHT. This is the ratio of per capita income of Hispanics divided by the per capita income of whites. Close the table when finished.

Tutorial 4-4

Join tables

Often you will need to display data on your map that is not directly stored with a map layer. For example, you might obtain data from other departments in your organization, purchase commercially available data, or download data from the Internet. If this data is stored in a table such as an Excel or comma-separated-value table and has geocodes such as census tract numbers matching your tract map layer, you can import it into a file geodatabase and join it to your geographic features for display on your map. Next, you will join the CensusTractData table to the polygon Tracts feature class. The same steps work if your map layer is a shapefile or map layer in another format supported by ArcMap.

1 In the ArcMap table of contents, right-click the Tracts layer, click Joins and Relates, then click Join.

2 Make the selections shown in the graphic on the right.

3 Click OK, Yes.

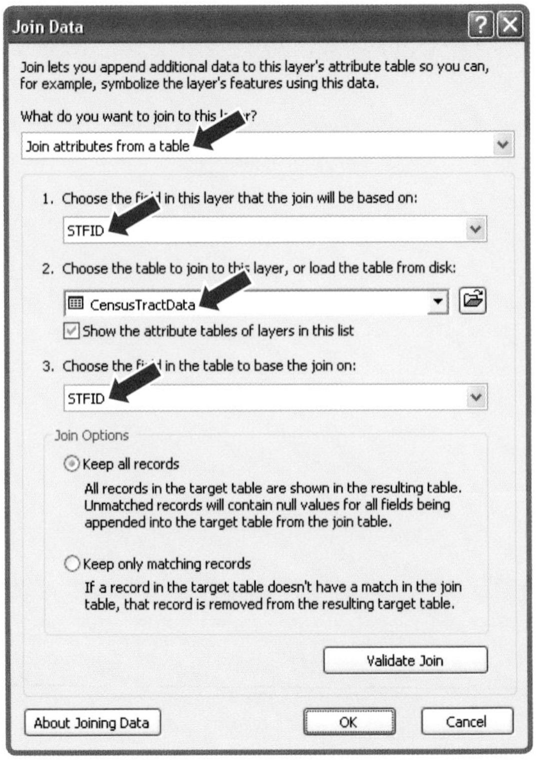

4 Right-click the Tracts layer in the table of contents, click Open Attribute Table, scroll to the right in the table, and verify that ArcMap joined the CensusTractData table to the Attributes of Tracts table.

Tracts

Shape_Length	Shape_Area	OBJECTID_1 *	OBJECTID	GEO_ID	POPTOT	POPWHITE	POPBLACK
3.397847	0.272108	1	1	14000US04013010100	4120	4071	0
0.432454	0.009786	2	2	14000US04013020201	829	29	7
0.80614	0.021225	3	3	14000US04013020202	6403	1212	20
0.060034	0.000225	4	4	14000US04013030302	4499	3711	67
0.048449	0.000123	5	5	14000US04013030303	5533	4308	104
0.064359	0.000258	6	6	14000US04013030304	6523	5822	161

5 Leave the Tracts table open.

4-1
4-2
4-3
4-4
4-5
4-6
A4-1
A4-2

Tutorial 4-5

Create centroid coordinates in a table

The centroid of a polygon is the point at which the polygon would balance on a pencil point if it were cut out of cardboard. Together, polygons and their centroids give you the ability to display two attributes of the same map layer, one as a choropleth map and the other as a size-graduated point marker map. ArcMap has an algorithm that calculates and adds centroid coordinates to your attribute table, thereby allowing you to create a new point layer.

Add x,y coordinates to a polygon attribute table

1 In the Tracts table, click the drop-down arrow of the Table Options button and click Add Field.

2 Type **X** as the Name, select Double for Type, and click OK.

3 Repeat steps 1 and 2 except call the new field **Y**.

4 Scroll to the right in the Tracts table; right-click the Tracts.X column heading; click Calculate Geometry, Yes; examine the Calculate Geometry window; and click OK. The attribute name is fully qualified to Tracts.X to indicate that of the joined set of tables in the display. ArcMap identifies the X attribute as being in the Tracts table.

5 Repeat step 4 except right-click Tracts.Y and select Y Coordinate of Centroid.

Tracts

PCINCHIS	PCINCWHT	STFID *	RHatWht	RHisWht	Tracts.X	Tracts.Y
13752	23969	04013020201	0.735742	0.573741	-111.664435	33.641264
7363	16897	04013020202	0.454519	0.435758	-111.793494	33.519152
14301	18324	04013030302	0.289456	0.780452	-112.107299	33.647254
6812	14373	04013030303	0.559521	0.473944	-112.091176	33.643751
18271	21426	04013030304	0.620041	0.852749	-112.073926	33.647386
12381	20416	04013030307	0.795846	0.606436	-112.091226	33.651027

Export a table

When you export joined tables as a table, you get all the attributes of the joined tables stored as one table permanently. Then there are several possible uses for the resultant table, one of which is to use it to make a new point layer based on the centroid coordinates.

1 In the Tracts table, click the drop-down arrow of the Table Options button ▾ and click Export.

2 In the Output table field of the Export Data window, click the browse button; change the Save as type to File Geodatabase tables; browse to \ESRIPress\GIST1\MyExercises\Chapter4\; double-click MaricopaCountyFiles.gdb; change Name to **TractCentroids**; and click Save, OK, and Yes. Open the table to see that it has all of the columns of both joined tables. Then close the table.

3 Close the Tracts attribute table.

Create a feature class from an XY table

1 On the main menu, click Windows, Catalog. This opens a version of ArcCatalog as a window in ArcMap, thus providing quick access to GIS utility programs.

2 Expand the \ESRIPress\GIST1\ folder connection to MyExercises, Chapter4, and contents of the MaricopaCountyFiles.gdb file geodatabase.

3 Right-click TractCentoids, click Create FeatureClass, and click From XY Table.

4 In the Create Feature Class from XY Table window, click the Coordinate System of Input Coordinates button, click Import, double-click Tracts in MaricopaCountyFiles.gdb, and click OK. The coordinate system's geographic spherical coordinates are the same as of the Tracts layer, so the simplest option is to import the system specification from Tracts.

5 Click the browse button for Output, change the Save As type to File and Personal Geodatabase feature classes, browse to MaricopaCountyFiles.gdb, change the Name to **CensusTractCentroids**, click Save, and click OK.

6 Close the Catalog window in ArcMap.

7 Click the Add Data button, browse to MaricopaCountyFiles.gdb, and double-click CensusTractCentroids.

8 Open the Cities attribute table, click the row selector for Phoenix to select that record and polygon on the map, and close the table.

9 **Right-click Cities in the TOC, click Selection, and Zoom to Selected Features.** Now you can get a better look at the centroids point layer you just created.

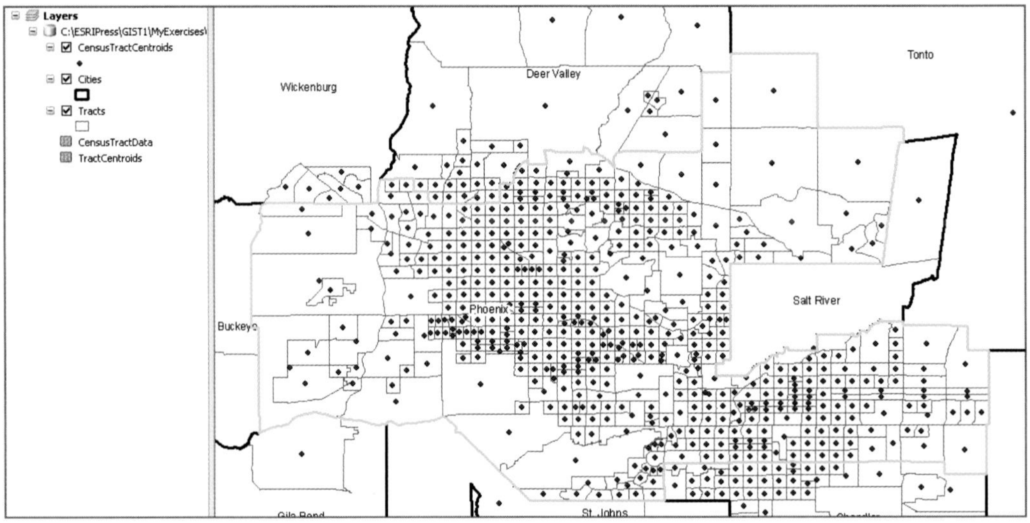

Symbolize a choropleth and centroid map

Here you will symbolize a map using both the tract polygons and centroids. Let's see how per capita income compares with percentage of total population that is Hispanic.

1 **Right-click the Tracts layer in the TOC and Properties, and click the Symbology tab.**

2 **In the Show panel, click Quantities and Graduated Colors. Under Fields, select POPHISP for the Value field, change Normalization to POPTOT, click Classify, change Method to Quantile, and click OK.**

3 **In the Symbology tab, click the Label column heading to the right of the Symbol and Range headings, click Format Labels, click the Numeric Category, change the number of decimal places to 2, and click OK twice.**

4 **Right-click CensusTractCentroids in the TOC, click Properties, and click the Symbology tab.**

5 **In the Show panel, click Quantities and Graduated Symbols. Under Fields, change Value to PCINCTOT.**

6 **Change the Symbolize Size range to 2 to 10.**

7 **Click the Template button, choose Circle 2, and click OK.**

8 **In the Classification panel, change the number of classes to 4, click Classify, change Method to Quantile, and click OK.**

9 **In the Symbology tab, click the Label column heading, click Format Labels, click the Numeric Category, change the number of decimal places to 0, click OK twice, and save your map document.** Now you can plainly see that Phoenix has areas with concentrations of Hispanic population, and those areas tend to be low income.

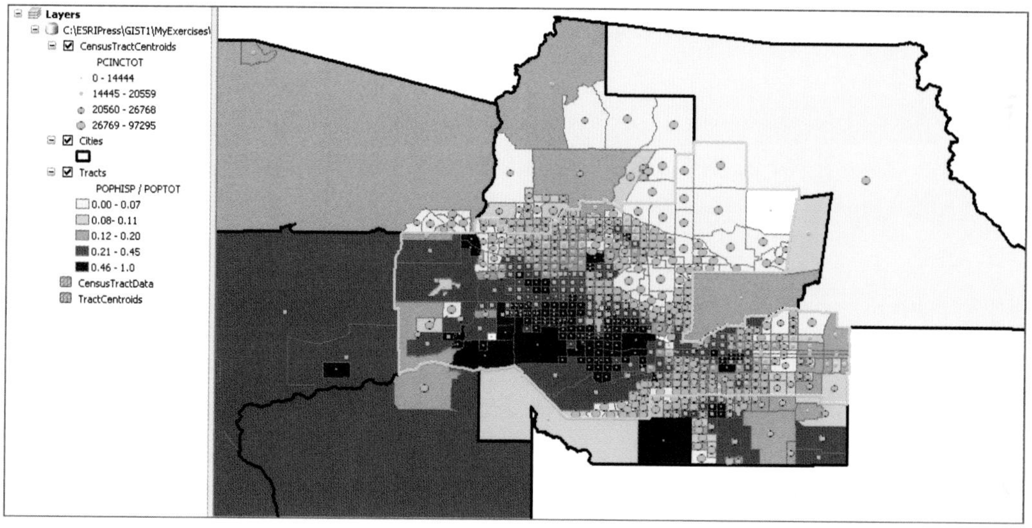

Tutorial 4-6

Aggregate data

The next part of this tutorial has you count, or aggregate, points within police administrative areas called car beats and then display the results on a map. A car beat is the patrol area of a single police car. The end result will be a choropleth map of car beats displaying the number of crime-prone businesses of a certain kind: eating and drinking places. There are preliminary tasks before aggregating the data. First you need to assign each business the identifier for the car beat in which it resides. This requires the unique GIS functionality of a spatial join, using the polygon map layer for car beats as the input data. Then you have to join a code table to the businesses feature class and use the code descriptions to select the subset of businesses to aggregate. Finally, there are aggregation steps.

Spatially join point and polygon layers

1 In ArcMap, open Tutorial4-6.mxd from the \ESRIPress\GIST1\Maps\ folder. The map that opens displays police car beats in Rochester, New York, as polygons, and all businesses as points.

2 Right-click the Businesses layer in the TOC, click Joins and Relates, and click Join.

3 Type or make the selections as shown in the image.

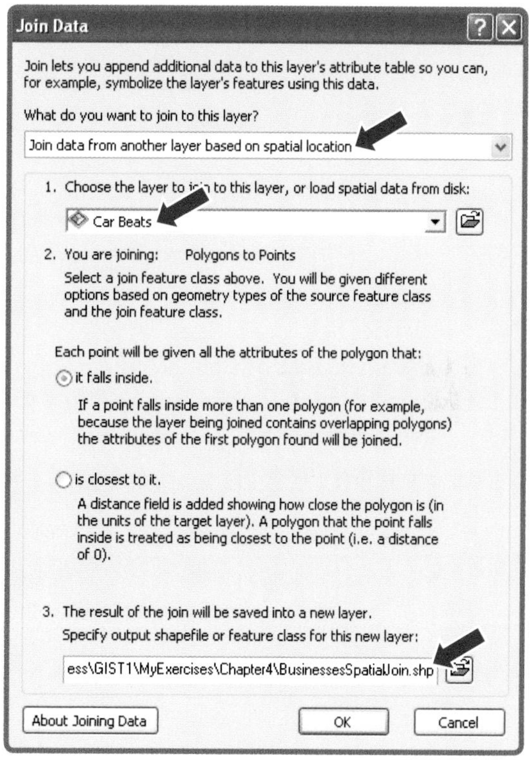

4 Click OK. ArcMap creates the new BusinessesSpatialJoin shapefile and adds it to your map document.

5 Right-click Businesses in the TOC and click Remove.

Examine tables to join

The next task is to join a code table to the BusinessesSpatialJoin shapefile so that you have its code descriptions available for processing.

1 **Right-click BusinessesSpatialJoin in the TOC, click Open Attribute Table, and drag the right side of the SIC column header to the right to make the column wider.** The SIC attribute is the Standard Industrial Code, a U.S. Census Bureau classification for private-sector businesses. Note that SIC is a text value, because the values are left aligned. Note also in the right side of the table that each business has BEAT as an attribute from the spatial join. Later, BEAT will be the basis for counting businesses by car beat.

2 **In the TOC, click the List by Source button, right-click the SIC table in the TOC, and click Open.** This is a code table with a definition for each SICCODE value, also with text data type for the code. The code values are hierarchical with the highest level having two digits (01=Agriculture), three digits being the next level (011=Cash Grains), and the finest level at four digits (0111=Wheat). Next, you will join this table to the Businesses table, so that you can see the nature of each business, but of course only matching four-digit codes will join. The join is called a one-to-many join because a single SIC code is used many times for each business of that type.

3 Close both tables.

Join a code table to a feature attribute table

1 Right-click the BusinessesSpatialJoin layer in the table of contents, click Joins and Relates, and click Join.

2 Type or make the selections shown in the graphic, but do not click OK.

3 Click the Validate Join button. You get a report that 9,253 of 9,325 business records successfully join.

4 Click Close in the Join Validation window and click OK.

5 Open the BusinessesSpatialJoin attribute table, scroll to the right and see the joined SIC code descriptions, then close the table.

6 Save your map document.

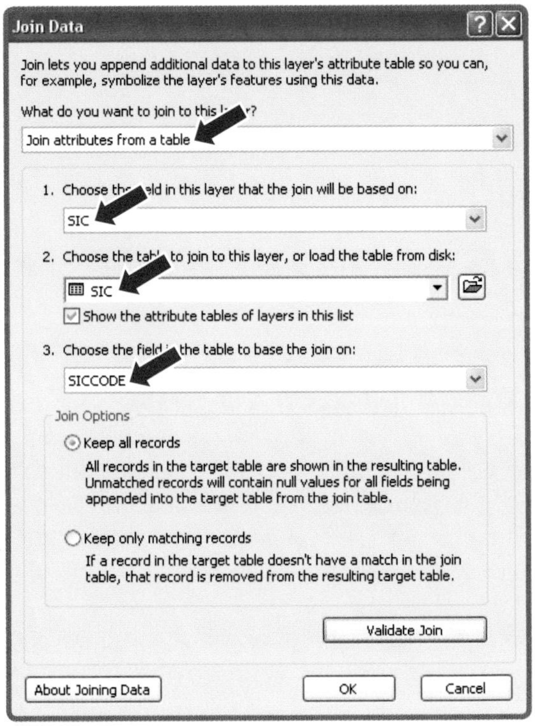

Select and export a subset of a feature class

The subset of eating and drinking places desired have SIC code values 5812 and 5813.

1 On the main menu, click Selection, Select by Attributes.

2 In the Select by Attributes dialog box, select BusinessesSpatialJoin from the Layer drop-down list.

3 Scroll down in the list of fields, double-click SIC.SICCODE, click the = button, click the Get Unique Values button, scroll down and double-click '5812', click the Or button, double-click SIC.SICCODE, click the = button, and double-click '5813' in the Unique Values list. Your "Where" expression must read SIC.SICCODE = '5812' OR SIC.SICCODE = '5813'. If not, you can edit it directly in the expression box.

4 Click the Verify button. You should get a message that the expression was successfully verified. If not, look for an error, fix it by editing the expression directly, and verify it until you get it correct.

5 Click OK, Apply, and Close. ArcMap highlights all of the eating and drinking places with the selection point marker. At the lower right of the ArcMap window it should say that 457 features are selected.

6 Open the BusinessesSpatialJoin attribute table, click the Show selected records button, scroll to the right and verify that the selected records are for eating or drinking places, and then close the table.

7 Right-click BusinessesSpatialJoin in the TOC and click Data, Export Data.

8 In the Export Data window, click the browse button; change the Save as type to Shapefile; browse to \ESRIPress\GIST1\MyExercises\Chapter4\; type **EatingDrinkingPlaces.shp** as the Name; click Save, OK, and Yes.

9 Click Selection on the main menu, click Clear Selected Features, and then turn BusinessesSpatialJoin off in the TOC while leaving EatingDrinkingPlaces on.

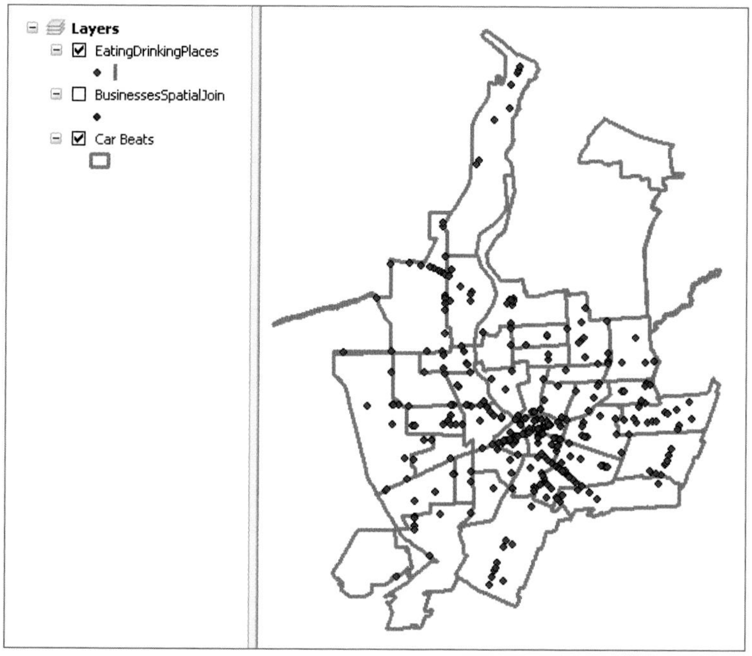

YOUR TURN

Extract a second set of businesses, those having to do with automobiles or motor vehicles. You can use the following query expression: SIC.SICDESCR LIKE '%auto%' OR SIC.SICDESCR LIKE '%Motor vehicle%'. This expression uses wild card characters, "%" in ArcMap, that stand for zero, one, or more characters to identify desired records. So '%auto%' will retrieve values such as 'General automotive repair shops'. The text values in expressions are case sensitive, which is why the expression above capitalizes Motor. Save the output as **\ESRIPress\GIST1\MyExercises\Chapter4\AutoMotorVehiclePlaces.shp**.

Count points by polygon ID

Now you can count the number of eating and drinking places by car beat.

1 **Open the attribute table of EatingDrinkingPlaces, scroll to the right, right-click the column heading of the BEAT column, and click Summarize.**

2 **For item 3 in the Summarize dialog box, specify the output as \ESRIPress\GIST1\MyExercises\Chapter4\EatingDrinkingCount.dbf; click OK; and click Yes.**

3 **Close the EatingDrinkingPlaces table.**

4 **At the top of the TOC, check that the List By Source button is clicked.**

5 **Right-click the EatingDrinkingCount table in the table of contents and click Open.** The Count_BEAT field contains the total number of retail business points in each car beat polygon.

OID	BEAT	Count_BEAT
0		2
1	224	20
2	231	5
3	232	16
4	233	4
5	234	28
6	236	3

6 **Close the table.**

Join a count table to a polygon map

1 **Right-click Car Beats in the TOC, click Joins and Relates, and click Join.**

2 **Type or make the selections shown as shown in the image.**

3 **Click OK.**

4 **Open the Car Beat attribute table, scroll to the right, verify that each beat has a Count_Beat value, and close the table.**

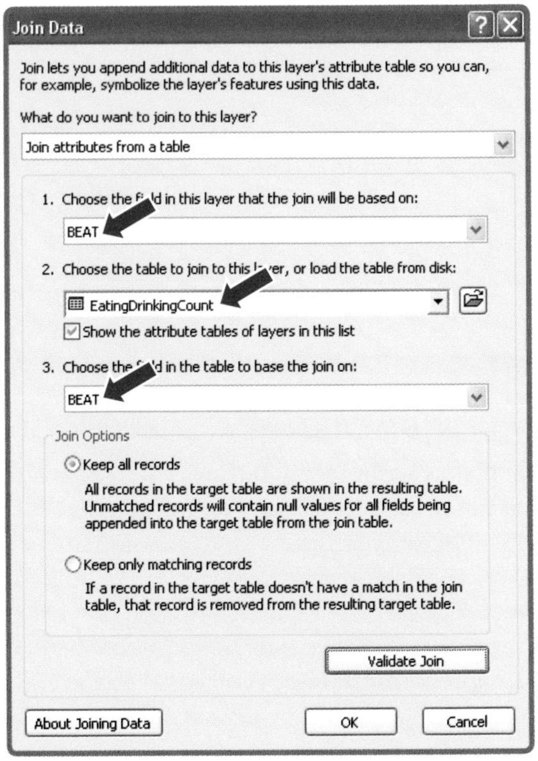

Symbolize the choropleth map

With counts of eating and drinking places now joined to the car beats polygon layer, you are ready to create a car beat choropleth map. The resulting map will provide a good means for scanning the entire city for areas with high concentrations of crime-prone establishments.

1 Right-click Car Beats in the table of contents. Click Properties, the Labels tab, and the Label Features in this layer check box.

2 Click the General tab and change the Layer Name to **Number of Eating and Drinking Places**.

3 Click the Symbology tab, Quantities, and Graduated colors.

4 Change the Value field to Count_BEAT, choose a monochromatic color ramp, and click Classify.

5 In the Classification dialog box, choose 7 classes, set the Method drop-down list to Quantile, and click OK twice.

6 Turn off all layers except Number of Eating and Drinking Places.

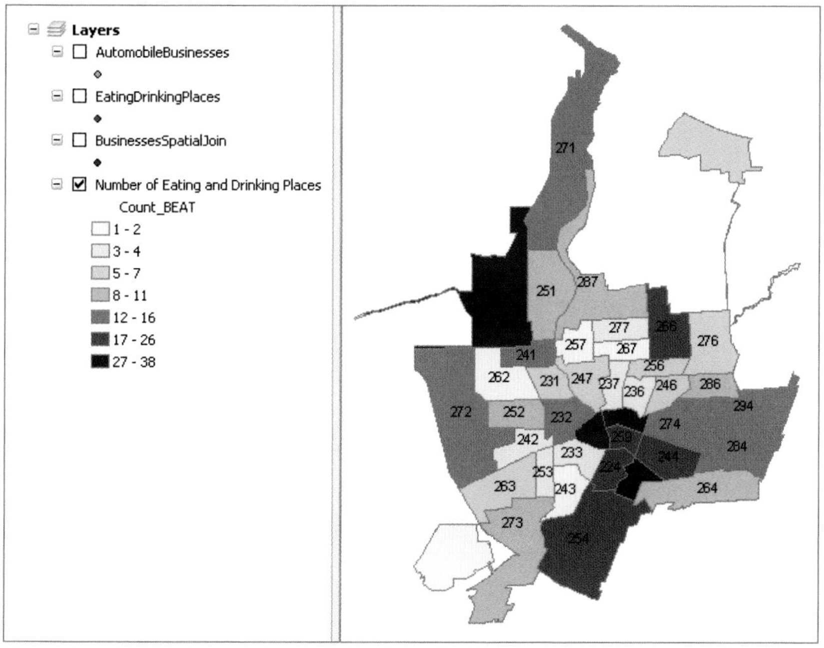

7 Save your map document and exit ArcMap.

4-1
4-2
4-3
4-4
4-5
4-6
A4-1
A4-2

Assignment 4-1

Compare municipal tax compositions

Public finance experts prefer that municipalities obtain a larger share of taxes from earned income rather than real estate (or property) taxes. They reason that real estate taxes are unfair because they place a larger burden on low- and fixed-income families. For example, property values and real estate taxes increase over time, but retired people generally have fixed incomes. So over time, retired people use a larger proportion of their income for real estate taxes. It is fairer to tax earned income—whether it is fixed or increasing over time.

As a guide for municipalities to be fairer to low- and fixed-income inhabitants, create two maps of municipalities in Allegheny County, Pennsylvania—one with percentage of tax collections that are from real estate and the other with percentage of tax collections that are from earned income.

Start with the following:

Revenue table:

- \ESRIPress\GIST1\Data\DataFiles\Revenue.xls—Excel table of 2004 municipal revenue data from the Pennsylvania Department of Economic Development. Attributes of this table include:

  NAME—municipality name and primary key

  SHORTNAME—municipality name used for labeling

  TAX—total tax revenues

  REAL—tax revenues from real estate

  INCOMETAX—tax revenues from earned income

Allegheny County map layers:

- \ESRIPress\GIST1\Data\AlleghenyCounty.gdb\Munic—polygon layer of Allegheny County municipalities

  NAME—municipality name and primary key

- \ESRIPress\GIST1\Data\AlleghenyCounty.gdb\Rivers—polygon layer of three major rivers

Create a file geodatabase

Create a new file geodatabase stored as **\ESRIPress\GIST1\MyAssignments\Chapter4\Assignment4-1YourName.gdb**. Import the above table and map layers into the geodatabase.

Create a map document

Create a map document called **\ESRIPress\GIST1\MyAssignments\Chapter4\Assignment4-1 YourName.mxd** that has two data frames. Both data frames need the two map layers and table from the file geodatabase with the table joined to the municipality map. The data frames differ only by the attribute used to symbolize a choropleth map of municipalities:

- One data frame displays REAL normalized by TAX
- The other data frame displays INCOMETAX normalized by TAX

Use quantiles with five categories for a numerical scale. Create a map layout with portrait page orientation and the two data frames, two legends, title, and other map elements that you choose. Label the municipalities. Use guidelines and design the layout carefully. Quite often, maps need to be placed in reports and other documents, so you must export them as graphics. Export your layout as a JPEG image, **\ESRIPress\GIST1\MyAssignments\Chapter4\Assignment4-1YourName.jpg**. Insert the image in a Word document, **\ESRIPress\GIST1\MyAssignments\Chapter4\Assignment4-1YourName.doc**, saved in the same folder. Include a brief description of patterns found in the maps.

Hints

Start with a single data frame. Add the table and map layers to the data frame, and join the table to the municipality map layer. Label the municipalities. Then copy and paste the data frame (right-click it) to create the second data frame.

You can edit map legends by converting them to graphics. For a legend in layout view, do the following:

- Zoom to a legend by right-clicking it and clicking Zoom to Selected Elements.
- Right-click the legend and click Convert to Graphics.
- Right-click the legend and click Ungroup. You can further ungroup paint chips and labels in the legend, if needed, to edit labels.
- For example, double-click a text box, such as one that is too wide. Place the insertion point where you would like to split the line into two, press the Enter key, and click OK.
- For another example, click a text box that you wish to delete and press the delete key.
- When finished editing, select all elements of a legend, right-click them and click Group to turn the legend back into a single graphic.

Note: You will find patterns in the two maps. To help you with interpretation, you should know that the poorer municipalities are along the rivers, in the old industrial parts of the county. The wealthier, suburban municipalities are in the northern and southern parts of the county. Does it look like taxation is fair?

WHAT TO TURN IN

If your work is to be graded, turn in the following files:

File geodatabase: \ESRIPress\GIST1\MyAssignments\Chapter4\Assignment4-1YourName.gdb

ArcMap document: \ESRIPress\GIST1\MyAssignments\Chapter4\Assignment4-1YourName.mxd

Word document: \ESRIPress\GIST1\MyAssignments\Chapter4\Assignment4-1YourName.doc

Image file: \ESRIPress\GIST1\MyAssignments\Chapter4\Assignment4-1YourName.jpg

If instructed to do so, instead of the above individual files, turn in a compressed file, **Assignment4-1YourName.zip**, with all files included. Do not include path information in the compressed file.

Assignment 4-2

Compare youth population and total school enrollment

In an earlier exercise, you studied the schools in the city of Pittsburgh by enrollment using a point feature map. Another way to study the same data is to spatially join the school points to a polygon layer (e.g., census tracts), and then sum the number of students in each polygon. After a few more steps, the end result is a choropleth map symbolizing census tracts with the newly summarized school data.

Start with the following:

- \ESRIPress\GIST1\Data\DataFiles\PghTracts.shp—polygon layer of Pittsburgh census tracts, 2000
- \ESRIPress\GIST1\Data\DataFiles\Schools.shp—point layer of all schools with student enrollment data

Create a file geodatabase

In ArcCatalog, create a file geodatabase **\ESRIPress\GIST1\MyAssignments\Chapter4\Assignment4-2YourName.gdb** with the above layers imported.

Create a map document

In ArcMap, create a map document saved as **\ESRIPress\GIST1\MyAssignments\Chapter4\Assignment4-2YourName.mxd** with a copy of each of the above layers added from the file geodatabase. Carry out a spatial overlay of schools with tracts using the tip provided below, creating a new layer called TractSchoolJoin in Assignment4-2YourName.gdb. Symbolize this layer as a choropleth map using five classes and quantiles with SUM_ENROLL created in the spatial join process. Symbolize the original tract layer with Quantities, Graduated symbols for AGE_5_17, also with five classes and quantiles. Create an 8.5-by-11-inch landscape layout with map, legend, and heading. Export the layout to a JPEG image file, **\ESRIPress\GIST1\MyAssignments\Chapter4\Assignment4-2.jpg**, with resolution 150 dpi, and insert it into a Word document, **\ESRIPress\GIST1\MyAssignments\Chapter4\Assignment4-2YourName.doc**, in the same folder.

Spatial join tip

Below is a shortcut for spatially joining points to polygons that automatically counts and summarizes data.

1. Right-click the PghTracts layer, click Joins and Relates, and click Join.
2. Spatially join the Schools point layer to the PghTracts layer.
3. Click Sum in the join dialog box. This will sum all of the numerical fields in the Schools point layer, including ENROLL.

4. Save the new layer in your file geodatabase as **TractSchoolJoin**.
5. Open the attribute table of the new TractSchoolJoin layer and examine the fields that were created by joining the points to the polygons. Of particular interest will be the fields "Count_", which is the number of schools (points) in each census polygon, and "Sum_ENROLL", which is the sum of students enrolled in each school. Some fields will be <null> because those census tracts have no schools or student enrollment.

STUDY QUESTION

1. Does it appear that schools are well located relative to the youth population?

WHAT TO TURN IN

If your work is to be graded, turn in the following files:

File geodatabase: \ESRIPress\GIST1\MyAssignments\Chapter4\Assignment4-2YourName.gdb

ArcMap document: \ESRIPress\GIST1\MyAssignments\Chapter4\Assignment4-2YourName.mxd

Word document: \ESRIPress\GIST1\MyAssignments\Chapter4\Assignment4-2YourName.doc

Image file: \ESRIPress\GIST1\MyAssignments\Chapter4\Assignment4-2YourName.jpg

If instructed to do so, instead of the above individual files, turn in a compressed file, **Assignment4-2YourName.zip**, with all files included. Do not include path information in the compressed file.

5

Spatial data

There are vast collections of spatial data available from government agencies. You can readily download much of this data for free on the Internet, but before doing so it is helpful to get some background and see some of the major forms of this data. Spatial data is complex, with both vector and raster formats available in many file formats and with several attending characteristics such as coordinate system, feature or cell attribute properties, and intended map scale for application. This chapter provides a hands-on introduction to spatial data and then has you download or use samples from some of the major governmental suppliers.

Learning objectives

- *Examine metadata*
- *Work with map projections*
- *Learn about vector data formats*
- *Explore sources of vector maps*
- *Download and process tabular data*
- *Explore sources of raster maps*

Tutorial 5-1

Examine metadata

Spatial data needs much documentation for interpretation and proper use. "Metadata" is the term for such documentation. It describes the context, content, and structure of GIS data. In ArcGIS Desktop, data providers use ArcCatalog to create and view metadata.

Open a map document

1 **In ArcMap, open Tutorial5-1.mxd from the \ESRIPress\GIST1\Maps\ folder.** You can see that the Pennsylvania Counties boundary lines are crudely drawn compared to Allegheny County Municipalities. The metadata for Pennsylvania Counties explains why.

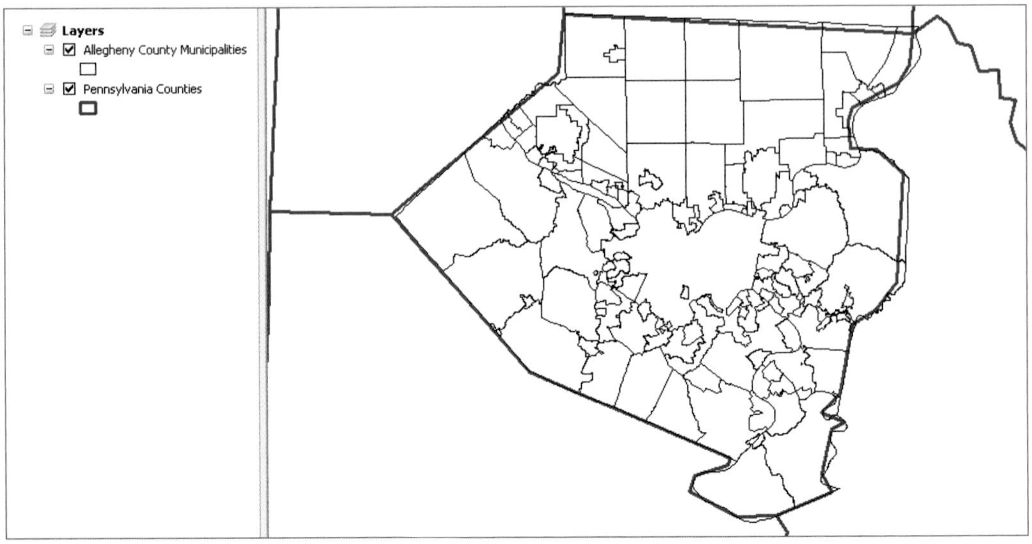

2 **Zoom in to the northwestern boundary of Allegheny County for a closer look at the differences between the resolution of the two map layers.**

Open a metadata file

1 **Open ArcCatalog. Locate and select the Pennsylvania Counties layer in the Catalog tree. Click the Description tab. Go to Customize and select ArcMap Options, Metadata tab. Change the Metadata style to North American Profile of ISO191152003.**

2 **Click FGDC Metadata and the Data Quality link. Read the Positional Accuracy entry.** You learn that the source map was small scale, 1:25M. Features will be crude when zoomed to the scale of Tutorial5-1.mxd. Moreover, the data providers simplified the map's detail through generalization, which removes vertices while preserving shape. This decreases map size and increases processing speed. So Pennsylvania Counties is not appropriate for use at the scale of individual counties as in the current map document.

POSITIONAL ACCURACY
HORIZONTAL POSITIONAL ACCURACY
HORIZONTAL POSITIONAL ACCURACY REPORT
The geospatial part of this data set was extracted from the ArcUSA 1:25M database, then generalized. Generalizing reduces positional accuracy and the tolerance was not recorded so the positional accuracy exceeds that of the ArcUSA data set. The positional accuracy of the ArcUSA 1:25M data set is 1792 meters which is based on generalizing-to a tolerance of 500 meters-the 1:2,000,000-scale USGS Digital Line Graph (DLG) source data.

3 **Scroll to the bottom of the Metadata window and click Entities and Attributes.** Scrolling through this section, you will see metadata about each attribute.

ATTRIBUTE
ATTRIBUTE LABEL FIPS
ATTRIBUTE DEFINITION
The combined state and county FIPS code. County FIPS codes begin with 001 for each state—use the combined code (five-digit number) to uniquely identify a county in the country.
ATTRIBUTE DEFINITION SOURCE Department of Commerce, National Institute of Standards and Technology
ATTRIBUTE DOMAIN VALUES
CODESET DOMAIN
CODESET NAME Federal Information Processing Standards
CODESET SOURCE National Institute of Standards and Technology

ATTRIBUTE
ATTRIBUTE LABEL POP2000
ATTRIBUTE DEFINITION
The 2000 population of the county.
ATTRIBUTE DEFINITION SOURCE Department of Commerce, Census Bureau
ATTRIBUTE DOMAIN VALUES
UNREPRESENTABLE DOMAIN
Populations for the features.

4 **Close the Metadata window.**

YOUR TURN

Explore some of the other metadata for Pennsylvania Counties, including spatial reference information on the coordinate system of the map layer. This is relevant to the next topic of this chapter. You will see that the map layer has geographic coordinates.

Tutorial 5-2

Work with map projections

There are two types of coordinate systems—geographic and projected. Geographic coordinate systems use latitude and longitude coordinates for locations on the surface of a sphere while projected coordinate systems use a mathematical conversion to transform latitude and longitude coordinates to a flat surface.

Set world projections

Because the earth is nearly spherical and maps are flat, GIS applications require that a mathematical formulation be applied to the earth to represent it on a flat surface. This is called a map projection, and it causes distortions in some combination of distance, area, shape, or direction. ArcMap has more than 100 projections from which you may choose. Typically, though, only a few projections are suitable for your data.

1 **In ArcMap, open Tutorial5-2.mxd from the \ESRIPress\GIST1\Maps\ folder.**

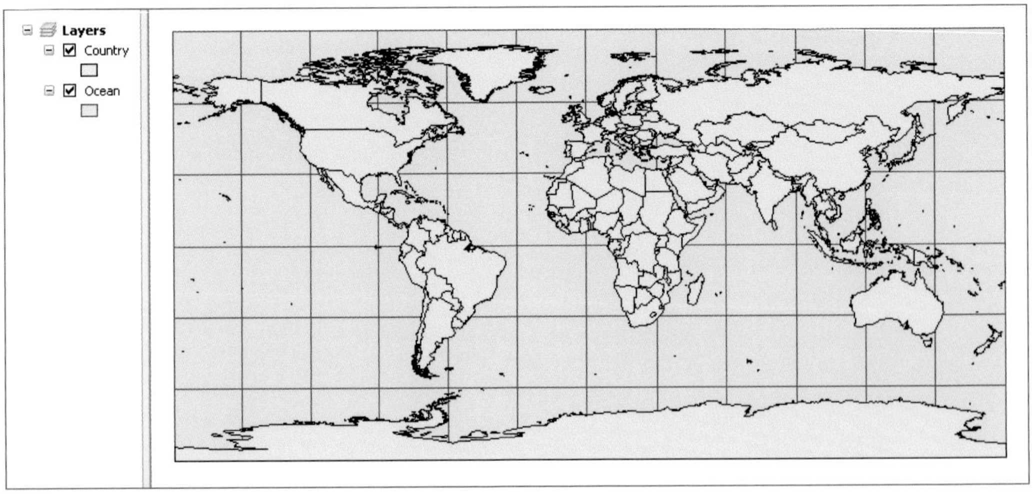

2 **Place your cursor over the westernmost point of Africa and read the coordinates on the bottom of the ArcMap window (approximately –16.6, 21.6 decimal degrees).** The map and data frame are in geographic coordinates, decimal degrees, which are angles of rotation of earth's radius from the prime meridian on the equator. These coordinates are not intended for viewing on a flat surface such as your computer screen, so there are huge distortions; for example, the north and south pole points are the horizontal lines bounding the top and bottom of the map. Instead of geographic coordinates, you should use one of the projected coordinate systems appropriate for viewing flat maps of the world. ArcMap has several and can easily project the map on the fly.

Change the map's projection to Mercator

1. **Right-click the Layers data frame, click Properties, then click the Coordinate System tab.**

2. **In the Select a coordinate system panel, expand the Predefined folder, Projected Coordinate Systems folder, and the World folder.**

3. **Scroll down the coordinate systems, click Mercator (world), and click OK.**

4. **Zoom to full extent.** The purpose of the Mercator projection is for navigation because straight lines on the projection are accurate compass bearings. This projection greatly distorts areas near the polar regions and distorts distances along all lines except the equator. The Mercator projection is a conformal projection, meaning that it preserves small shapes and angular relationships.

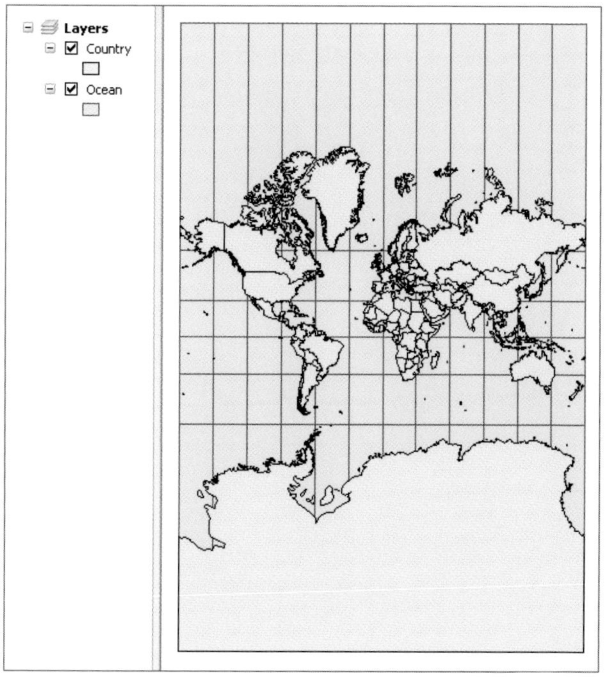

YOUR TURN

Repeat the four steps of the previous exercise, but this time select the Hammer-Aitoff projection in the third step. This projection is nearly the opposite of the Mercator. The Hammer-Aitoff is good for use on a world map, being an equal-area projection that preserves area. However, it distorts direction and distance. Repeat the steps again, this time choosing the Robinson projection. This projection minimizes distortions of many kinds, striking a balance between conformal and equal-area projections. Do not save changes to the map document.

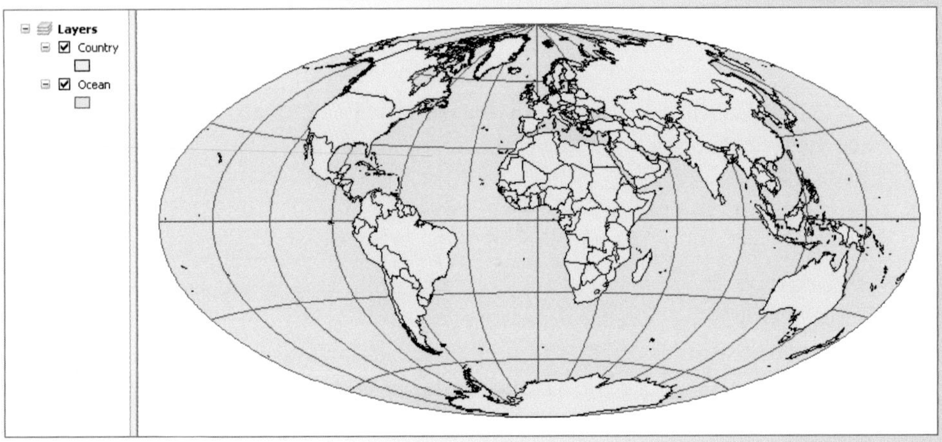

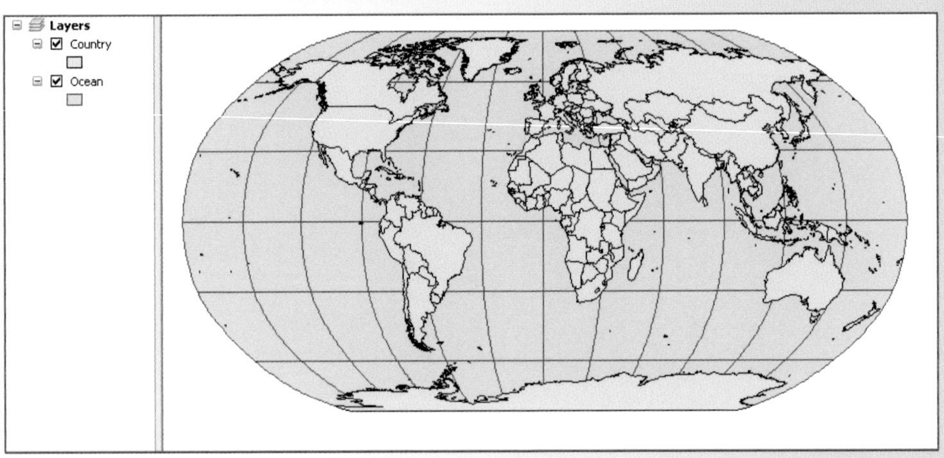

Set projections of the United States

Next, you will get some experience with projections commonly used for maps of the continental United States. Some projections are standard for organizations. For example, Albers equal area is the standard projection of both the U.S. Geological Survey and the U.S. Census Bureau.

1 **In ArcMap, open Tutorial5-3.mxd from the \ESRIPress\GIST1\Maps\ folder.** Initially, the map display is in geographic coordinates.

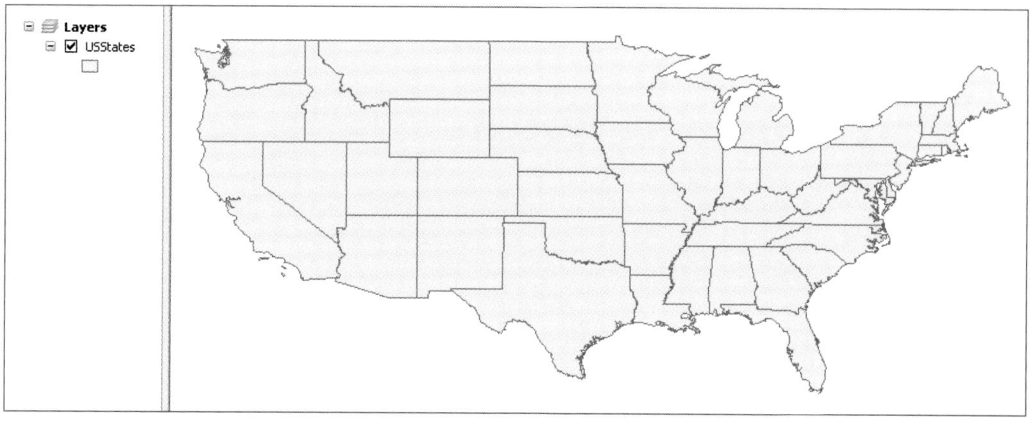

2 **Right-click the Layers data frame, click Properties, then click the Coordinate System tab.**

3 **In the Select a coordinate system panel, expand Predefined, Projected Coordinate Systems, Continental, North America.**

4 **Click North America Albers Equal Area Conic, then click OK.**

5 **Zoom to full extent.**

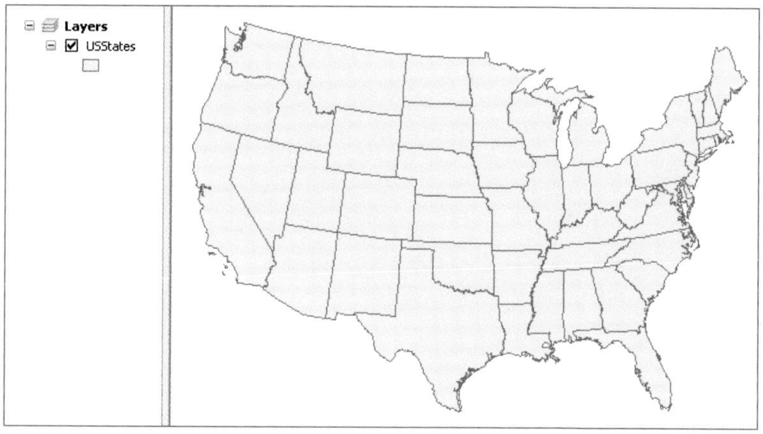

YOUR TURN

Experiment by applying a few other projections to the U.S. map such as North America Equidistant. As long as you stay in the correct group—Continental, North America—all of the projections look similar. The conclusion is that the smaller the part of the world that you need to project, the less distortion. There remains much distortion at the scale of a continent, but much less so than for the entire world. By the time you get to a part of a state, such as Allegheny County, practically no distortion is left, as you will see next. Do not save changes to your map document.

State plane coordinate system

The state plane coordinate system is a series of projections. It divides the 50 U.S. states, Puerto Rico, and the U.S. Virgin Islands into more than 124 numbered sections referred to as zones, each with its own finely tuned projection. Used mostly by local government agencies such as counties, municipalities, and cities, the state plane coordinate system is for large-scale mapping in the United States. The U.S. Coast and Geodetic Survey developed it in the 1930s to provide a common reference system for surveyors and mapmakers. The first step in using a state plane projection is to look up the correct zone for your area.

1 **Start your Web browser, go to `www.ngs.noaa.gov/TOOLS/spc.shtml`, and click the Find Zone link.**

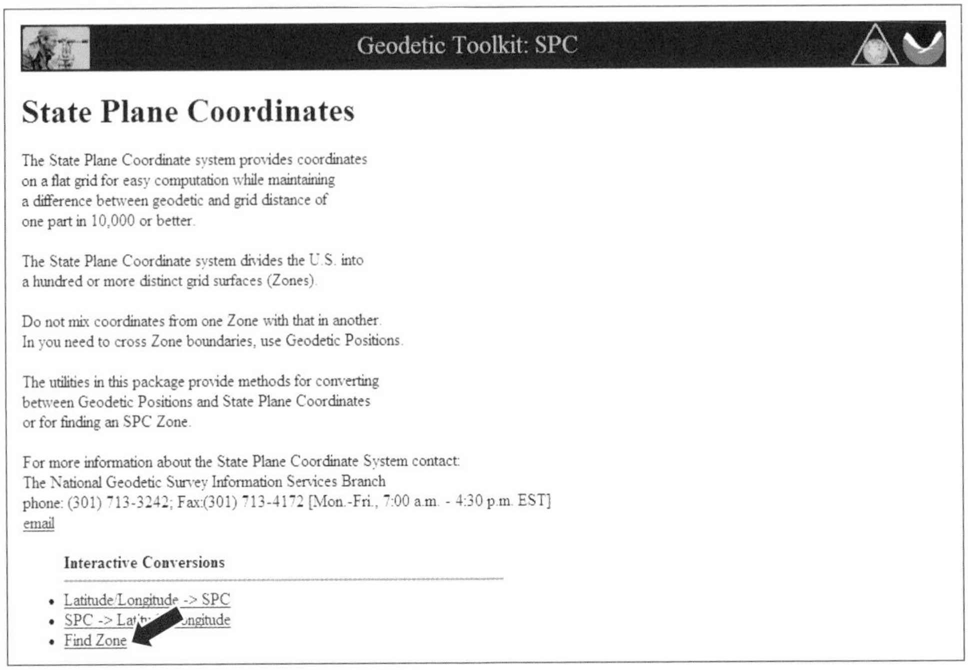

2 In the resulting Web page, with the By County option button clicked, click Begin.

3 Select Pennsylvania, click Submit, then click Allegheny and Submit. Pennsylvania's Allegheny County is in State Plane Zone 3702.

4 Close your browser.

Add a state plane projected layer to a map document

As a default, the first map layer that you add to a map document sets the projection for the data frame. If all of your map layers have projection data included, you should have no problem combining maps with different projections. ArcMap will reproject all map layers to the data frame's projection on the fly. First, you will add a layer with a state plane projection and then a layer with geographic coordinates.

1 In a new empty map, click the Add Data button, browse to \ESRIPress\GIST1\Data\AlleghenyCounty.gdb, and add Munic to the map. The coordinates appearing in the lower right corner of the display now appear in state plane units (feet). The origin of these coordinates (0,0) is at the lower left corner of Pennsylvania.

2 Change the Munic symbology to a hollow fill, black outline.

3 Click the Add Data button, browse to \ESRIPress\GIST1\Data\AlleghenyCounty.gdb, and add Tracts to the map.

4 Change the Tracts symbology to a hollow fill, light gray (20%) outline.

5 Make sure the List By Drawing Order button is selected in the TOC. Move the Munic layer to the top of the TOC.

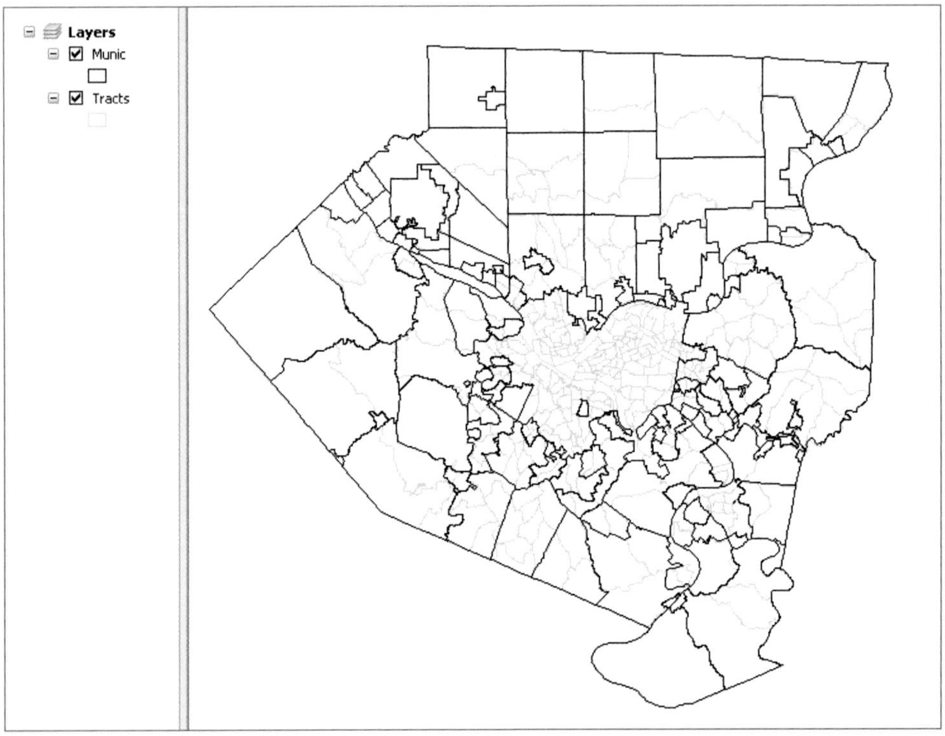

6 Right-click Munic in the TOC, click Properties and the Source tab, and note the coordinate system in the Data Source panel. This layer has state plane coordinates.

7 **Repeat step 6 except for Tracts.** This layer has geographic coordinates. Both layers appear in the data frame using state plane coordinates because you added Munic first and it has state plane coordinates. Notice that map coordinates are in feet, which correspond to state plane.

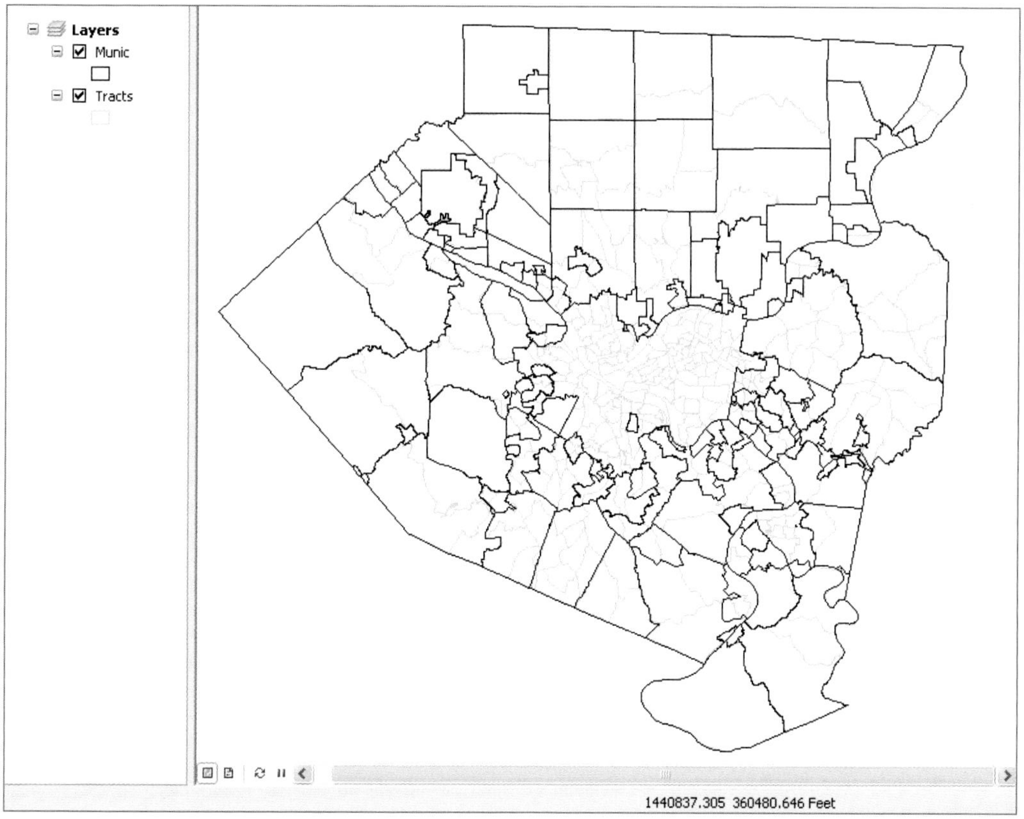

8 **Leave your map document open.**

UTM coordinate system

The U.S. military developed the universal transverse Mercator (UTM) coordinate system in the late 1940s. It includes 60 longitudinal zones defined by meridians that are 6° wide. ArcGIS has UTM projections available for the northern and southern hemisphere of each zone. These projections, like state plane, are good for areas about the size of a state (or smaller) and have the advantage of covering the entire world.

1 **Start your Web browser, go to `www.dmap.co.uk/utmworld.htm`, and determine the zone for western Pennsylvania.** You should find that western Pennsylvania is in zone 17 north.

2 **Close your browser.**

5-1 5-2 5-3 5-4 5-5 5-6 A5-1 A5-2

3 **In the TOC, right-click the Layers data frame, and click Properties.**

4 **Click the Coordinate System tab. Expand Predefined, Projected Coordinate Systems, UTM, and NAD 1983. Click NAD 1983 UTM Zone 17N and OK.** The coordinate system and map appearance change accordingly. Notice that the coordinates are now in meters. UTM is a metric system and thus uses meters. If your display remains in feet, open the data frame properties and set the display units to meters.

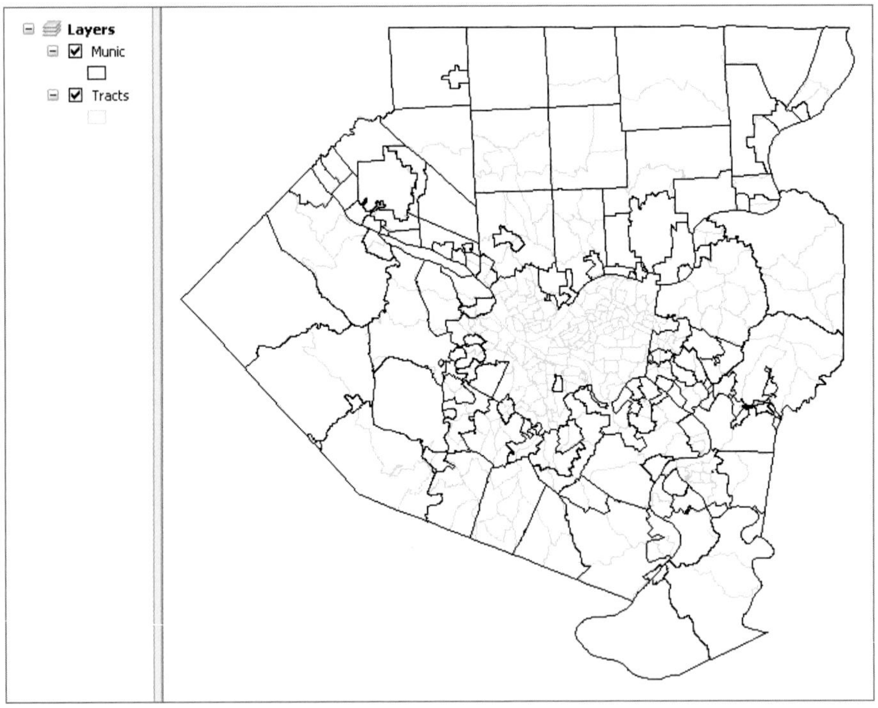

5 **Do not save your map document.**

Assign a projection to a shapefile

All GIS layers should have a projection defined, but sometimes you will receive a shapefile or other GIS layer that does not have a projection data included in its data, so you will need to assign this yourself. Note that the needed projection data is not metadata but data that is part of the functional part of the map layer. A common coordinate system for North American files is the Geographic Coordinate System, North American Datum 1983 (NAD 1983) projection, which is used by organizations such as the U.S. Census and is the coordinate system of the shapefile that you are about to use.

1 **In a new blank map, click the Add Data button, browse to \ESRIPress\GIST1\Data\DataFiles\, and add AlleghenyCountyTracts.shp to the map.** You get a warning that this layer has an unknown spatial reference (no projection data included in its files), and that while you can display the layer, you cannot edit it. Click OK.

2 **Right-click AlleghenyCountyTracts in the TOC, click Properties, and click the Source tab.** Notice in the Data Source panel that the coordinate system is unknown.

3 **Click OK, then on the main menu, click Windows and Catalog.**

4 **In Catalog, navigate to the \ESRIPress\GIST1\Data\DataFiles\ folder; right-click AlleghenyCountyTracts.shp; click Properties; and click the XY Coordinate System tab.**

5 **Click Select, double-click Geographic Coordinate Systems, double-click North America, click NAD 1983.prj, and click Add.** Of course you must know the correct coordinate system to assign to the map layer from external information, metadata, or from other sources. If you assign the wrong coordinates, your map will not display correctly in your map document.

6 **Click OK.**

7 **Repeat step 2 to see that the layer now has its coordinate system data included.**

8 **Leave your map document open.**

Tutorial 5-3

Learn about vector data formats

This tutorial reviews several file formats commonly found for vector spatial data, other than the file geodatabase covered in chapter 4. Included are ESRI shapefiles and coverages as well as computer aided design (CAD) files, XY event files, and other tabular data formats.

Examine a shapefile

Many spatial data suppliers use the shapefile data format for vector map layers because it is so simple. Shapefiles appeared about the same time that personal computers became popular. A shapefile consists of at least three files: a SHP file, DBF file, and a SHX file. Each of these files uses the shapefile's name but with the different file types. The SHP file stores the geometry of the features, the DBF file stores the attribute table, and the SHX file stores an index of the spatial geometry for quick processing. Next, you will examine AlleghenyCountyTracts.shp in more detail.

1 **Examine AlleghenyCountyTracts.shp in Catalog.** It appears as an entry in one line with an icon representing a polygon map layer. ArcCatalog treats the several files as a unit and provides utilities such as renaming the shapefile in one location. In fact, as you will see next, there are several files that make up the shapefile layer.

2 **Open a My Computer window and navigate to \ESRIPress\GIST1\Data\DataFiles\\.** Now you can see that there are seven files for the shapefile, including the projection (.prj) file that you created above when you added a spatial reference for the layer's coordinate system.

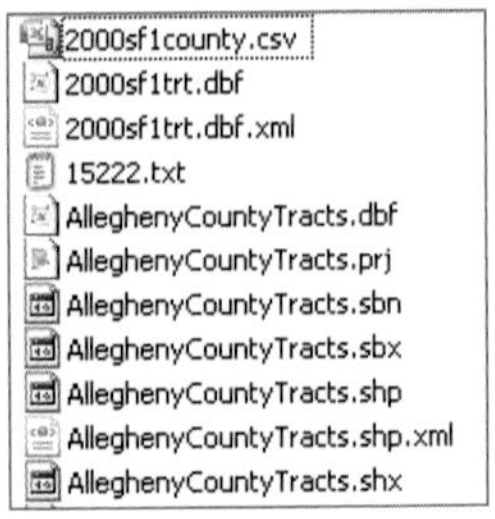

3 **Close the My Computer window.**

Coverages

The coverage is an old ESRI spatial data format from times when personal computers did not even exist. Coverages typically store one or more feature classes that are related. For example, in a cadastral (landownership) dataset it is common for a coverage to store the parcel boundaries as polygons and the parcel lines making up the polygons as arcs (lines). You can add coverage data to ArcMap and use it for analysis and presentation, but you cannot edit coverage data with ArcMap. When browsing data within Windows Explorer, coverages appear as folders containing several files. Below you can see the four coverages in the EastLiberty folder with the contents of the Building coverage appearing in the right-hand side of Windows Explorer. Building has 18 files.

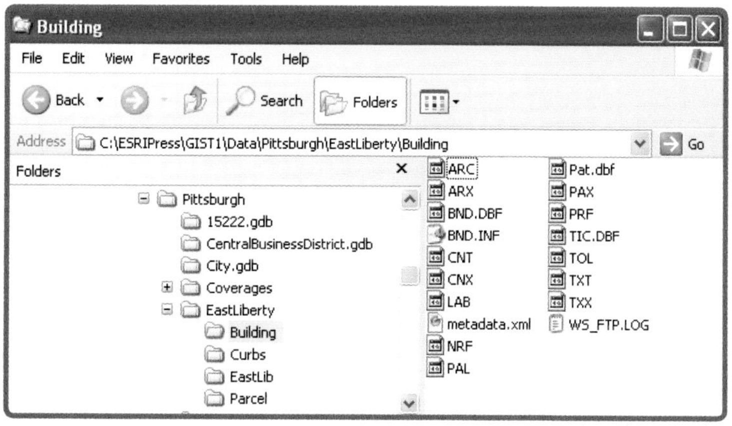

Add a coverage to ArcMap

1 **In a new blank map, click the Add Data button, browse to \ESRIPress\GIST1\Data\Pittsburgh\EastLiberty\, and double-click Building.**

2 **Click the Polygon layer icon, and click Add.** A coverage behaves like any other vector layer in ArcMap. It has the same appearance and has an attribute table.

YOUR TURN

Add Curb arcs and parcel polygon coverages to your map for the East Liberty neighborhood. Examine the attribute table for the Parcel layer.

Convert a coverage to a shapefile

If you need to edit the attribute tables or geometry of a coverage, you must export it first to the shapefile or file geodatabase format in ArcMap.

1 **In the TOC, right-click the Building Polygon layer, click Data, and click Export Data.**

2 **Browse to \ESRIPress\GIST1\MyExercises\Chapter5\ and save the output shapefile as Building.shp.**

3 **Click OK, then click Yes to add the exported data to the map as a layer.** Now you could edit the polygons in Building.shp and add the missing spatial reference for the layer. You will learn about editing shapefiles in chapter 6.

4 **Do not save your map document.**

CAD files

Many organizations have CAD (computer-aided design) files, drawings that you can display in ArcMap in their native format. ArcMap can add CAD files in one of two CAD formats: as native AutoCAD (.dwg) or as Drawing Exchange Files (.dxf) that most CAD software can create. When viewed in Catalog, a CAD dataset appears with a light blue icon. An AutoCAD file is much like a coverage in that it has different kinds of vector features in the same file. You can see CampusMap.dwg in ArcCatalog in the image on the right.

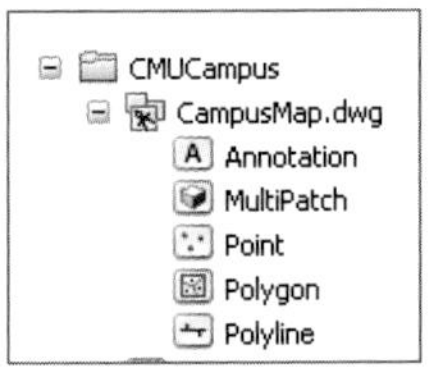

Add a CAD file as a layer for display

1 **In a new blank map, click the Add Data button, browse to \ESRIPress\GIST1\Data\CMUCampus\, click the CampusMap.dwg icon, and Add.** The following map of the Carnegie Mellon University campus appears in ArcMap. It contains many feature types, including lines, polygons, and text. This map is for display only.

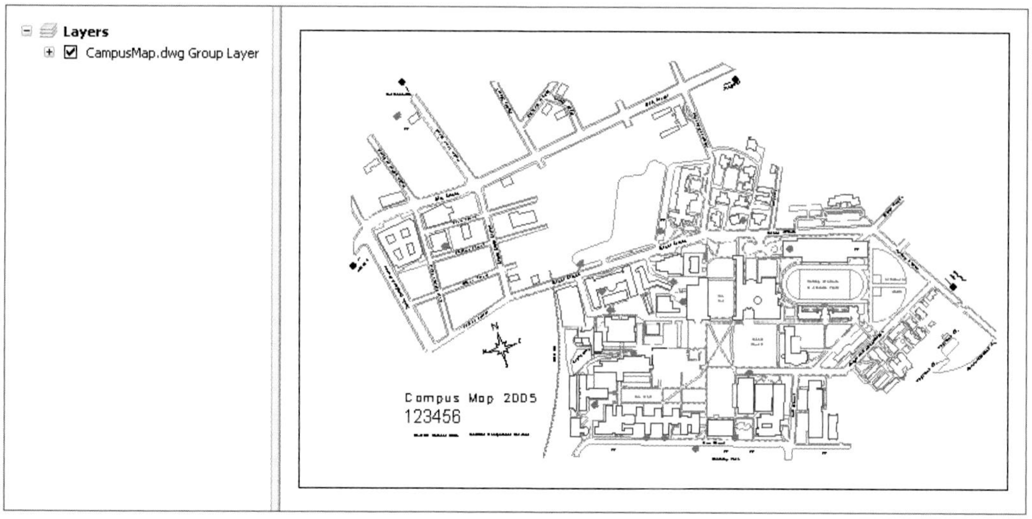

2 **Remove the CampusMap.dwg layer from the TOC.**

3 **Click the Add Data button, double-click the CampusMap.dwg icon, and double click the Polygon icon.** ArcMap adds the polyline feature class from the CAD dataset. You can select the features in this layer and change their display properties, but you cannot edit the layer.

4 **In the TOC, double-click the CampusMap.dwg Polygon layer.**

5 **Click the Drawing Layers tab.** Notice that you can turn Layers on and off. You can also symbolize the layers using the Symbology tab.

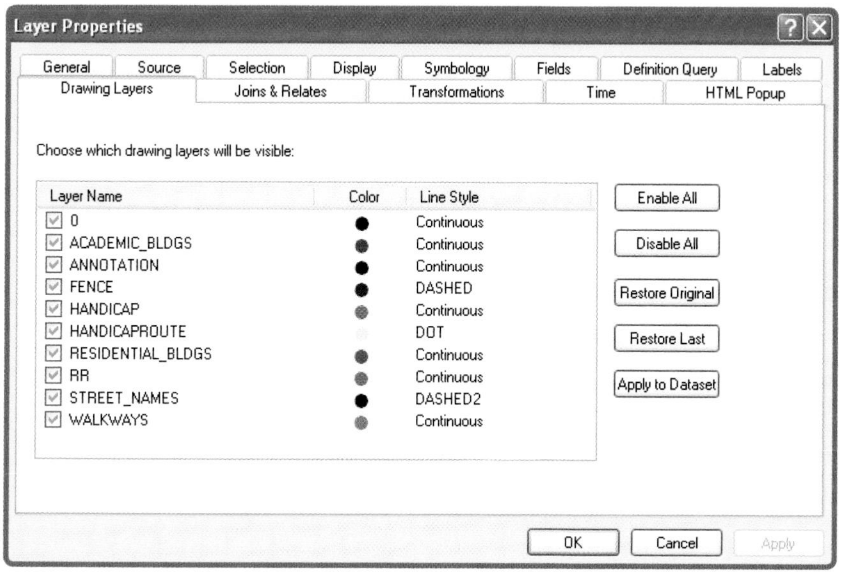

6 **Do not save your map document.**

Export shapefiles to CAD

Sometimes you may need to deliver shapefile data to a person working with CAD software. Using the Export tools in ArcCatalog, you can export shapefiles to AutoCAD (.dwg) or Drawing Exchange Files (.dxf) formats, which can then be opened by most commercial CAD applications.

1 **Create a new blank map and click the Add Data button.**

2 **Browse to \ESRIPress\GIST1\Data\Pittsburgh\CentralBusinessDistrict.gdb, click CBDStreets, and click Add.**

3 **In the TOC, right-click the CBDStreets layer, and click Data and Export to CAD.**

4 **Complete the dialog box as follows:**

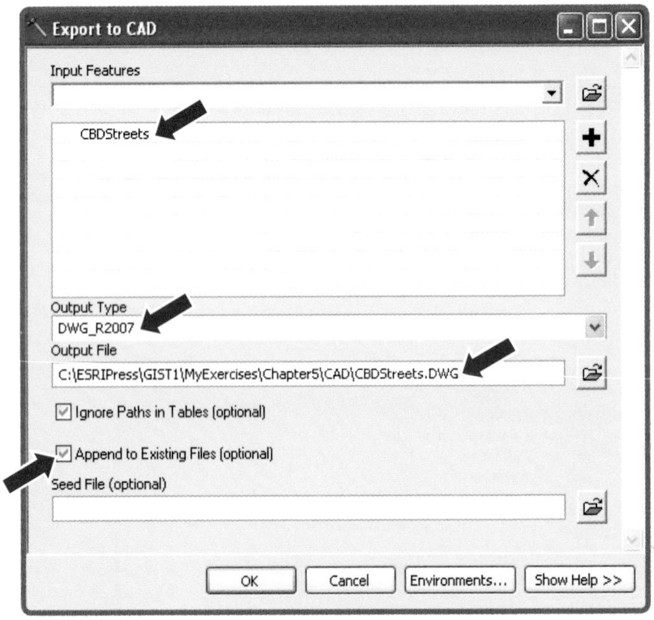

XY event files

It is possible to get point data in tabular form with columns for x- and y-coordinates. A good source of such data is from a Global Positioning System (GPS) device. Many Web sites include coordinates for point features along with other attributes.

1 **In a new blank map, click the Add Data button, browse to \ESRIPress\GIST1\Data\DataFiles\, click Earthquakes.dbf, and click Add.**

2 **Also add CACounties from \ESRIPress\GIST1\Data\UnitedStates.gdb\.** This layer is also in geographic coordinates, so the XY data on earthquakes will display with the counties.

3 **Right-click Earthquakes, click Open, and scroll to the right to see the X and Y fields.** The x- and y-coordinates here are latitude and longitude values for earthquake locations. You could simply display the earthquakes directly from this table. For many purposes, however, it would be better to have a map layer. So, next, you will create a shapefile from the table using Catalog.

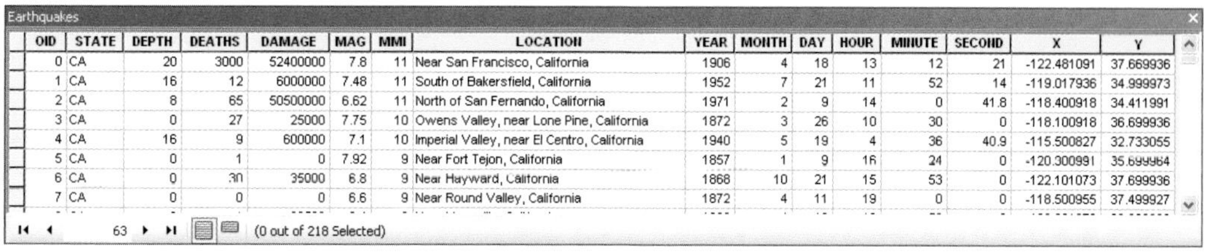
Earthquakes

OID	STATE	DEPTH	DEATHS	DAMAGE	MAG	MMI	LOCATION	YEAR	MONTH	DAY	HOUR	MINUTE	SECOND	X	Y
0	CA	20	3000	52400000	7.8	11	Near San Francisco, California	1906	4	18	13	12	21	-122.481091	37.669936
1	CA	16	12	6000000	7.48	11	South of Bakersfield, California	1952	7	21	11	52	14	-119.017936	34.999973
2	CA	8	65	50500000	6.62	11	North of San Fernando, California	1971	2	9	14	0	41.8	-118.400918	34.411991
3	CA	0	27	25000	7.75	10	Owens Valley, near Lone Pine, California	1872	3	26	10	30	0	-118.100918	36.699936
4	CA	16	9	600000	7.1	10	Imperial Valley, near El Centro, California	1940	5	19	4	36	40.9	-115.500827	32.733055
5	CA	0	1	0	7.92	9	Near Fort Tejon, California	1857	1	9	16	24	0	-120.300991	35.699964
6	CA	0	30	35000	6.8	9	Near Hayward, California	1868	10	21	15	53	0	-122.101073	37.699936
7	CA	0	0	0	6.6	9	Near Round Valley, California	1872	4	11	19	0	0	-118.500955	37.499927

63 (0 out of 218 Selected)

4 **Close the Attributes of Earthquakes table.**

5 **Click Windows and Catalog.**

6 **In Catalog, navigate to the Data Files folder, right-click Earthquakes.dbf, click Create Feature Class, and click From XY Table.**

7 **Type or make selections as follows. Assign the coordinate system GCS_North_American_1983.**

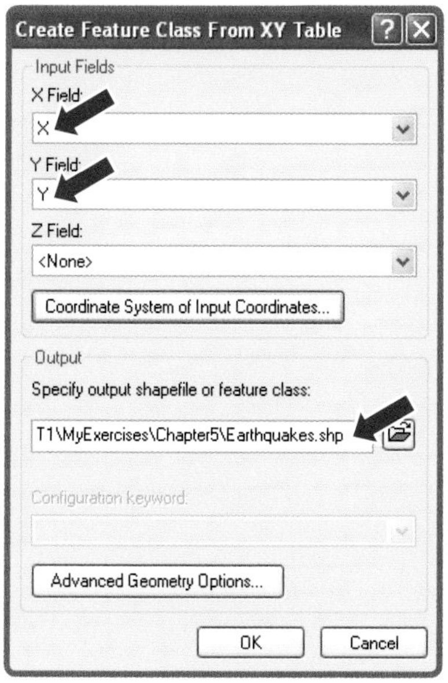

8 **Click OK.**

9 **Click the Add Data button, browse to \ESRIPress\GIST1\MyExercises\Chapter5\, click Earthquakes.shp, and click Add.** ArcMap displays the earthquake points from the new shapefile.

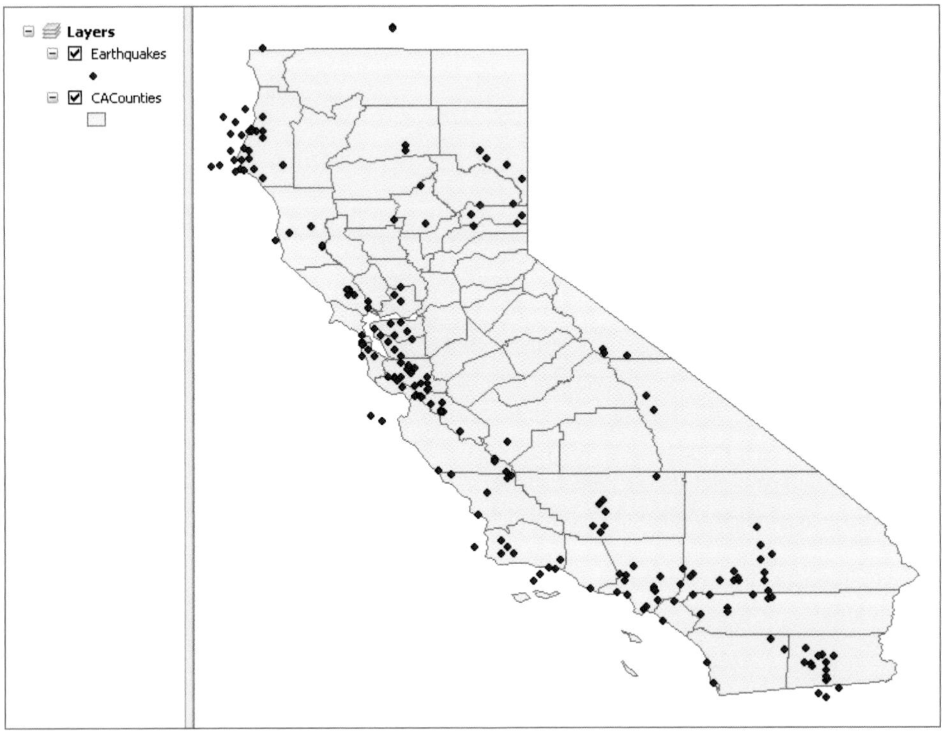

10 Do not save your map document.

Tutorial 5-4

Explore sources of vector maps

There is a large and ever-expanding collection of spatial data—including vector and raster map layers and data tables with geocodes—available for free download from the Internet or for direct use as map layers in map documents. The advantage of downloading spatial data is that you can modify it as you wish. The advantage of Web services is that you can use any selection from a vast collection of data without acquisition and storage on your own computer. You add map layers directly to your ArcGIS map documents from Web servers without downloading. Government agencies provide much of this spatial data as do GIS vendors such as ESRI. In this chapter you will learn about and use a few of the major spatial data suppliers. Note that if you have difficulty downloading the data files for this chapter, they are available in the \ESRIPress\GIST1\MyExercises\FinishedExercises\Chapter5\ folder.

ESRI Web site

ESRI maintains a Web site that is a useful resource for obtaining spatial data. You should visit this site often to learn what is new in the GIS community, find supporting articles, and access data.

1 **Open a Web browser and go to `www.esri.com`.** The content of this home page varies often, so your page will be different from the one that follows. Of particular importance to you on the main navigation bar are buttons for products (including free data), access to ESRI's Support Center, and access to online GIS periodicals *(ArcNews, ArcUser,* and *ArcWatch)* for users (you should read these free online publications often!).

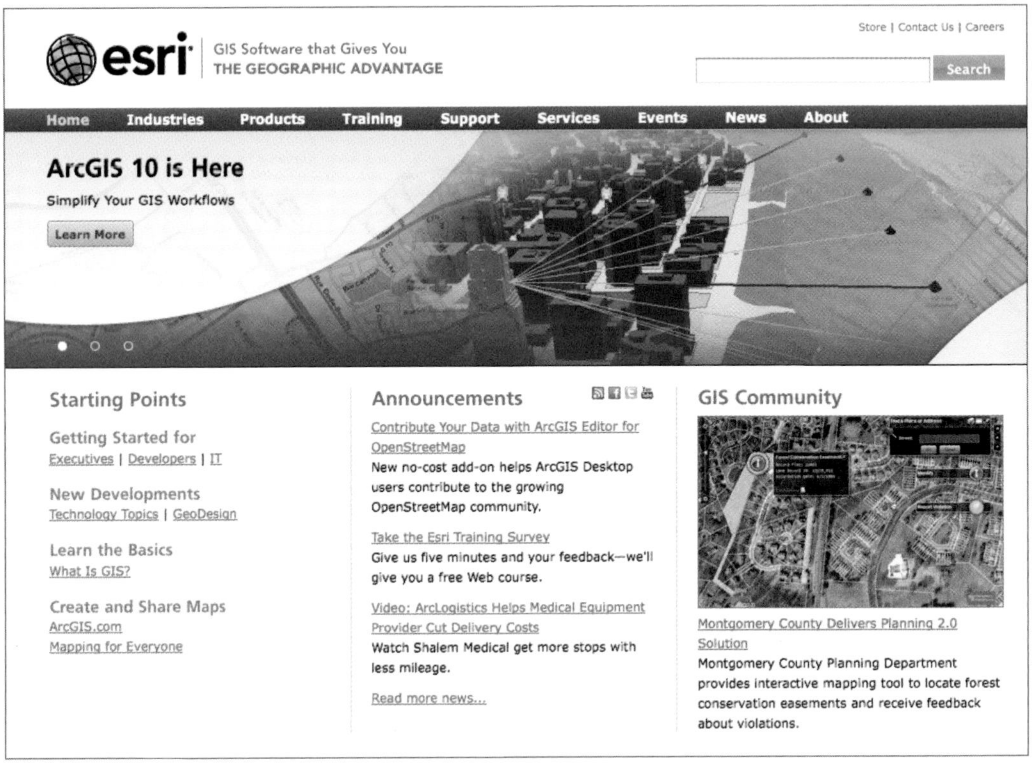

2 Click Products, Data, and Free Data. This Web page provides links to ESRI's Web map services as well as two portals for spatial data. First, you will download and use spatial data from the Census 2000 TIGER/Line data site. Later, you will use some ArcGIS online services and will download data from a site accessible through the geodata.gov portal.

Download Census TIGER/Line Data from ESRI's Web site

The Web site that you are about to use provides commonly used census map layers with good selections of census data attributes. Also, this is a good site for downloading TIGER/Line roads (streets).

1 Click the Census 2000 TIGER/Line Data link (or go to `www.esri.com/data/download/census2000-tigerline/index.html`).

2 On the Census 2000 TIGER /Line Data page, click the Preview and Download link on the left.

3 Click the state of Illinois (IL) on the map.

4 From the Select by County drop-down list, select Cook, then click Submit Selection.

5 **Select Census Tracts 2000 in the Available data layers list.**

Available data layers	File Size
☐ Block Groups 1990	726.9 KB
☐ Block Groups 2000	698.1 KB
☐ CMSA/MSA Polygons 2000	14.3 KB
☐ Census 2000 Collection Blocks	3.0 MB
☐ Census Blocks 1990	5.9 MB
☐ Census Blocks 2000	3.5 MB
☐ Census Tracts 1990	382.6 KB
☑ Census Tracts 2000	392.2 KB
☐ Congressional Districts - 106th	84.1 KB
☐ Congressional Districts - Current	84.0 KB
☐ County 1990	14.4 KB
☐ County 2000	14.3 KB

6 **Scroll down to see the Available Statewide Layers and select Census Tract Demographics (SF1) from the list.**

Available Statewide Layers	File Size
Census Block Demographics (PL94)	35.6 MB
☐ Census Block Demographics (SF1)	12.6 MB
☐ Census Block Group Demographics (SF1)	738.2 KB
☐ Census County Demographics (PL94)	26.0 KB
☐ Census County Demographics (SF1)	11.4 KB
☐ Census Place Demographics (PL94)	218.4 KB
☐ Census Place Demographics (SF1)	109.4 KB
☐ Census State Demographics (PL94)	1.6 KB
☐ Census State Demographics (SF1)	665.0 bytes
☐ Census Tract Demographics (PL94)	535.7 KB
☑ Census Tract Demographics (SF1)	251.8 KB

Proceed to Download

7 **Click Proceed to Download, Download File.** A File Download Web page opens when your file is ready.

8 **Click Save to save the file to the \ESRIPress\GIST1\MyExercises\Chapter5\ folder.**

9 **Close any open windows associated with your Web browser.**

Extract files

1 **Use your zip program to extract the zipped files to the \ESRIPress\GIST1\MyExercises\Chapter5\ folder.** This takes three steps. When you unzip the downloaded file you get two more zipped files that you also need to unzip in the same folder to yield a total of four files: tgr17000sf1trt.dbf (SF1 database), tgr17031tr00.dbf, tgr17031tr00.shp, and tgr17031tr00.shx (shapefile of census tracts for Cook County).

2 **Start ArcMap and add the files to a new blank map.**

3 **Open the attribute tables for the tgr17000sf1trt.dbf table and the tgr17031trt00 layer and explore their contents.** Verify that both the shapefile and table include the matching primary key, STFID, making it easy to join the table to the map.

4 **Do not save your map document.**

YOUR TURN

Download census data for another state and county of your choice to \ESRIPress\GIST1\MyExercises\Chapter5\. Include Line Features – Roads to get TIGER/Line roads.

Download U.S. Census Bureau Cartographic Boundary Files

The U.S. Census Bureau Web site, **www.census.gov**, is the resource for more detailed census data and TIGER basemaps. Advantages of this site are that you can get maps with the latest revisions as well as choose your own selection of census variables from the thousands available for download.

1 **Start your Web browser and go to www.census.gov.** The content of this home page varies over time, so your page may be different in appearance.

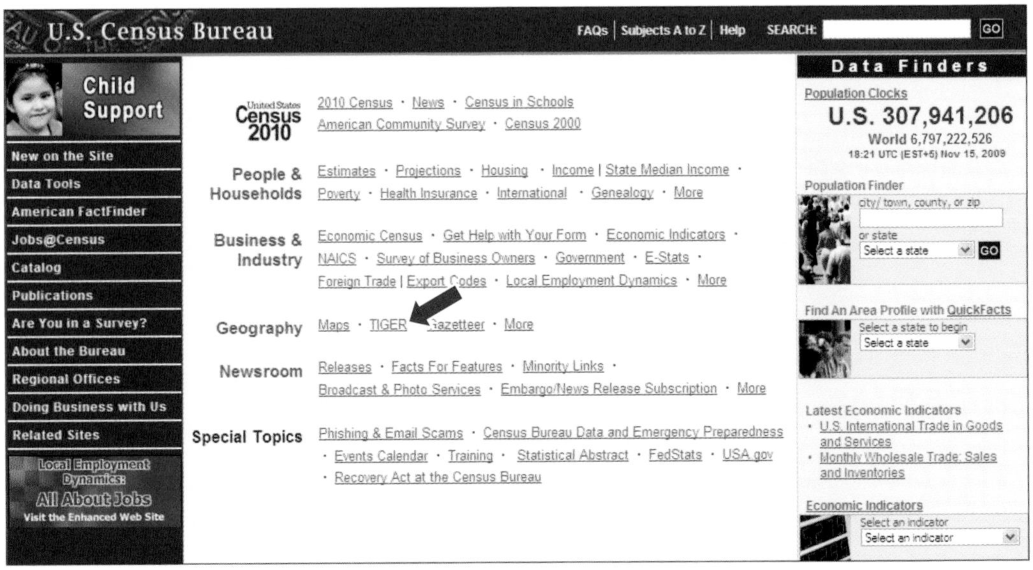

2 Click the TIGER link in the Geography section.

3 Click the link for 2009 TIGER/Line Shapefiles Main Page.

4 Click the Download the 2009TIGER/Line Shapefiles now link.

5 Select Illinois for the state, click submit, select Cook County for county, and click submit.

6 Click Census Tract (Census 2000), click Download Selected Files, click Open, and extract tl_2009_17031_tract000.zip to \ESRIPress\GIST1\MyExercises\Chapter5\.

Note: Some Web browsers may cite a security issue when downloading this file. You can also click the hyperlink for Census Tract (Census 2000) to download it.

7 Close your Web browser, open a My Computer window, and browse to \ESRIPress\GIST1\MyExercises\Chapter5\.

8 Double-click tl_2009_17031_tract00.zip and unzip the shapefile to \ESRIPress\GIST1\MyExercises\Chapter5\. That results in the shapefile, tl_2009_17031_tract00.shp, including all of its associated files.

9 In ArcMap, create a new blank map, add tl_2009_17031_tract00.shp, and examine the layer's attribute table. You will find many identifiers in the table but no census data per se. The identifier that you will use in the next tutorial is CTIDFP00 for tracts. Notice that CTIDFP00 aligns left (make the column wider to see this), indicating that it has text values instead of numeric, even though all characters are digits.

10 Close the attribute table and leave ArcMap open with the tracts map layer displayed.

Tutorial 5-5

Download and process tabular data

The U.S. Census Bureau's American FactFinder Web site is a good example of a source that provides tabular data with geocodes. Next, you will download a table with selected census variables for Cook County, Illinois, process them in Microsoft Excel, and finally join them to the tract map layer for display.

Download American FactFinder Data Tables

The American FactFinder is the U.S. Census Bureau site for downloading census data tables to join to census cartographic boundary maps.

You can download census variables of your choice from the Census short-form tables (SF 1), long-form tables (SF 3), or other tables.

1 Use your Web browser to go to `factfinder.census.gov`.

2 Click DOWNLOAD CENTER in the left panel.

3 From the DOWNLOAD CENTER page, click the Census 2000 Summary File 3 (SF 3) – Sample Data link.

4 In the Select a Geographic Summary Level panel, click All Census Tracts in a County (140).

5 Click the drop-down list for state and click Illinois.

6 Click the county drop-down list and click Cook County.

7 Click the Selected Detailed Tables Option button and click Go.

8 Scroll down, click P30 Means of Transportation to Work for Workers 16+ Years, and click Add. This single selection results in a table with several census variables on means of transportation to work.

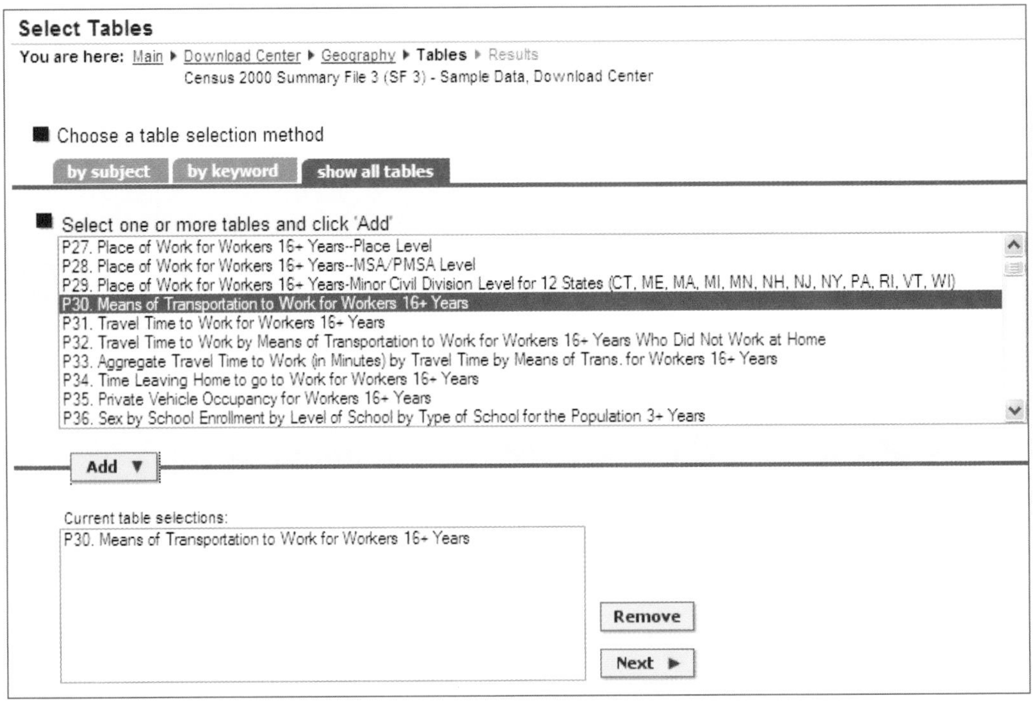

9 Click Next, Start Download. **Note:** This may take several minutes to download.

10 Save the resulting zipped file to \ESRIPress\GIST1\MyExercises\Chapter5\, then extract the zipped file to the same folder, resulting in the text file dc_dec_2000_sf3_u_data1.txt.

YOUR TURN

Download a few SF 1 census variables for Illinois census tracts.

Import text data into Microsoft Excel

The census data that you downloaded in the previous exercise as a text table needs some cleaning up using software such as Microsoft Excel before using in your GIS. First, you need to import the text table into Excel.

1 Open Microsoft Excel on your computer, click the Office button, click Open, select All Files (*.*) for Files of type, browse to \ESRIPress\GIST1\MyExercises\Chapter5\, and double-click dc_dec_2000_sf3_u_data1.txt.

2 In the Text Import Wizard, click the Delimited Option button, click Next, clear the Tab check box, click Other, type the "|" character (above the Enter key) in the text box to the right of the Other check box.

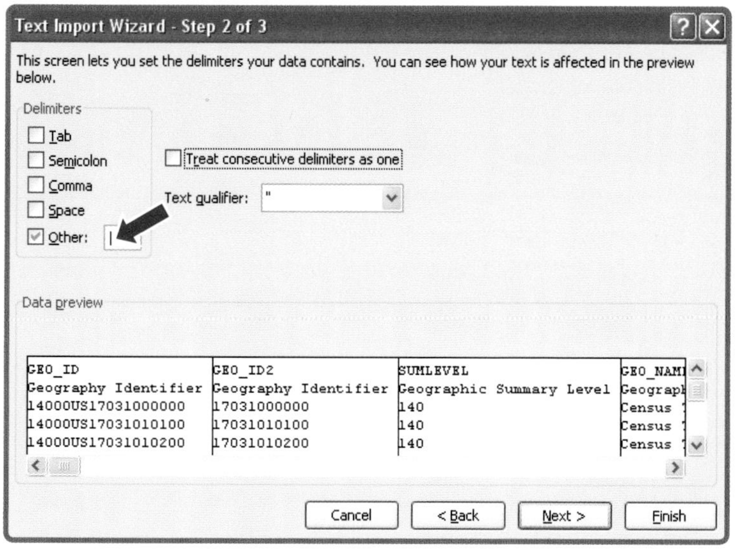

3 Click Finish.

4 Click inside the A column header cell, hold your mouse key down, and drag across to the right to the T column header to select columns A through T.

5 Position your cursor on the vertical boundary line between any two column header cells so that the cursor changes to a double-headed arrow, and double-click. Click anywhere in the table to clear the selection. Excel changes column widths to fit and display all data.

Assign self-documenting attribute names

The problem with the downloaded table is that instead of a single row at the top with attribute names, there are two rows at the top. The first has a cryptic name and the second has a definition for each attribute. You need to use the definitions in order to assign self-documenting names in the first row and then delete the second row. First, you can delete unneeded columns.

1 Right-click the column heading cell (A) for the row 1 value GEO_ID and click Delete.

2 **Likewise, delete columns that have row 1 values SUMLEVEL and GEO_NAME.** You should have a worksheet that appears as follows. If you made a mistake, click the Undo key until you get back to a good starting place.

	A	B	C	
1	GEO_ID2	P030001	P030002	P
2	Geography Identifier	Workers 16 years and over: Total	Workers 16 years and over: Means of transportation to work; Car; truck; or van	W
3	17031000000			
4	17031010100	2179	1220	
5	17031010200	4678	2655	
6	17031010300	3346	1682	

3 **Type new names in the row 1 cells as follows (press the tab key to move to the next cell):**

P030001	Workers
P030002	Vehicle
P030003	VehicleAlone
P030004	VehiclePooled
P030005	Public
P030006	Bus
P030007	Streetcar
P030008	Subway
P030009	Railroad
P030010	Ferryboat
P030011	Taxicab
P030012	Motorcycle
P030013	Bicycle
P030014	Walk
P030015	Other
P030016	Home

4 **Right-click the row selector for row 2 and click Delete.**

Change the identifier data type to text

The final problem is that the tract identifier in the tract map layer that you downloaded has text as its data type, while the matching GEO_ID2 column in the worksheet has the numeric data type. Text cells in Excel have a prefix character of a single quote that generally is not visible but is there. So you will add a single quote to each GEO_ID2 value.

1 **Click the A column selector cell to select that column.**

2 **With the Home tab selected in Excel, click the Find & Select button on the ribbon and click Replace.**

3 Type values as follows (note the single quote in the Replace with text box):

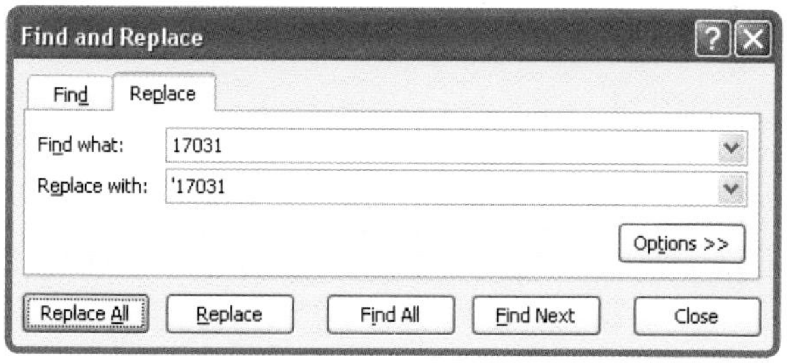

4 Click Replace All. All cells get notes, indicated by small green triangles, that numbers are stored as text.

5 Double-click the lower left tab with text dc_dec_2000_sf3_u_data1, and type **Tracts** to rename the tab.

6 Click the Office button and Save As, select Excel 97-2003 Workbook (*.xls) as the Save as type, change the File Name to **CookTracts.xls**, and save in \ESRIPress\GIST1\MyExercises\Chapter5\.

7 Close Excel.

YOUR TURN

Add the Tracts$ sheet from CookTracts.xls to the map document with the tract boundaries, join the table to the polygon layer, select an attribute of interest, and symbolize a choropleth map with quantile classification.

Tutorial 5-6

Explore sources of raster maps

While vector maps are discrete—consisting of points, lines connecting points, and polygons made up of lines—raster maps are continuous like photographs and use many of the same file formats as images on computers, including joint photographic experts group (.jpg) and tagged image file (.tif) formats. All raster maps are rectangular, consisting of rows and columns of cells known as pixels. Each pixel has an associated projected coordinate and attribute value such as altitude for elevation. Raster maps do not store each pixel's location explicitly but rather store data such as the coordinates of the northwest corner of the map, cell size (assuming square pixels), and the number of rows and columns from which a computer algorithm can calculate the coordinates of any cell. Raster maps can represent points, lines, and polygons as collections of turned-on pixels, but they are better for continuous phenomena such as elevation, land cover, and temperature. A key aspect of raster maps from your point of view is that they are very large files. So while you may store some important raster files on your computer, these kinds of maps are perhaps best obtained as map services available for display on your computer but stored elsewhere.

View raster maps for download

The sort of online viewer that you will use, "seamless" based on ESRI's ArcIMS or ArcGIS Server software, is common on Web sites. You zoom in to an extent that meets your needs, view layers of interest, and then download layers with that extent.

1 Open your Web browser and go to `http://seamless.usgs.gov/`. Make sure your pop-up blocker is turned off.

2 Click View & Download United States Data. The National Map Seamless Server opens with an interactive map for displaying and downloading map layers. You could download all of the raster map layers you are about to see, but will not do so because of the large file sizes that would result. Instead, you will download a raster map for a small area after viewing some of the layers available.

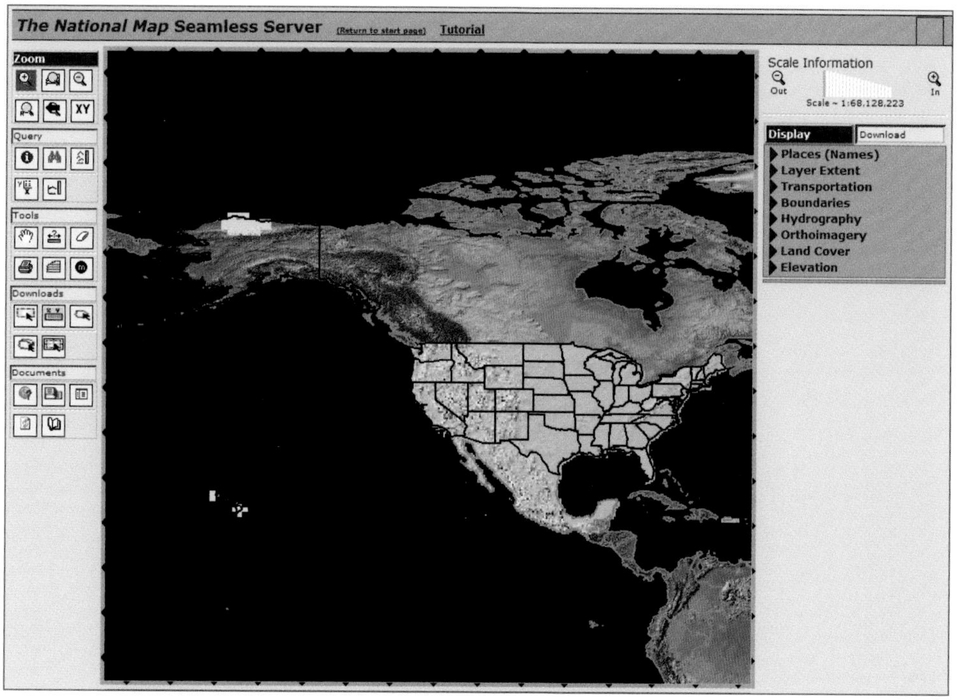

3 In the Zoom panel, click the Zoom to Region or Area button; click the drop-down arrow in the resulting Select a Locale field at the bottom of the map; and select Colorado Springs, CO. The viewer zooms to that locale displaying a major roads vector map and an elevation raster map.

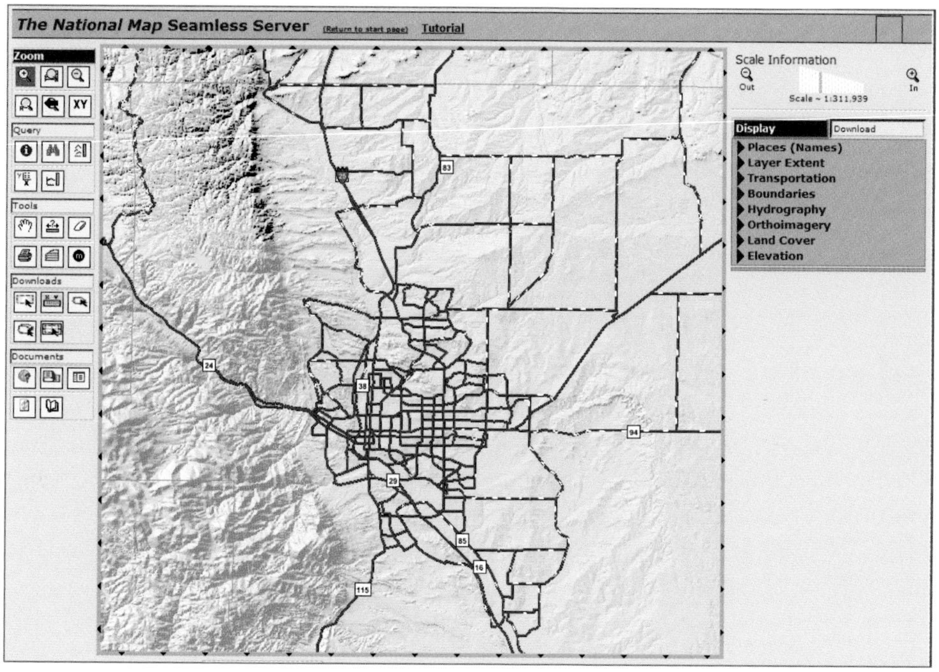

4 Under Display on the right side of the window, click Land Cover, NLCD 2001 Land Cover, and the Legend button. Those actions display a raster land-cover map and its legend. Notice the arrow pointing to a small enclosed area defined by major roads. Below you will zoom in to this area.

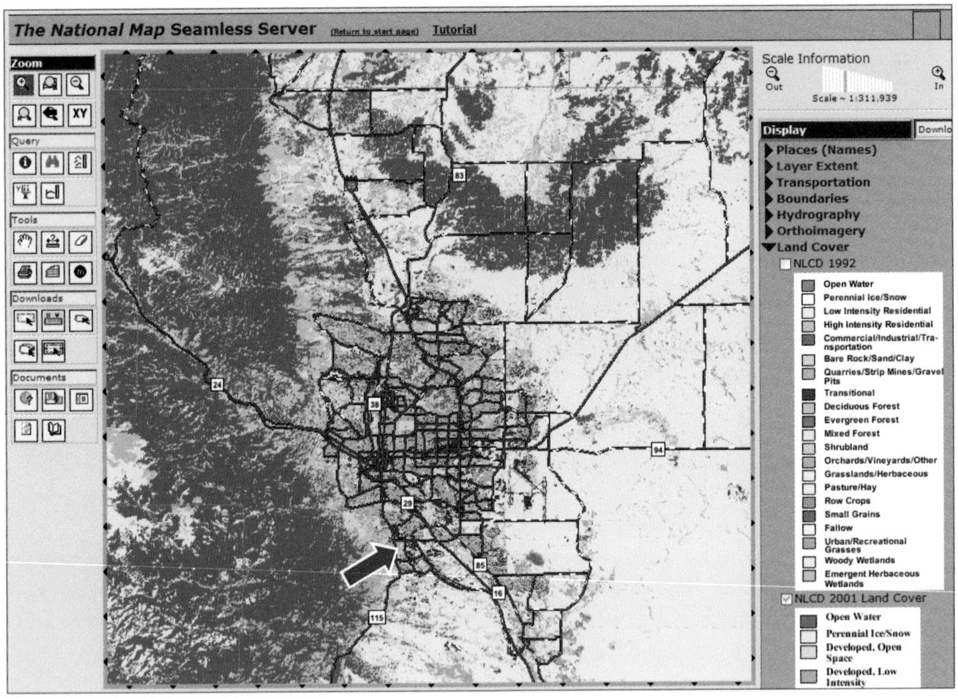

5 Click the Map Information and Meta Data button on the left under Documents. For the NLCD 2001 Land Cover row in the resulting window, click the Meta document icon. A window with detailed documentation opens on the land-cover layer.

6 Close the documentation window and the OGC Map Information window, leaving the Seamless Server window open.

7 Turn off the land-use layer. Click Land Cover under Display to close the land-cover list.

Download raster maps

Next, you need to zoom in to a small enough area so that the file you download does not have too big a file size.

1 Click the Zoom In button and drag a rectangle around the area indicated by the black arrow in the above map.

2 Click Orthoimagery under Display and turn on Colorado Springs, CO (Jun 2008).

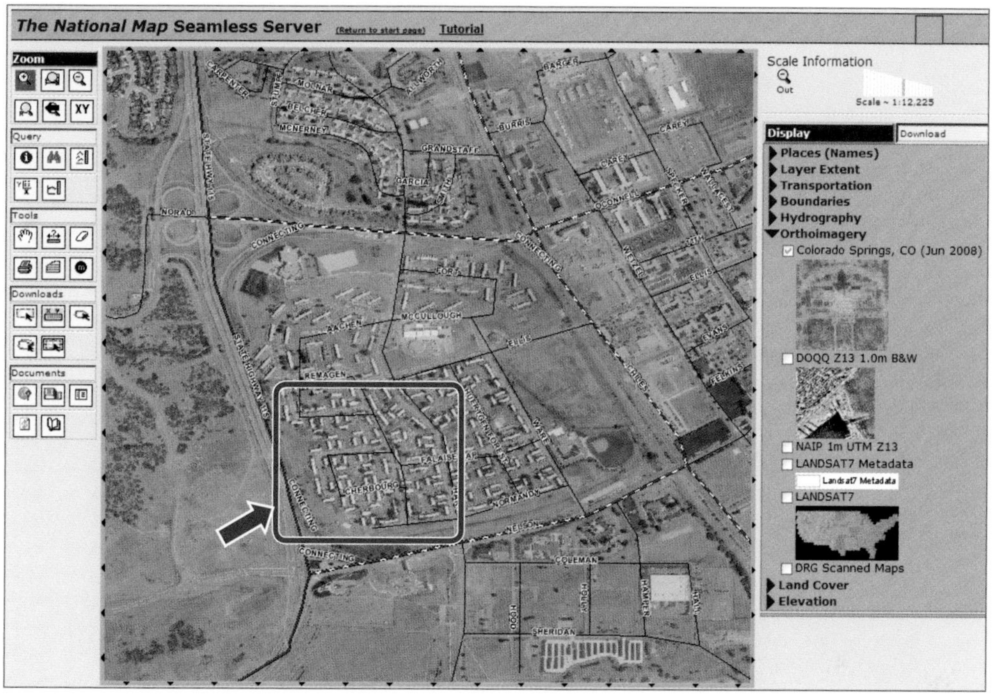

3 **Zoom further in to the area indicated by the arrow above.** This is a very high-quality aerial photo with pixel size of approximately 1 foot (electronic images are arrays of pixels where each pixel has a solid color and no boundary).

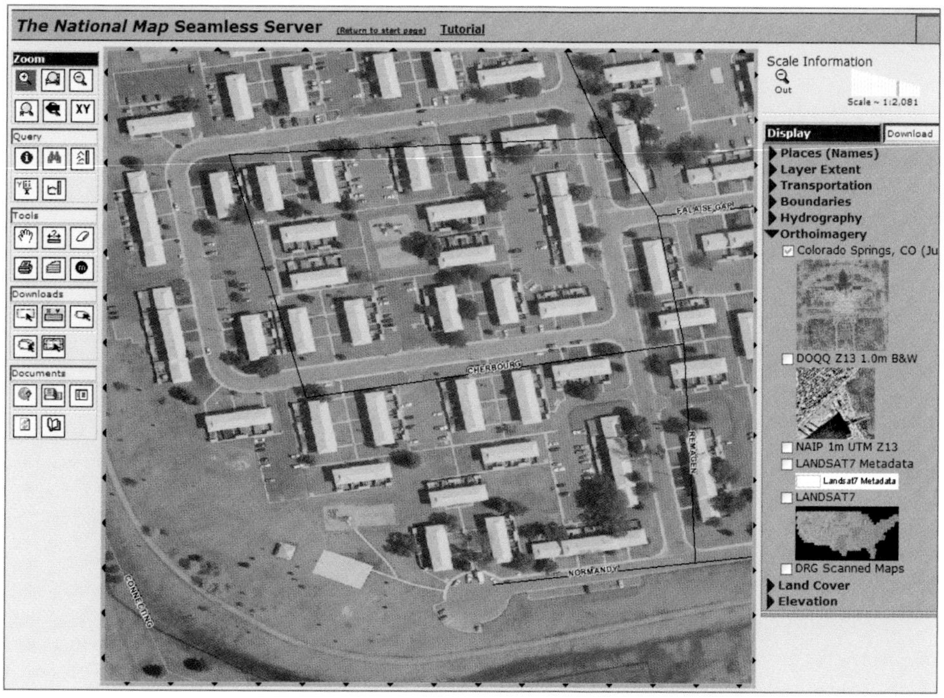

4 **Click the Download link to the right of Display, and click the Legend button to turn off the legend.**

5 **Click each line under Download, Structures through National Atlas. Turn off every layer except Colorado Springs, CO (Jun 2008).**

6 **Click the Define Rectangular Download Area button, drag a rectangle around the block bounded on the south by Cherbourg Street, and wait for the Seamless Server to respond.** The Seamless Server prepares a download of about 2 MB in size.

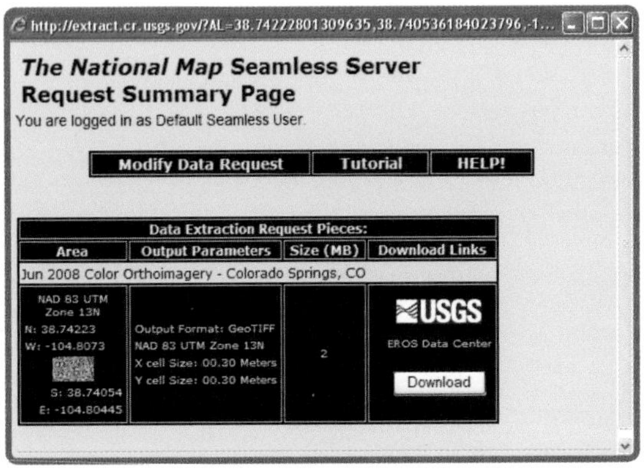

7 **Open the extracted file, extract to \ESRIPress\GIST1\MyExercises\Chapter5\, and close browser windows.**

YOUR TURN

Open a new map document in ArcMap, add your downloaded raster file (a TIFF file), and zoom in to the point where you can see the individual pixels. Do not save the map document. If you were to download vector layers for Colorado Springs, they would display correctly in your map document, overlaying the image but perhaps not fitting well due to inaccuracies inherent in TIGER maps.

Use a Web map service

ESRI provides a Web service, ArcGIS Online, as part of the ArcGIS package. You can add several different raster layers to a map document as services in addition to map layers from your computer or local area network. You can access ArcGIS Online directly from ArcMap's main menu.

1 **In ArcMap, create a new blank map.**

2 **Click File, Add Data, and Add Data; browse to \ESRIPress\GIST1\Data\UniteStates.gdb\; and click NYManhattanCounty and Add.** The Web service that you will use will get the map extent from the layer you just added.

3 **Symbolize the layer with a hollow fill and a red outline width 2.**

4 **Click File, Add Data, and Add Data from ArcGIS Online.** Your Web browser opens to the ArcGIS Online page.

5 **Type World Imagery in the search window and press Enter.**

6 **In the World Imagery panel, click the Add Data button and Close.** It takes a few moments for the map image to download to ArcMap.

7 **Zoom in to Central Park.** The map image that displays is from ESRI's server instead of your computer's hard disk.

8 **Repeat step 4, search for USA Topographic Maps, and click Add; wait for the map to load.** A topographic map of Manhattan and Central Park opens.

YOUR TURN

Try adding a few more maps from ArcGIS Online. Do not save your map document.

Assignment 5-1

Create a map of Maricopa County, Arizona, voting districts, schools, and voting-age population using data downloaded from the ESRI Web site

One community that uses GIS for elections is Maricopa County, Arizona, one of the nation's fastest-growing communities with 3.2 million residents and 1.5 million registered voters. Maricopa County uses GIS to ensure accurate voting boundaries, maintain voter lists, locate polling places, plan voting precincts, recruit poll workers, and deliver supplies.

In this assignment you will focus on skills needed to download data from an ESRI Web site and prepare the data to use in ArcGIS. For the Maricopa County voting GIS, you will download voting districts, streets, and census blocks for the purpose of building an interactive GIS to be used in selecting schools for use as polling sites. You will also download block-level census data from the ESRI Web site, join this data to the block map, and use it to display the spatial distribution of the voting-age population. An X,Y table shows schools and their geographical coordinates.

Start with the following:

Shapefiles
From the ESRI Census 2000 TIGER/Line Data Portal, download the following shapefiles and save them in \ESRIPress\GIST1\MyAssignments\Chapter5\:

- tgr04013blk00.shp—census block 2000 polygons
- tgr04013lkA.shp—line features (roads)
- tgr04013vot00.shp—voting districts 2000 polygons

Data tables
From the same ESRI Web site, download the following table and save it in \ESRIPress\GIST1\MyAssignments\Chapter5\:

- tgr04000sf1blk.dbf—census block demographics (SF1)

From \ESRIPress\GIST1\Data\MaricopaCounty\, use the following:

- CountySchools.dbf—the XY coordinates are for state plane projection for Maricopa County, Arizona

File geodatabase
Import all of the above shapefiles and tables into a new file geodatabase called **\ESRIPress\GIST1\MyAssignments\Chapter5\ Assignment5-1YourName.gdb**. Convert CountySchools.dbf from an XY table into a point feature class called **\ESRIPress\GIST1\MyAssignments\Chapter5\Assignment5-1YourName.gdb\MaricopaCountySchools** using Catalog.

Create an interactive GIS

Create a new map document called **\ESRIPress\GIST1\MyAssignments\Chapter5\Assignment5-1YourName.mxd**. Use scales to display detailed layers when zoomed in to 1:100,000 scale. At that scale, display labels for voting districts, schools, and streets. This provides a tool for analyzing potential voting places, voting district by voting district.

Look up the state plane zone for Maricopa County and use it for your map document's data frame. Add spatial reference data for the ESRI shapefiles: GCS_North_American_1983 (NAD 1983.prj). When you add the x,y data, edit the Coordinate System of Input Coordinates to use the correct state plane coordinates of Maricopa County Arizona. Add a field to block census data: Voters = [POP2000] – [AGE_UNDER5] – [AGE_5_17].

For very small-grain spatial data, such as provided by census blocks, a good approach is to use small, square point markers of the same size and with a monochromatic color ramp. Symbolize blocks using graduated symbols for the Voters attribute and use a "trick" to make all symbols the same size. Use size from 4 to 4 to get same size and then double-click each symbol to change color for the monochromatic ramp. Set the background color to No Color. The benefit of the "trick" is that ArcMap uses point markers instead of choropleth maps for the blocks.

Create an 11-by-8.5-inch landscape layout with map, legend, and title. Zoom in to a populated area of your map document with a map scale of 1:24,000. Export the layout as **\ESRIPress\GIST1\MyAssignments\Chapter5\Assignment5-1YourName.jpg**. Create a Word document, saved as **\ESRIPress\GIST1\MyAssignments\Chapter5\Assignment5-1YourName.doc**, that has a title, your name, your map layout image, and a paragraph suggesting schools to be used as polling places for observed voting districts.

WHAT TO TURN IN

If your work is to be graded, turn in the following files:

File geodatabase: \ESRIPress\GIST1\MyAssignments\Chapter5\Assignment5-1YourName.gdb

ArcMap document: \ESRIPress\GIST1\MyAssignments\Chapter5\Assignment5-1YourName.mxd

Word document: \ESRIPress\GIST1\MyAssignments\Chapter5\Assignment5-1YourName.doc

Image file: \ESRIPress\GIST1\MyAssignments\Chapter5\Assignment5-1YourName.jpg

If instructed to do so, instead of the above individual files, turn in a compressed file, **Assignment5-1YourName.zip**, with all files included. Do not include path information in the compressed file.

Assignment 5-2

Create a map of Pinellas County with census data displayed and Web service added

In this exercise you will focus on skills needed to download data from the U.S. Census Bureau's Web sites and prepare the data for use in ArcGIS. You will download block group (Census 2000) polygons for Pinellas County, Florida, and corresponding, selected census variables in a table that you will join to the map. You will symbolize a choropleth map and add some Web service map layers for further information on Pinellas County.

Start with the following:

Shapefile

- tl_2009_12103_bg00.shp—2009 TIGER/Line Shapefile of census block groups from the Census Bureau's TIGER site and saved to \ESRIPress\GIST1\MyAssignments\Chapter5\

Census table

- dc_dec_2000_sf3_u_data1.txt—downloaded from the Download Center on `factfinder.census.gov` using these settings: Census 2000 Summary File 3 (SF 3) - Sample Data, All Block Groups in a county, Florida, Pinellas County, H6 Occupancy Status saved in \ESRIPress\GIST1\MyAssignments\Chapter5\

Prepare data for use

Clean up and prepare the census data in Microsoft Excel using steps similar to those in the exercises. Create an Excel file called PinellasBlockGroups.xls with GEO_ID2 saved as text and renamed columns **TotalUnits**, **Occupied**, and **Vacant**.

Create a new map document called **\ESRIPress\GIST1\MyAssignments\Chapter5\Assignment5-2 YourName.gdb**. Import your Excel file and shapefile into it.

Create a map document

Create a new map document called **\ESRIPress\GIST1\MyAssignments\Chapter5\Assignment5-2 YourName.mxd**. Use ArcGIS Online to add World Imagery data to your map document. Add the block group layer and census table from your geodatabase to the map document. Join the table to the map layer and prepare a choropleth map of vacant housing units normalized using the total housing units and five quantiles. Change the transparency of the layer of the legend to 50 percent to see the world imagery below it.

Create an 11-by-17-inch portrait layout with map, layer, scale (miles) and title. In the layout, zoom in to a portion of the county with about 5 to 10 percent of the area (scale about 1:15,000) and relatively high vacancy rates. Export the layout as **\ESRIPress\GIST1\Chapter5\Assignment5-2 YourName.jpg**.

WHAT TO TURN IN

If your work is to be graded, turn in the following files:

File geodatabase: \ESRIPress\GIST1\MyAssignments\Chapter5\Assignment5-2YourName.gdb

ArcMap document: \ESRIPress\GIST1\MyAssignments\Chapter5\Assignment5-2YourName.mxd

Image file: \ESRIPress\GIST1\MyAssignments\Chapter5\Assignment5-2YourName.jpg

If instructed to do so, instead of the above individual files, turn in a compressed file, **Assignment5-2YourName.zip**, with all files included. Do not include path information in the compressed file.

6

Digitizing

This chapter shows you how to create and edit spatial data. You will learn how to digitize new vector features and add attribute data to a table. You will also adjust vector data spatially to make it align with a basemap layer.

Learning objectives

- *Digitize polygon features*
- *Use advanced edit tools*
- *Digitize point features*
- *Digitize line features*
- *Spatially adjust features*

Tutorial 6-1

Digitize polygon features

You will create a new polygon feature class and then add features to it using heads-up digitizing with your mouse.

Create a new polygon feature class

1 Start ArcCatalog.

2 In the Catalog tree, browse to \ESRIPress\GIST1\MyExercises\Chapter6\MidHill.gdb.

3 Right-click MidHill.gdb, click New, and click Feature Class.

4 In the Name field of the New Feature Class window, type **CommercialZone**.

5 For Type, select Polygon Features and click Next.

6 Expand Projected Coordinate Systems, State Plane, NAD 1983 (US Feet); click NAD 1983 StatePlane Pennsylvania South FIPS 3702 (US Feet), and click Next three times.

7 Type **ZoneNumber** as a new field, select Short Integer as the Data Type, and click Finish. The result is a new polygon feature class added to MidHill.gdb.

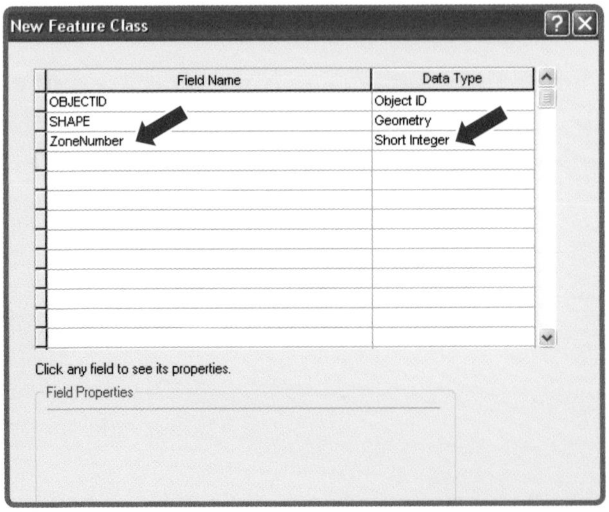

8 **Expand MidHill.gdb in the Catalog tree to see your new feature class, and then close ArcCatalog.**

Open a map document

1 **Start ArcMap, click browse for more, browse to \ESRIPress\GIST1\Maps\, and double-click Tutorial6-1.mxd.** The Tutorial6-1 map document opens in ArcMap, showing a map of the Middle Hill neighborhood of Pittsburgh, Pennsylvania. You will use the Commercial Properties and Street Centerlines layers as references for digitizing commercial zones.

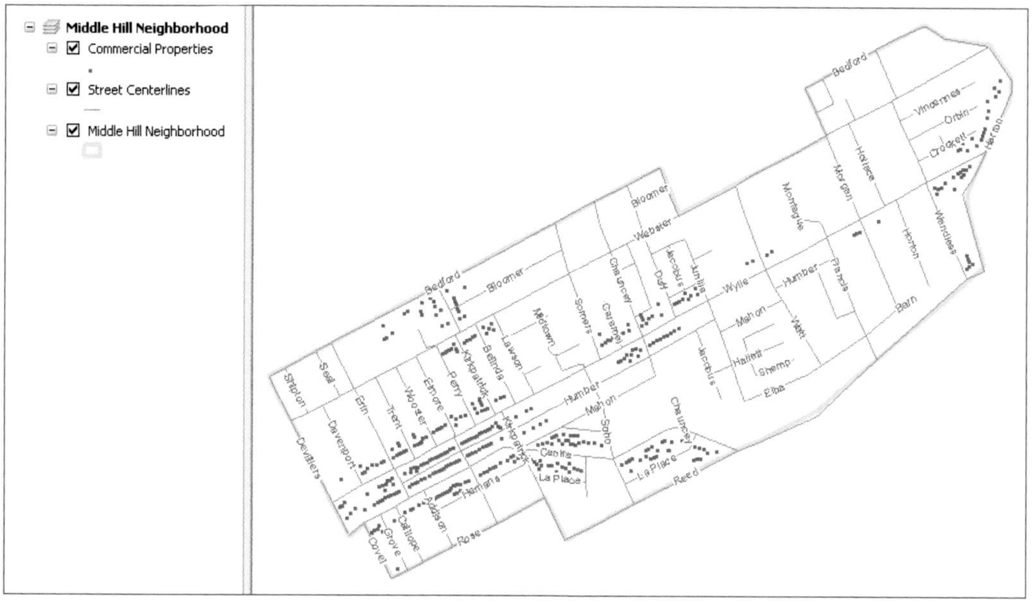

2 **Click the Add Data button.**

3 **Browse to \ESRIPress\GIST1\MyExercises\Chapter6\MidHill.gdb, click CommercialZone, and click Add.** This adds the CommercialZone layer to the map, although there are no features in it yet.

4 **Add CommercialZone from Midhill.gdb to your map document.** CommercialZone appears in the TOC, but of course nothing displays on the map because at this point there are no features in this new map layer. Next, you can digitize new features, starting with some practice polygons that will give you some experience but that you will not save.

Start editing with the Editor toolbar

1 On the main menu, click Customize, Toolbars, Editor. The Editor toolbar appears. You can move it or dock it anywhere in ArcMap. Dock it on top of the ArcMap window below the Standard toolbar.

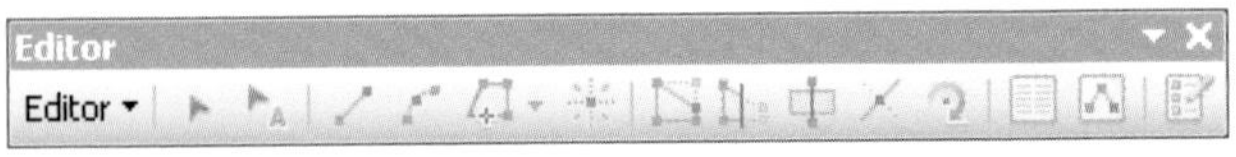

2 On the Editor toolbar, click Editor, Start Editing.

3 Click CommercialZone as the layer to edit and click OK. Start editing. Create Features and Construction tools panels appear on the right of the map. You can adjust these panels by dragging the boundary between them.

4 Click CommercialZone in the Create Features panel, then click Polygon in the Construction tools panel.

Practice digitizing a polygon

1 Click the List By Selection button in the TOC and make CommercialZone the only selectable layer.

2 Hide the Create Features panel. Zoom to the Middle Hill Neighborhood layer as shown below.

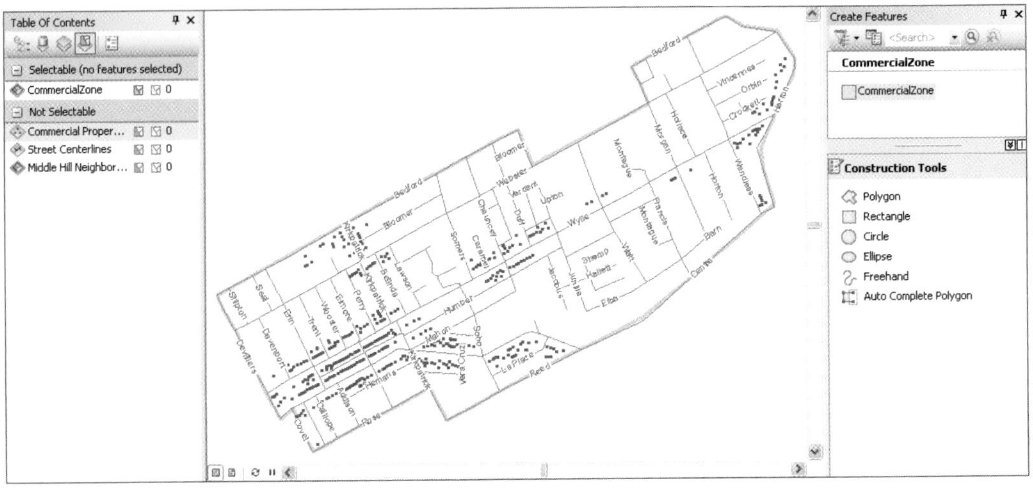

3 On the Editor toolbar, click the Straight Segment tool.

4 Position the crosshair cursor anywhere on the map and click to place a vertex.

5. Move your mouse and click a series of vertices one at a time to form a polygon (but do not double-click!). You will find that your new vertices snap to existing vertices of other features. You will learn how to turn this behavior on and off later.

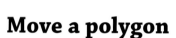

6. Double-click to place the last vertex.

Move a polygon

1. On the Editor toolbar, click the Edit tool.
2. Click and hold down the mouse button anywhere inside your new polygon.
3. Drag the polygon a small distance and release.

Delete a polygon

1. With the Edit tool still selected, click anywhere inside your new polygon.
2. Press the Delete key on the keyboard.

YOUR TURN

Practice creating new polygons using the Polygon, Rectangle, Circle, and Ellipse tools from the Construction tools panel. Delete your practice polygons when finished.

Edit polygon vertex points

Next, you will learn how to work with vertices. You will move, add, and delete vertices from a new polygon.

1. Click Bookmarks, Erin Street.
2. In the Construction tools panel, click Polygon.

3 **Click the Straight Segment tool and draw another new polygon feature as shown in the image, snapping to intersections.**

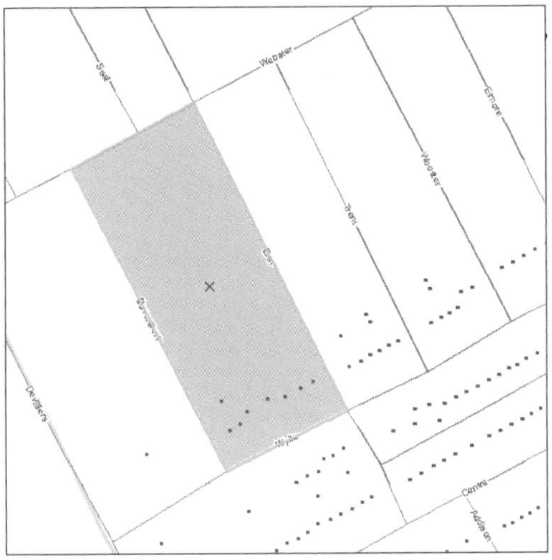

4 **Click the Edit tool.**

5 **Double-click the new polygon.** Grab handles, small squares, appear on the polygon at its vertex locations. Next, you will see that you can edit the shape of a feature by moving a vertex.

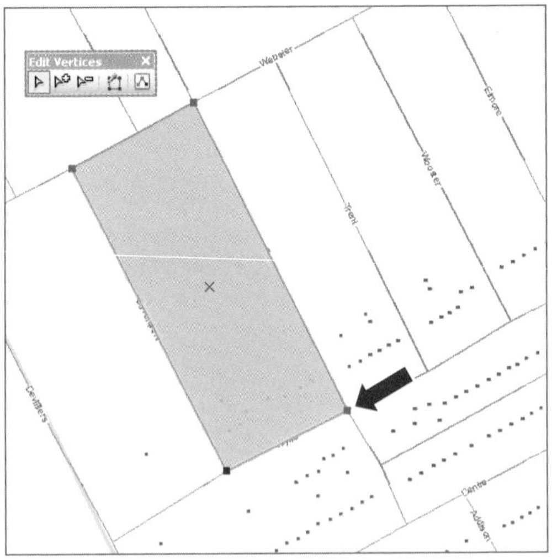

6 **Position the cursor over one of the vertices.**

7 **Click and drag the vertex somewhere nearby and release.** The polygon's shape changes correspondingly.

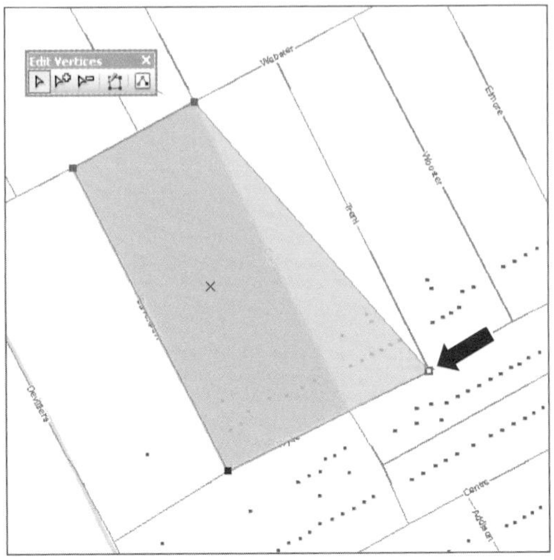

8 **Click anywhere on the map or polygon to confirm the new shape.**

Next, you will practice editing digitized polygons and learn how to add, delete, and move vertices.

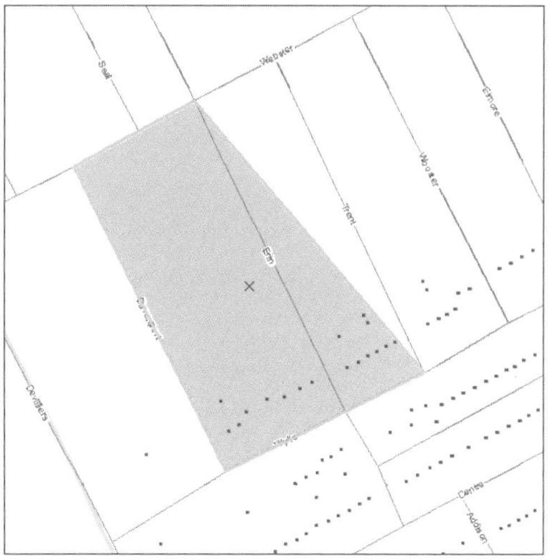

Add vertex points

1 **Double-click inside the polygon.** Grab handles appear on the polygon and the small Edit Vertices toolbar appears .

2 Click the Add Vertex tool .

3 **Move the mouse along the line between two vertices and click.** This adds a new vertex at the location of the cursor. Now you can move the new vertex to change the polygon's shape.

4 **Position the cursor over the new vertex, then click and drag the vertex to a new position and release.**

5 **Click anywhere on the map to confirm the new shape.**

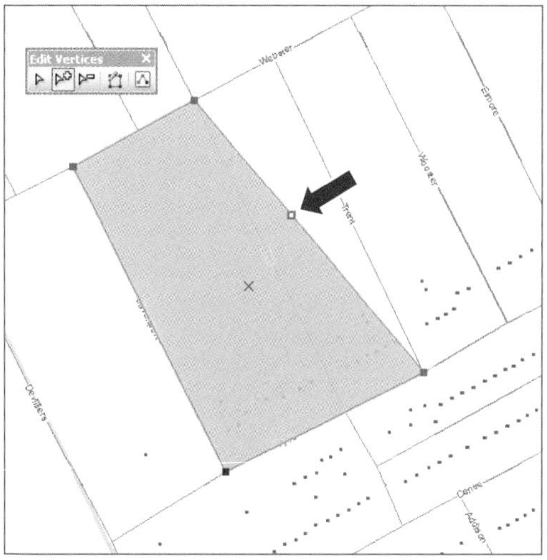

YOUR TURN

Practice adding a few more vertices and changing the shape of the polygon.

Delete vertex points

1 **Double-click inside the new polygon.**

2 Click the Delete Vertex tool .

3 **Place your mouse cursor over a new vertex point and click.**

4 Click anywhere on the map to confirm the new shape.

YOUR TURN

Practice changing the shape of the new polygon by moving, adding, and deleting vertices. When finished, delete the polygon.

Specify a segment angle and length

1 **In the Construction tools panel, click Polygon.**

2 **On the Editor toolbar, click the Straight Segment tool and digitize a starting point for a polygon segment.**

3 **Move your cursor to start drawing a line, right-click, and click Length.**

4 **Type 250 and press Enter.**

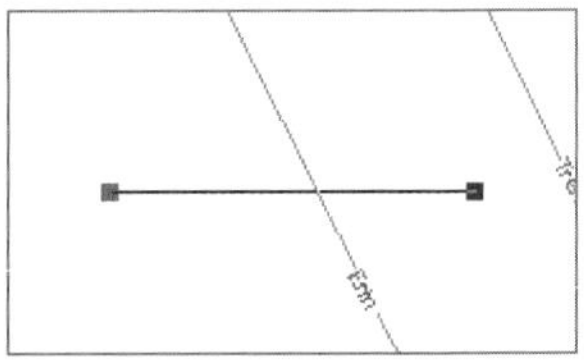

5 **Right-click, click Direction, type 0, and press Enter.** The line is 250 feet long and its direction is to the right (you measure angles counterclockwise with zero being east or to the right).

YOUR TURN

Digitize a few more points with specified segment lengths and angles, then double-click to finish the polygon. Delete the polygon when finished.

Digitize polygons

1 **Zoom to the cluster of commercial block centroids at the top left of the map.** The map to the right has the polygon drawn that you are about to digitize roughly. You will fine-tune the polygon in a second pass of digitizing, so you do not need much precision at first.

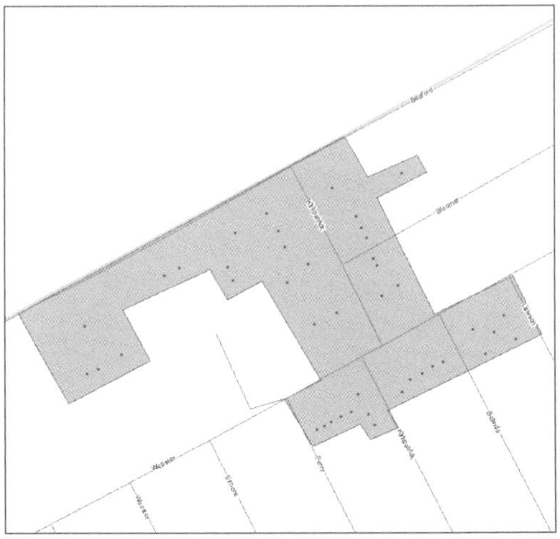

2 **In the Construction tools panel, click Polygon. On the Editor toolbar, click the Straight Segment tool and digitize the polygon seen on the previous page by clicking one vertex at a time and double-clicking to finish.** Wherever possible, use street centerlines as a guide for digitizing your lines.

YOUR TURN

Zoom in to a part of your new polygon and use the Add, Delete, and Move Vertex tools to refine the polygon's shape. Use the Pan tool on the Tools toolbar to move around your polygon's boundary and eventually refine all of it. You need to alternate between the Edit tool, confirming a change, and the Pan tool. Click the Full Extent button and then zoom in to a cluster of commercial points to digitize another polygon. Repeat until you have digitized all polygons seen below. When you complete the final polygon, click Editor and Save your edits.

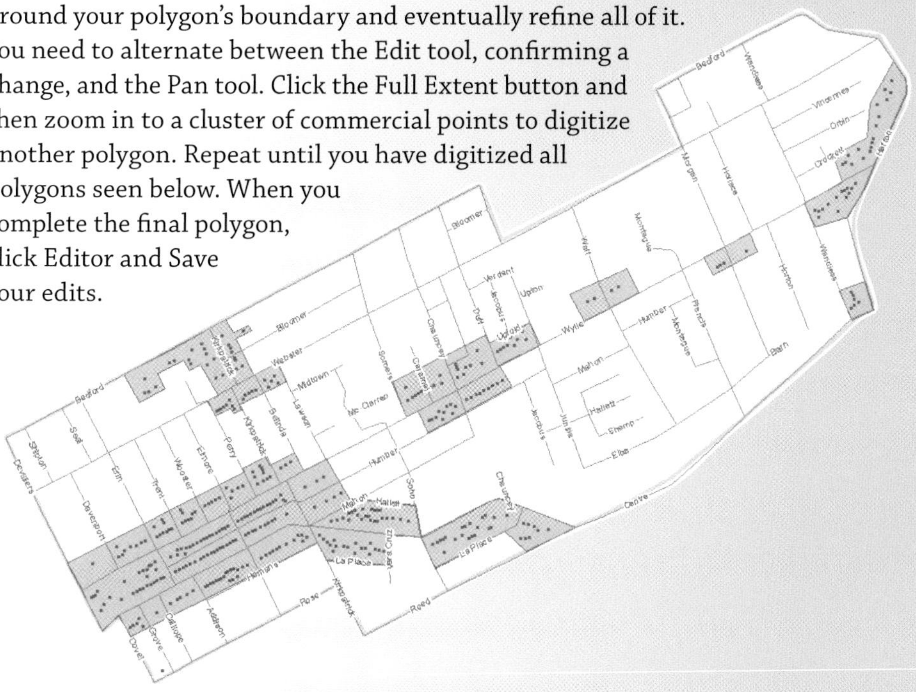

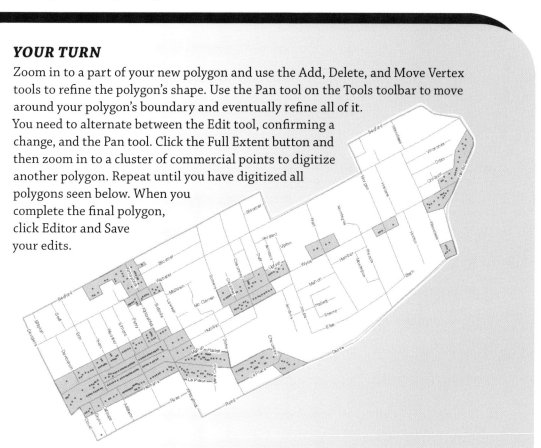

Edit feature attribute data

Now that you have digitized the commercial polygons, you will assign zone numbers to them.

1 **Open the CommercialZone attribute table.**

2 **Click in the first cell of the ZoneNumber field, type 1, and press Enter.**

3 **In sequential order, continue numbering the remaining cells in the ZoneNumber field.**

CommercialZone

OBJECTID	SHAPE *	SHAPE_Are	ZoneNumber
1	Polygon	167769.878	1
2	Polygon	928871.117	2
3	Polygon	137303.702	3
4	Polygon	251611.383	4
5	Polygon	70607.3583	5
6	Polygon	71823.2920	6
7	Polygon	30271.1874	7
8	Polygon	180852.108	8

4 Click Editor, Stop Editing, and Yes to save your edits.

5 Close the attribute table.

Label the commercial zones

1 Turn off the Commercial Properties layer.

2 In the TOC, right-click the CommercialZone layer and click Properties.

3 Select the Labels tab and type or make selections as follows:

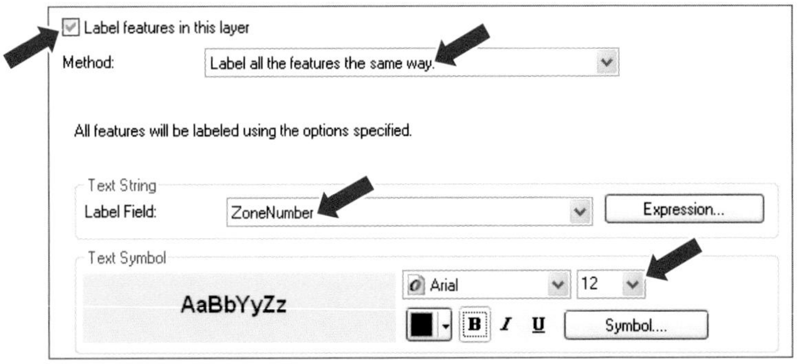

4 Click OK. Your label numbers may not match those below, depending on your order of digitizing.

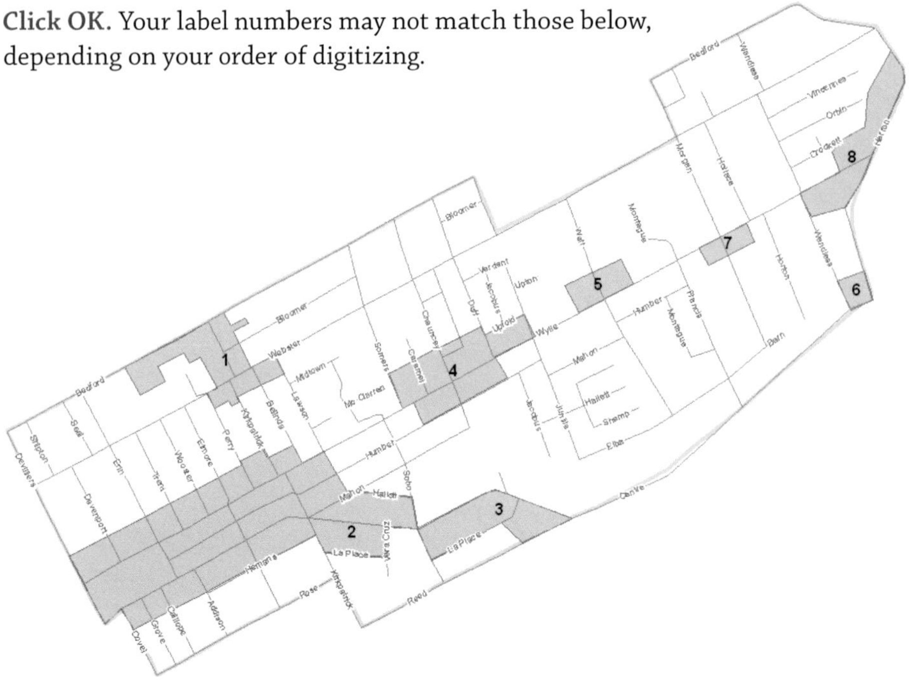

5 Save the map document as **\ESRIPress\GIST1\MyExercises\Chapter6\Tutorial6-1.mxd** and leave it open.

YOUR TURN

In ArcMap, click Windows > Catalog and use Catalog with steps similar to those at the start of this chapter to create a new polygon feature class in \ESRIPress\GIST1\MyExercises\Chapter6\MidHill.gdb called **Practice**. Use the same coordinate system used at the start of the chapter and do not create any new attributes. You will need the new feature class in the next section. When finished, close the Catalog window.

Tutorial 6-2

Use advanced edit tools

There are several advanced editing tools. Here you will try the Snapping, Trace, Generalize, Smooth, and Cut Polygons tools, all of which affect the shape of digitized polygons.

Set Snapping tools

ArcMap automatically snaps to all layers in a map document. There may be too many features, and turning off snapping will allow for easier edits.

1 Remove CommercialZone from the TOC. Click Bookmarks, Erin Street.

2 Turn on the Commercial Properties layer.

3 Click Editor, Start Editing, the Practice icon, and OK.

4 Click Practice in the Create Features panel.

5 Click Editor, Snapping, Snapping Toolbar.

6 Click the Point, Vertex, and Edge Snapping tools to turn them off, and leave End Snapping on. 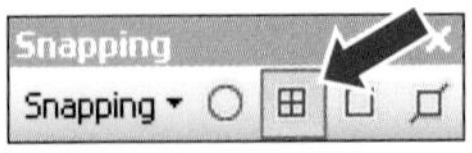

7 Click the street endpoint as shown in the image to snap to it.

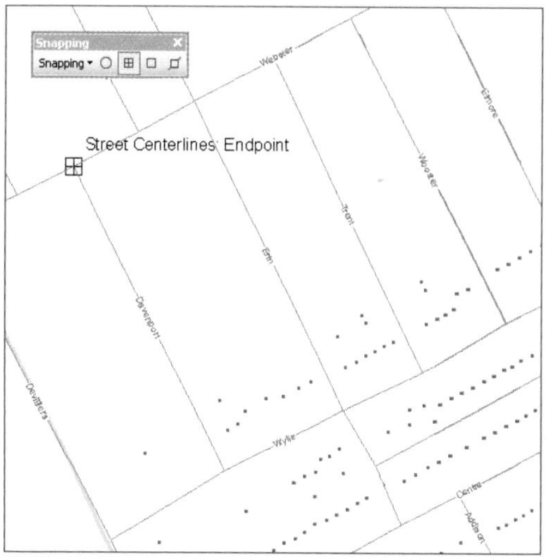

8 Continue clicking street endpoints to create a polygon that encloses two blocks as shown in the image.

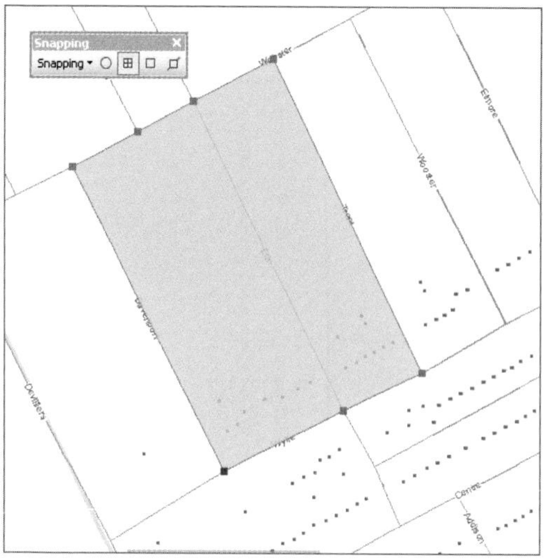

YOUR TURN

Turn on Edge and Point snapping. Practice creating polygons snapping to these features.

Trace tool

Tracing is a quick way to create new segments that follow the shapes of other features. Tracing is particularly useful when the features you want to follow have curves or complicated shapes, because snapping is more difficult in those cases.

1 **Delete any existing polygons.**

2 **Click Bookmarks, LaPlace Street.**

3 **Click Polygon as the Construction tool if it is not already selected.**

4 **On the Editor toolbar, click the Trace tool** . You may have to click the list arrow on the fifth button to the right on the Editor toolbar to access the Trace tool.

5 **Click the intersection of Soho and Centre streets and drag your mouse to the right, click the first vertex encountered, trace without clicking to complete the polygon, double-clicking to finish.**

You can click the Undo button to start over as needed. The resultant polygon nicely follows straight and curved segments.

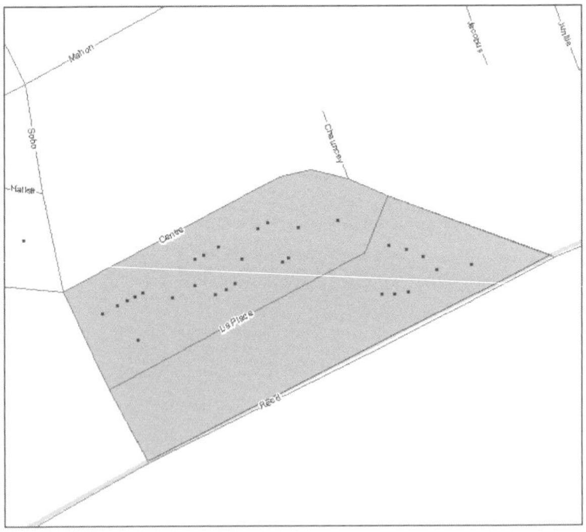

Generalize tool

Generalizing creates features for use at small scales with less detail while preserving basic shapes. For example, the U.S. Census generalizes many of its cartographic boundary files.

1 **On the Editor toolbar, click Editor, More Editing Tools, and Advanced Editing.** The Advanced Editing toolbar appears.

2 **Click the Edit tool** **and double-click inside the new traced polygon.**

3 On the Advanced Editing toolbar, click the Generalize tool, type a Maximum allowable offset of **100**, and click OK. The result is a polygon with fewer vertices, no two of which have a line segment between them less than 100 feet. You can click the Undo button to try a different offset.

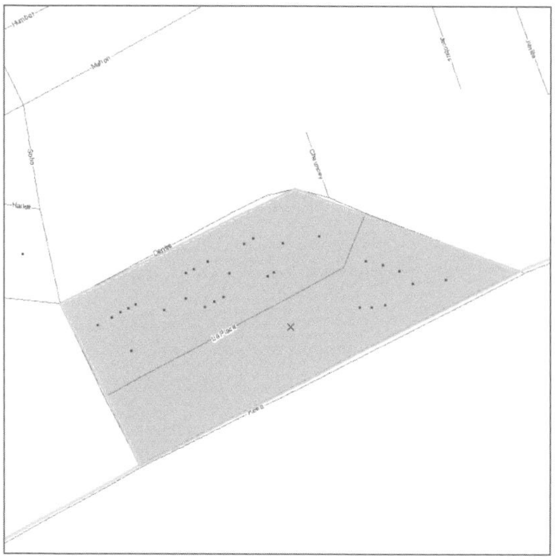

Smooth tool

The Smooth tool is the opposite of generalize. This tool smooths sharp angles in polygon outlines to improve aesthetic or cartographic quality.

1 Zoom out and pan the map to an area outside of the Middle Hill neighborhood.

2 Digitize a new polygon with 20 to 40 vertices.

3 Click the Edit tool and click inside the polygon.

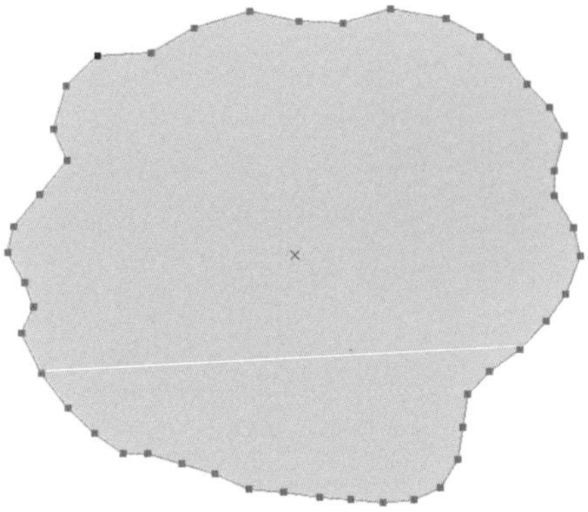

4 Click the Smooth tool on the Advanced Editing toolbar, type a maximum allowable offset of **10**, and click OK. This adds many shape vertices to create smooth curves between the polygon's vertices.

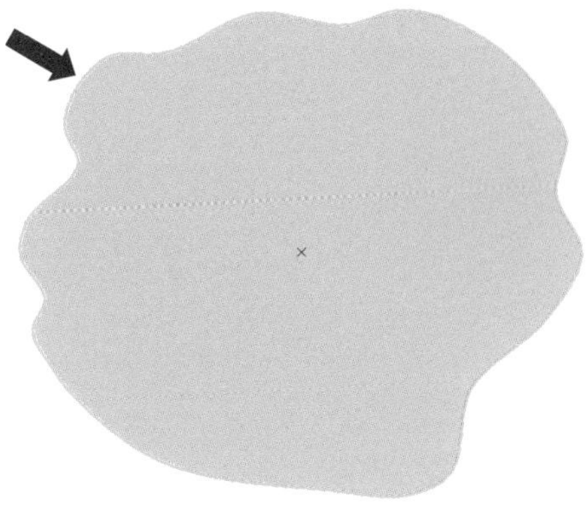

5 Close the Advanced Editing toolbar.

6 Click the Edit tool, click inside the new polygon, and press the Delete key.

Cut Polygons tool

The Cut Polygons tool creates two polygons from one original polygon. Note that you must click on the outside segment of the polygon you wish to cut.

1 **Click Bookmarks, Erin Street**

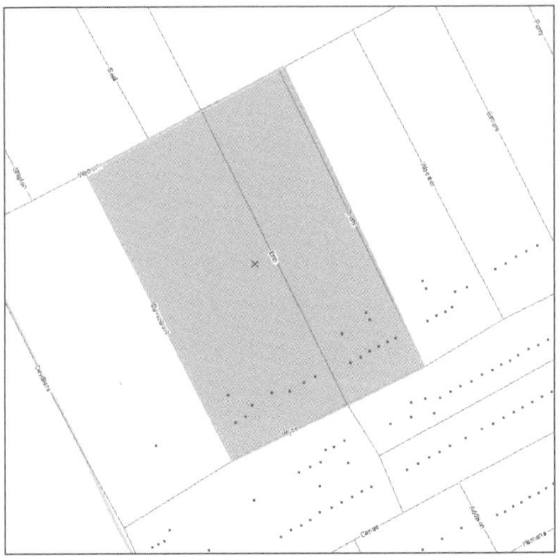

2 **If not already drawn, digitize a new polygon from the intersections of Webster and Davenport, Webster and Trent, Wylie and Trent, and Wylie and Davenport as shown in the image.**

3 **On the Editor Toolbar, click the Cut Polygons tool.**

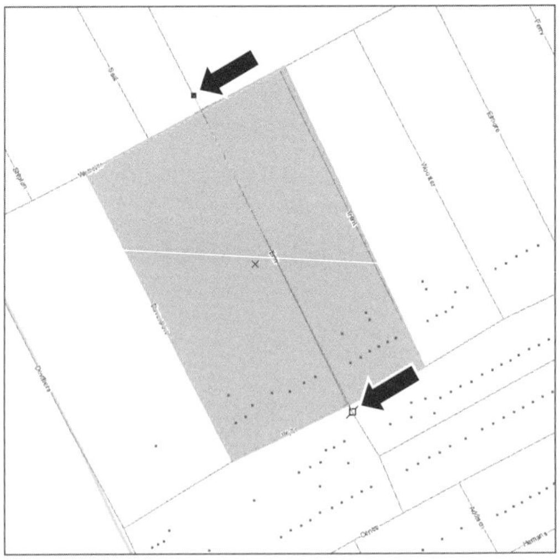

4 **Click Erin Street just above Webster, then double-click Erin Street just below Wylie as shown in the image.** The result is two new polygons.

5 **Stop Editing without saving any changes.**

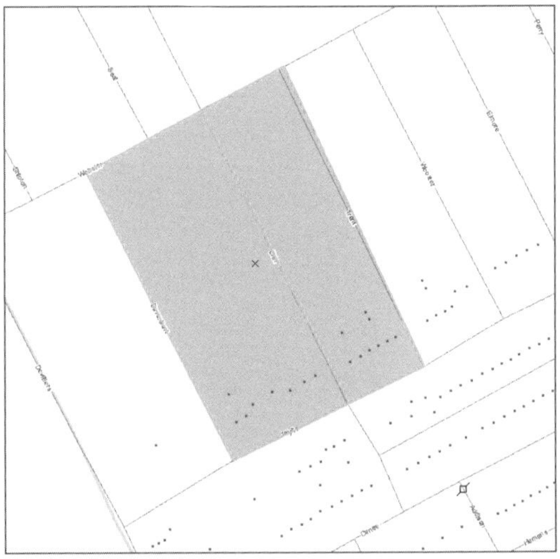

Tutorial 6-3

Digitize point features

Many governmental agencies use GIS for homeland security. Two GIS layers common to emergency preparedness applications are evacuation routes and shelter facilities. In this exercise, you will digitize shelter locations as points.

Create a point feature class for evacuation shelters

1. Click Windows, Catalog.
2. In the Catalog tree, browse to \ESRIPress\GIST1\MyExercises\Chapter6\MidHill.gdb.
3. Right-click MidHill.gdb, click New, and click Feature Class.
4. In the Name field, type **EvacShelter**.
5. In the Alias field, type **Evacuation Shelters**.
6. Select Point Features for Type and click Next.
7. Click Import, browse to the \ESRIPress\GIST1\Data\Pittsburgh\Midhill.gdb folder, click CommercialProperties, and Add.
8. Click Next three times, and click Finish.
9. Close Catalog.

Add evacuation shelter points

1. In ArcMap, click File and Open, browse to \ESRIPress\GIST1\Maps\, and open Tutorial6-3.mxd.
2. Click the Add Data button, navigate to \ESRIPress\GIST1\MyExercises\Chapter6\MidHill.gdb, and add EvacShelter to the map.

3 In the TOC, click the legend symbol for the Evacuation Shelters layer; change the symbol to Square 2, the color to Mars Red, and the size to 10; and click OK.

4 From the Editor toolbar, click Editor, Start Editing.

5 Click Evacuation Shelters as the layer to edit and click OK.

6 Click Evacuation Shelters in the Create Features panel.

YOUR TURN

The red squares in the map below are the shelter locations. Using the Point tool, click the corresponding locations in your map to add the shelter points to the EvacShelter layer. When finished, click Editor, Stop Editing. Click Yes to save edits to EvacShelter.

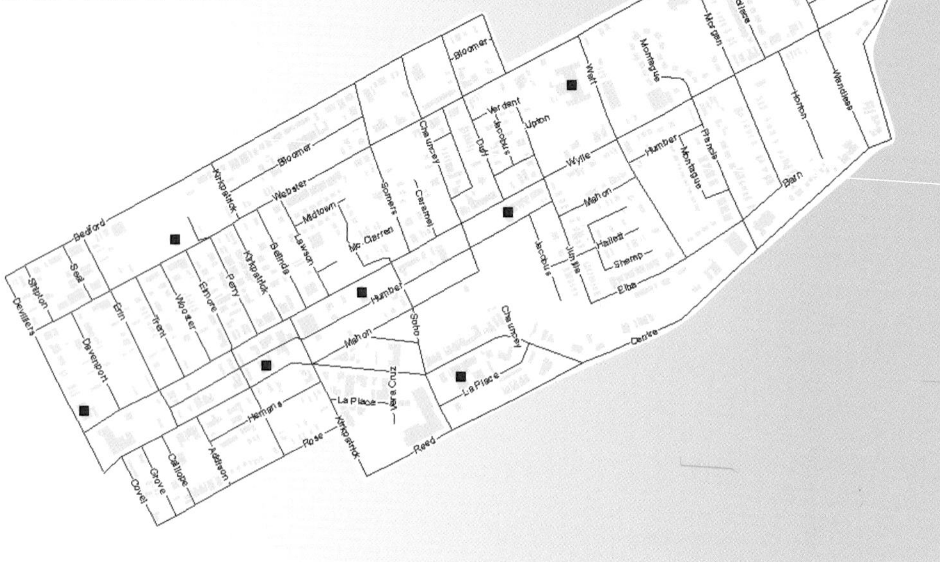

Add a field to the EvacShelter table

You did not add an attribute table field when you created the feature class, but you can add it now.

1 In the TOC, right-click the Evacuation Shelters layer.

2 Click Open Attribute Table.

3 In the EvacShelters table, click the Table Options button and Add Field.

4 In the Name field, type **ID** and click OK.

5 Repeat steps 3 and 4 except name the field ShelterName and make its type text.

Edit EvacShelter records

1 From the Editor toolbar, click Editor, Start Editing.

2 Click Evacuation Shelters as the layer to edit, and click OK.

3 In the Evacuation Shelters table, click the small gray box to the left of the first record in the table. This highlights the record in the table and the related feature in the map.

4 Using the table and map provided below as your input, add the ID and Name attributes to the selected record, then repeat the process for the remaining records.

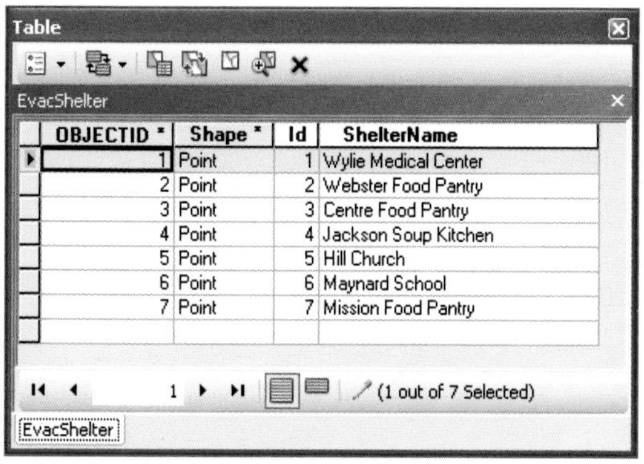
Table
EvacShelter

OBJECTID *	Shape *	Id	ShelterName
1	Point	1	Wylie Medical Center
2	Point	2	Webster Food Pantry
3	Point	3	Centre Food Pantry
4	Point	4	Jackson Soup Kitchen
5	Point	5	Hill Church
6	Point	6	Maynard School
7	Point	7	Mission Food Pantry

(1 out of 7 Selected)
EvacShelter

5 On the Edit menu, click Stop Editing. Click Yes to save your edits. Close the table, but leave ArcMap open.

Label map with shelter name

1 In the TOC, right-click the Evacuation Shelter layer and click Properties.

2 Click the Labels tab.

3 Type or make selections as follows:

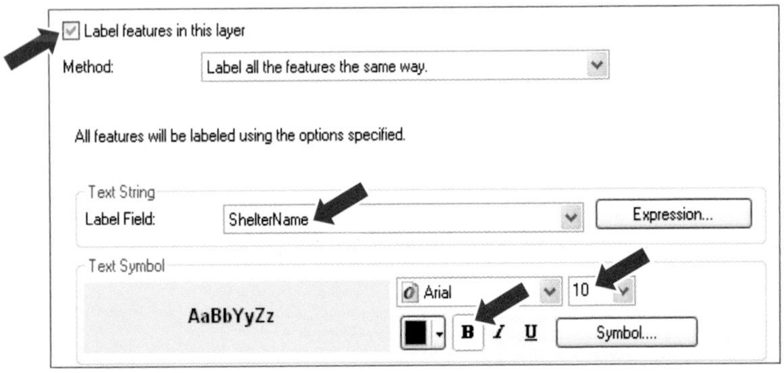

4 Click OK. The resulting map shows emergency planners where shelters are located.

5 Save your map document as **\ESRIPress\GIST1\MyExercises\Chapter6 \Tutorial6-3.mxd** and leave it open.

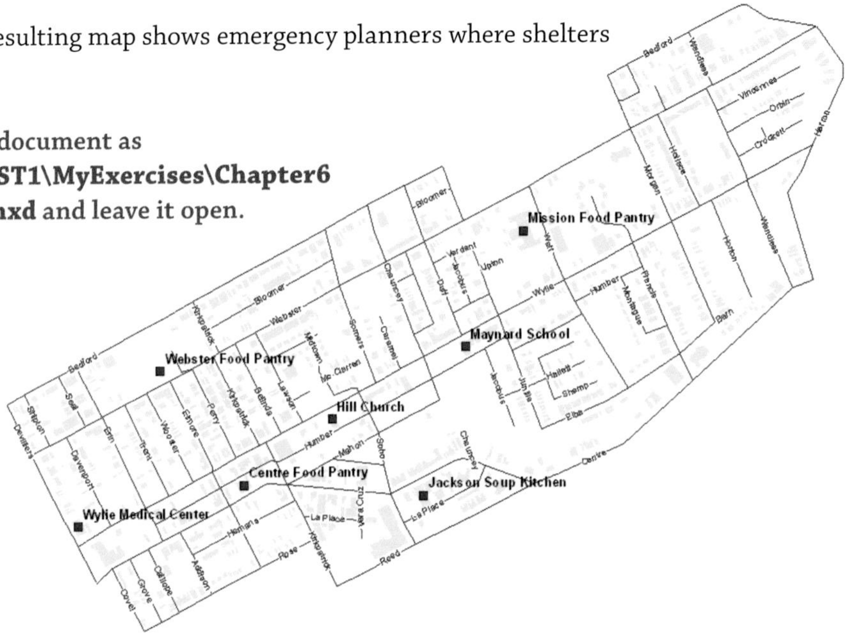

Tutorial 6-4

Digitize line features

Now you are ready to digitize an evacuation route to and from the evacuation shelters. In this exercise you will create a line shapefile for this scenario.

Create a line shapefile for an evacuation route

1 Click Windows, Catalog.

2 In the Catalog tree, browse to \ESRIPress\GIST1\MyExercises\Chapter6.

3 Right-click the Chapter6 folder, click New, and click Shapefile.

4 In the Name field, type **EvacRoute** and select Polyline for Feature Type.

5 Click Edit and Import; browse to \ESRIPress\GIST1\Data\Pittsburgh\MidHill.gdb; and click Streets, Add, and OK twice.

6 Close the Catalog window.

Add and symbolize the evacuation route shapefile

1 In ArcMap, add EvacRoute.shp to the map.

2 In the TOC, click EvacRoute's line symbol to open the Symbol Selector.

3 In the search text box at the top of the Symbol Selector window, type **Arrow Right Middle** and click the search button ; click the resulting symbol; change the Color to Cretean Blue and Width to 2; and click OK.

Prepare area for digitizing and start editing

1 Zoom to the western half of the Middle Hill neighborhood as shown below.

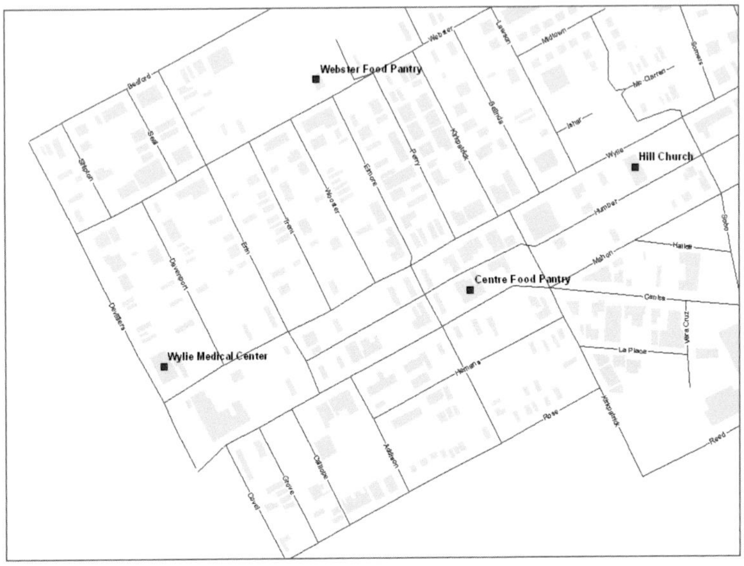

2 On the Editor toolbar, click Editor, Start Editing.

3 Click the EvacRoute layer and OK.

4 Click EvacRoute in the Create Features panel.

Digitize by snapping to features

You will snap digitized lines to features to make sure that line segments connect where they should. Make sure that Endpoint snap is on so you snap to the endpoint of the street segment.

1 Click the Straight Segment tool .

2 Click Wylie Medical Center to choose the route's starting point.

3 Move the cursor to the nearest street below the shelter, right-click, and click Perpendicular from the context menu.

4 **Click Wylie Avenue to place a vertex there.** ArcMap will force the line to be perpendicular with the Wylie Avenue street centerline.

5 Move the cursor to the first intersection on the street centerline (Wylie and Davenport) and click.

6 Continue snapping to street intersections along the evacuation route shown in the image and double-click to finish the route at Webster Food Pantry.

Save your edits and the map

1 From the Editor toolbar, click Editor, Stop Editing.

2 Click Yes to save your edits.

3 Save the map document as **\ESRIPress\GIST1\MyExercises\Chapter6\Tutorial6-4.mxd**.

3 In the Transparency field, type **20**. Click OK.

4 Repeat steps 1–3 for 26_45.tif.

Add a building outline

1 Click the Add Data button.

2 Browse to the \ESRIPress\GIST1\Data\CMUCampus\ folder and add HBH.shp to the map.

3 In the TOC, rename the HBH layer **Hamburg Hall**.

4 Change Hamburg Hall's symbol color to Mars Red and its symbol width to 1.5.

5 Zoom to the map extent. The Hamburg Hall layer, originated as a CAD drawing, is not in proper alignment or scale with the buildings shown on the aerial photos. Next, you will adjust the building layer so that it properly aligns with the aerial photo.

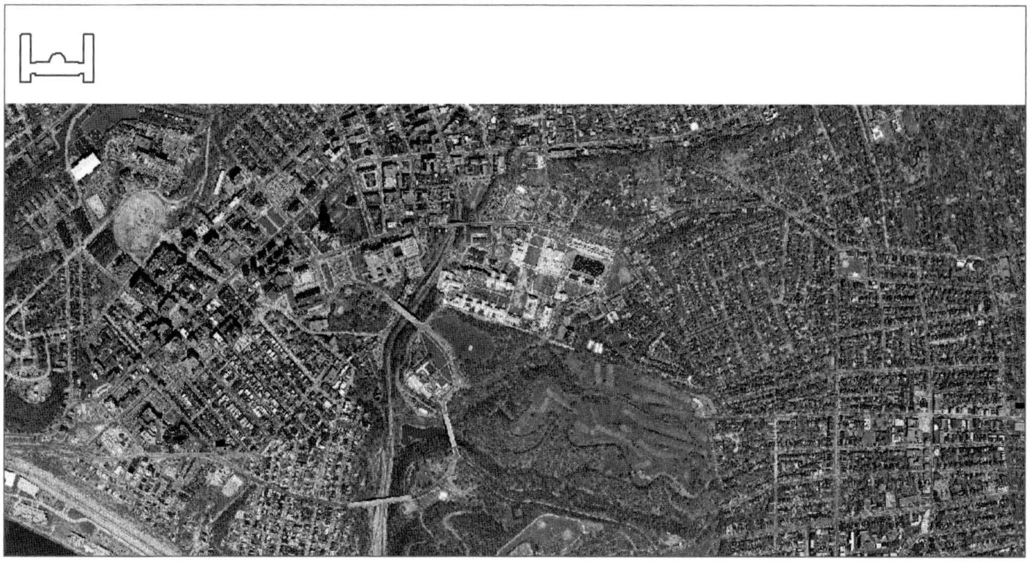

Move the building

1 On the Editor toolbar, click Editor, Start Editing. Make sure the Create Features layer is Hamburg Hall (HBH).

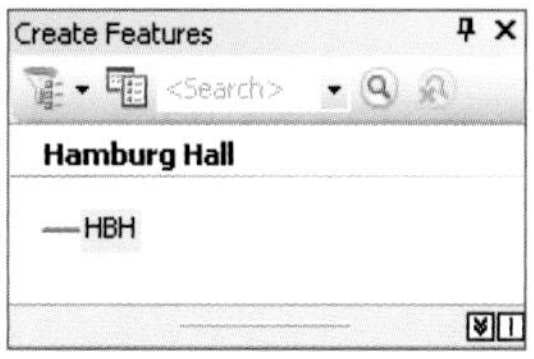

2 Click the Edit tool .

3 Click the outline of the Hamburg Hall feature.

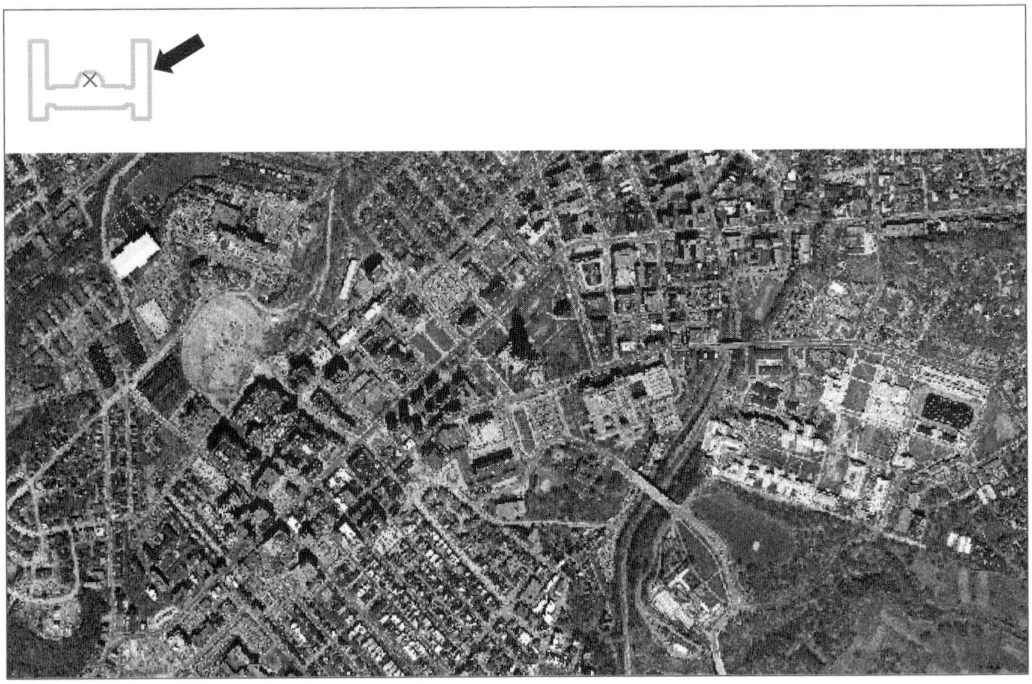

4 Place your mouse cursor directly over the outline of the Hamburg Hall feature so that the cursor icon changes to a four-headed arrow.

5 Click and drag the Hamburg Hall feature to the following location on the photo.

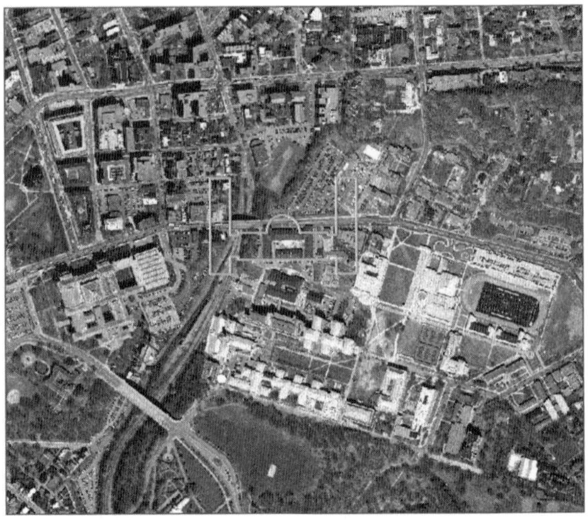

6 Zoom in to the building feature, as shown in the image.

Now you can better see Hamburg Hall in the photo. The building layer is too large and is upside down.

Rotate the building

1 Click the Edit tool and click the outline of the Hamburg Hall shapefile.

2 On the Editor toolbar, click the Rotate tool.

3 Click the lower-right grab handle of Hamburg Hall and, while holding down the mouse button, rotate it 180 degrees, as shown in the image.

Add displacement links

To align the feature with the aerial photo, you will use a transformation tool.

1 **Click Editor, More Editing Tools, and Spatial Adjustment.** This opens the Spatial Adjustment toolbar.

2 **Click Spatial Adjustment, Adjustment Methods, Transformation - Similarity.**

3 **Click the New Displacement Link tool .**

4 **Click the upper left corner of the Hamburg Hall building feature.**

5 **Click the corresponding location on the aerial photo.**

6 Continue adding displacement links to the building feature and the aerial photo, as shown in the image.

Edit displacement links

If you select the wrong position on the building or map, you can use the edit displacement tools to adjust your picks.

1 From the Spatial Adjustments toolbar, click the Select Elements tool.

2 Click one of the displacement links.

3 Click the Modify Link tool.

4 Click and drag the link to a new position.

5 Drag the link back to its original location.

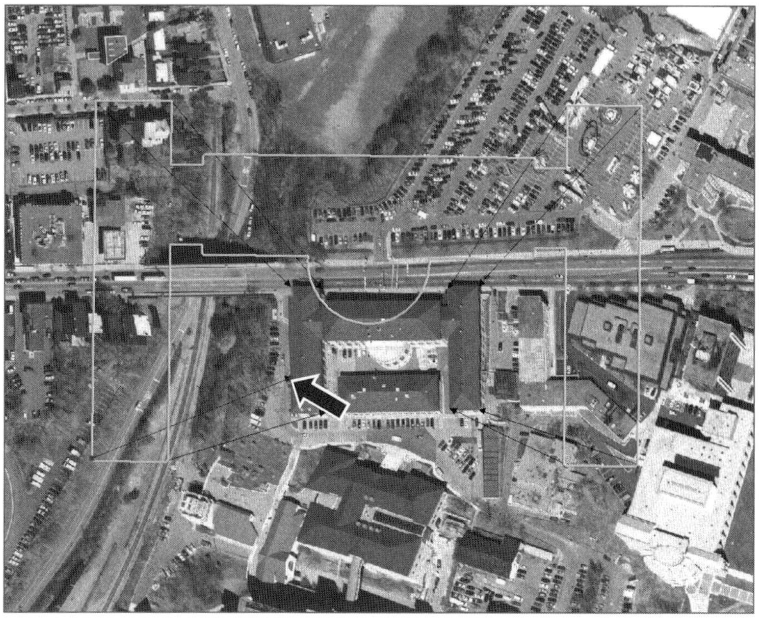

YOUR TURN

Zoom to Hamburg Hall in the aerial photo and use the Modify Links tool to more precisely move the displacement links to the corners of the building.

Adjust the building

1. **From the Spatial Adjustment toolbar, click Spatial Adjustment, Adjust.**

2. **Stop editing and save your edits.** ArcMap scales down the Hamburg Hall feature to match the geometry of the feature in the aerial photo. If the resulting match is not very good, select the Hamburg Hall feature, redefine new displacement links, and run the Adjust command again.

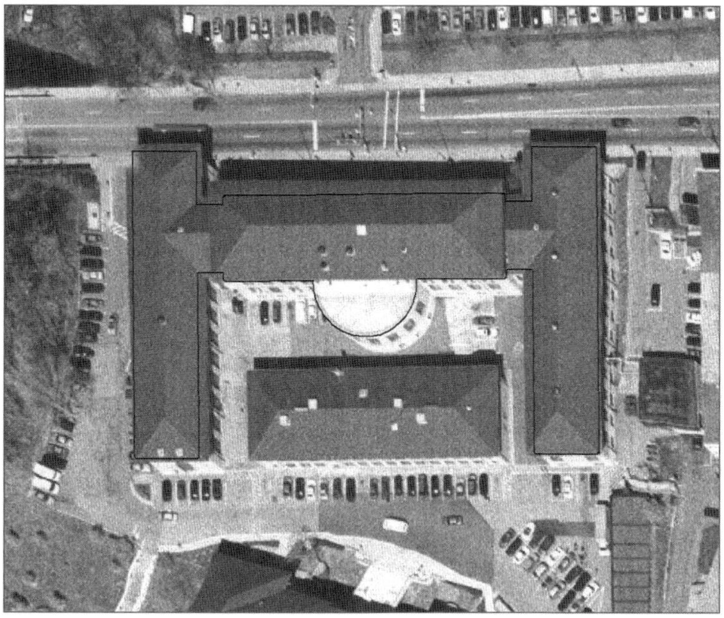

3. Save your map as **\ESRIPress\GIST1\MyExercises\Chapter6\Tutorial6-5.mxd**.

Assignment 6-1

Digitize police beats

Community-oriented police officers are responsible for preventing crime and solving underlying community problems related to crime. Among other activities, these officers walk "beats," which are small networks of streets in specified areas. Often the beats are designed in cooperation with community leaders who help set policing priorities. Beats change as problems are solved and priorities change. Hence, it is good to have the capability to digitize and modify police beats.

In this assignment, you will digitize two new polyline police beats for the city of Pittsburgh Zone 2 Police District based on street centerlines that make up these beats.

Start with the following:

- \ESRIPress\GIST1\Data\Pittsburgh\Zone2.gdb\streets—TIGER streets for Zone 2 Police District
- \ESRIPress\GIST1\Data\Pittsburgh\Zone2.gdb\zone2—polygon layer for boundary of Zone 2 Police District

Create a police beat map

In ArcCatalog, create a new file geodatabase called **\ESRIPress\GIST1\MyAssignments\Chapter6\Assignment6-1YourName.gdb** that includes two newly digitized line feature classes for police beats 1 and 2 imported into it. See the guidelines on the next page for what streets should make up the beats. Call these line feature classes Beat1 and Beat2.

In ArcMap, create a new map document called **\ESRIPress\GIST1\MyAssignments\Chapter6\Assignment6-1YourName.mxd** with a layout showing a map with an overview of the Police Zone 2 outline with existing streets and the newly digitized beats and maps zoomed in to beats 1 and 2. Show the beats with thick line widths and bright, distinctive colors; and show streets as lighter "ground" features. In the overview map, label the beats "Beat #1" and "Beat #2," and label the streets in the detailed maps. Include a scale bar in feet. See hints for digitizing.

Export your map to a PDF file called **\ESRIPress\GIST1\MyAssignments\Chapter6\Assignment 6-1YourName.pdf**.

Street centerline guides for Beat #1

- 1 through 199—17th St (four segments)
- 1 through 99—18th St (two segments)
- 1 through 199—19th St (one segment)
- 1 through 199—20th St (four segments)
- 1 through 99—Colville St (one segment)
- 1700 through 1999—Liberty Ave (one segment)
- 1700 through 1999—Penn Ave (three segments)

- 1700 through 1999—Smallman St (four segments)
- 1700 through 1999—Spring Way (one segment)

Street centerline guides for Beat #2

- 100 through 299—7th St (two segments)
- 1 through 299—8th St (three segments)
- 100 through 299—9th St (four segments)
- 800 through 899—Exchange Way (one segment)
- 700 through 899—Ft Duquesne Blvd (three segments)
- 700 through 899—Liberty Ave (three segments)
- 100 through 199—Maddock Pl (one segment)
- 700 through 899—Penn Ave (three segments)

Hint

Open the feature attribute table for the streets. Move the table so you can see both the table and the streets on the map. Sort the table by field 'NAME' and make multiple selections for a given beat in the table by simultaneously holding down the Ctrl key and clicking rows corresponding to the beat's street segments. The streets layer is a TIGER file map with TIGER-style address number data, so look for street number ranges in the following fields: L_F_ADD, L_T_ADD, R_F_ADD, and R_T_ADD. With all streets for a beat selected, digitize lines for every street making up beats in the new line layers (Beat1 and Beat2). Use various tools and techniques found in chapter 6.

WHAT TO TURN IN

If your work is to be graded, turn in the following files:

File geodatabase: \ESRIPress\GIST1\MyAssignments\Chapter6\Assignment6-1YourName.gdb

ArcMap document: \ESRIPress\GIST1\MyAssignments\Chapter6\Assignment6-1YourName.mxd

Exported map: \ESRIPress\GIST1\MyAssignments\Chapter6\Assignment6-1YourName.pdf

If instructed to do so, instead of the above individual files, turn in a compressed file, **Assignment6-1YourName.zip**, with all files included. Do not include path information in the compressed file.

Assignment 6-2

Use GIS to track campus information

GIS is a good tool to create "way finding" information maps. These maps can be used in many organizations that have large campuses or complicated buildings (airports, hospitals, office parks, colleges, and universities). For example, Carnegie Mellon University's campus can be confusing, especially to new students and visitors.

In this exercise, you will create a GIS campus map of parking, bus stops, and academic buildings by spatially adjusting buildings to an aerial photo map of the campus. You will digitize features showing bus stop and parking lot locations. Additional layers could be for routes around the campus that lead you to various buildings.

Start with the following:

- \ESRIPress\GIST1\Data\CMUCampus\25_45.tif and 26_45.tif—digital orthographics of CMU campus provided by Southwestern Pennsylvania Commission, Pennsylvania State Plane South NAD 1983 projection.

Note: These aerial photos are 80 MB in size. Do not include them in your geodatabase, just point to them in the map.

- \ESRIPress\GIST1\Data\CMUCampus\CampusMap.dwg—CAD drawing of CMU campus provided by the CMU facilities management department

Create a campus map

In ArcCatalog, create a new file geodatabase called **\ESRIPress\GIST1\MyAssignments\Chapter6\Assignment6-2YourName.gdb** with new feature classes Parking (polygons) and BusStops (points). Import campus map polygons for academic buildings only into the geodatabase (see Hints). Assign the NAD_1983_StatePlane_Pennsylvania_South_FIPS_3702_Feet projection to all feature classes. Do not import the aerial images but simply add them to your map from their original location.

In ArcMap, create a new map document called **\ESRIPress\GIST1\MyAssignments\Chapter6\Assignment6-2YourName.mxd** with a layout that shows the aerial photos of the CMU campus and academic buildings from the CampusMap drawing spatially adjusted to match the buildings in the aerial photo.

Digitize new polygons and points showing parking lots and bus stops. Digitize four separate parking lot polygon features in or around campus. You can see the parking lot locations because there are cars in the parking lots in the aerial photo. Digitize four bus stops at various street locations (you decide). Show the parking lots as semitransparent polygons so you can see the parking lots and cars in the aerial photo. Set the transparency of the aerial photos to 25 percent to better see the new features.

Export the finished map to a JPEG file called **\ESRIPress\GIST1\MyAssignments\Chapter6\Assignment6-2YourName.jpg**.

Hints

- Add the CMU campus CAD drawing as polyline features. In the TOC, right-click CampusMap.dwg Polyline and click Properties. Click the Drawing Layers tab and leave only Academic_Bldgs turned on. Export the buildings as a new polyline feature called **AcademicBldgs** to the Assignment6-2YourName geodatabase.
- Use ArcMap's editing tools to move the buildings closer to the aerial image. Then use spatial adjustment tools and zooming functions to adjust the buildings to the aerial photo. Continue using editing tools, such as Move and Rotate, to adjust the buildings according to the aerial photo below. Note that some buildings might be missing from the aerial photo, as these were built after the photo was taken.
- It is tricky to transform the academic buildings to the raster map. If you get an approximate transformation that has some mismatches, that is acceptable. Do the following steps: Zoom to the full extent to see the raster images and academic buildings. Using the Edit tool, select all of the academic buildings in the AcademicBldgs layer. Drag the buildings adjacent to the raster images. Select buildings on the four corners of campus and use points on this layer that match locations on the raster map image. On the Spatial Adjustment toolbar, use the New Displacement Link to roughly draw four lines from the buildings layer to the raster image. Zoom in to a point on the buildings map, click the corresponding link with the Select Elements tool on the Spatial Adjustment toolbar, click the Modify Link button, and move the link endpoint to be more precise. Do the same on the raster image side of the link. Repeat for the other three links.
- Use vertex edit functions to fine tune building corners for at least one building.

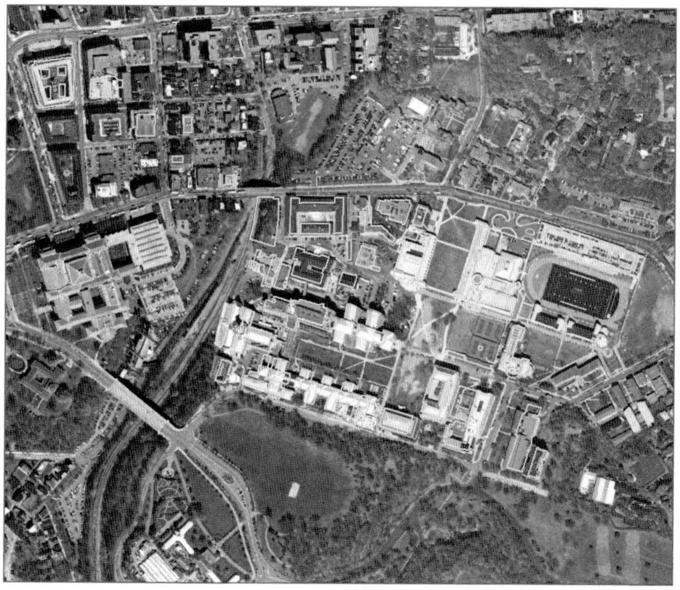

WHAT TO TURN IN

If your work is to be graded, turn in the following files:

File geodatabase: \ESRIPress\GIST1\MyAssignments\Chapter6\Assignment6-2YourName.gdb

ArcMap document: \ESRIPress\GIST1\MyAssignments\Chapter6\Assignment6-2YourName.mxd

Image file: \ESRIPress\GIST1\MyAssignments\Chapter6\Assignment6-2YourName.jpg

If instructed to do so, instead of the above individual files, turn in a compressed file, **Assignment6-2YourName.zip**, with all files included. Do not include path information in the compressed file.

7

Geocoding

Geocoding is the process used to plot address data—such as 123 Oak Street, Pittsburgh, PA 15213—as points on a map. You can geocode addresses to different levels such as ZIP Codes or streets, depending on the type of address data you have or wish to map. In this chapter, you will learn to geocode using source tables of address data and location reference data of TIGER street centerlines and ZIP Code polygons obtained from the U.S. Census Bureau. You will also learn how to fix errors in both source and reference data you use for geocoding.

Learning objectives

- *Geocode data by ZIP Code*
- *Geocode data by street address*
- *Correct source addresses using interactive rematch*
- *Correct street reference layer addresses*
- *Use an alias table*

Tutorial 7-1

Geocode data by ZIP Code

Geocoding to ZIP Codes is a common practice for many organizations because this data is often available in client and other databases. Furthermore, for marketing and planning, it is often sufficient to study the spatial distribution of clients by ZIP Code. ZIP Code areas lack an underlying design principle except for facilitating delivery of mail, so interpretation of results is sometimes limited. In these tutorials, you will match attendees for an art event sponsored by an arts organization in Pittsburgh, Pennsylvania, called FLUX. The event planners of FLUX would like to know where function attendees reside for planning and future marketing activities.

Open and examine the starting map document

1 **Start ArcMap and open \ESRIPress\GIST1\Maps\Tutorial7-1.mxd.** The map document includes the two needed inputs for geocoding, the ZIP Code map of Pennsylvania, and the data table Attendees.

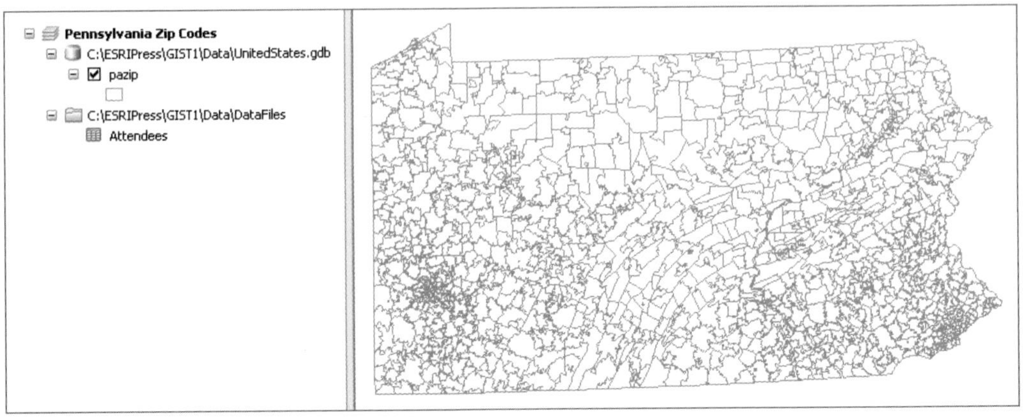

2 Click File and Save As, browse to \ESRIPress\GIST1\MyExercises\Chapter7\, and click Save.

3 **Right-click the Attendees table in the TOC and click Open.** The table contains the addresses, including 5-digit ZIP Code, and ages of all attendees of two recent FLUX events. Notice that two of the first six attendees are out of state and thus will not geocode with the Pennsylvania location reference map. As a supplement, before closing the table, let's tabulate how many such records there are. That will inform the performance assessment of geocoding within Pennsylvania.

OBJECTID *	Date_	Address	City	State	ZIP_Code	Age
1	20090629	2415 1ST AVE	SACRAMENTO	CA	95818	27
2	20090629	224 NORTH ST	STEUBENVILLE	OH	43952	32
3	20090629	PO BOX 622 535 4TH ST	MARIANNA	PA	15345	32
4	20090629	5126 JANIE DRIVE	PITTSBURGH	PA	15227	34
5	20090629	305 AVENUE A	PITTSBURGH	PA	15221	40
6	20090629	1431 CRESSON ST	PITTSBURGH	PA	15221	26

4 **At the top of the table, click the Select by Attributes button, double-click "State" in the top panel, click the = button, click Get Unique Values, scroll down and double-click 'PA', and click Apply and Close.** Information at the bottom of the table indicates that 1,124 out of 1,265 records (89 percent) are for Pennsylvania, so that is the maximum number of points that could be geocoded. If there are missing or incorrect ZIP Codes for Pennsylvania, the number geocoded will be less.

5 **Click the Clear Selection button and close the table.**

Create an address locator for ZIP Codes

The geocoding process requires several settings and parameters. Rather than have you specify them interactively each time you geocode data, ArcMap has you save settings in a reusable file, called an address locator. Included in the settings is a pointer to the reference data you will use to geocode the attendee data, the PAZip (Pennsylvania ZIP Codes) layer that is currently in your map document.

1 **In the Catalog window, navigate to \ESRIPress\GIST1\MyExercises\Chapter7\, right-click the Chapter 7 folder, and click New and Address Locator.**

2 **In the Create Address Locator window, click the browse button for the Address Locator Style, scroll down, click US Address – ZIP 5-Digit, click OK, and ignore the warning icon and message.**

3 **Click the browse button for Reference Data, browse to \ESRIPress\GIST1\Data\, double-click UnitedStates.gdb, and click PAZip and Add.**

4 Click the browse button for Output Address Locator, browse to \ESRIPress\GIST1\MyExercises\Chapter7, double-click Geocoding.gdb, type **PAZipCodes** for name, and click Save and OK.

5 Click OK and wait until Catalog informs you that the address locator is created. PAZipCodes appears as an address locator under Geocoding.gdb in the Catalog top panel.

6 Hide the Catalog window.

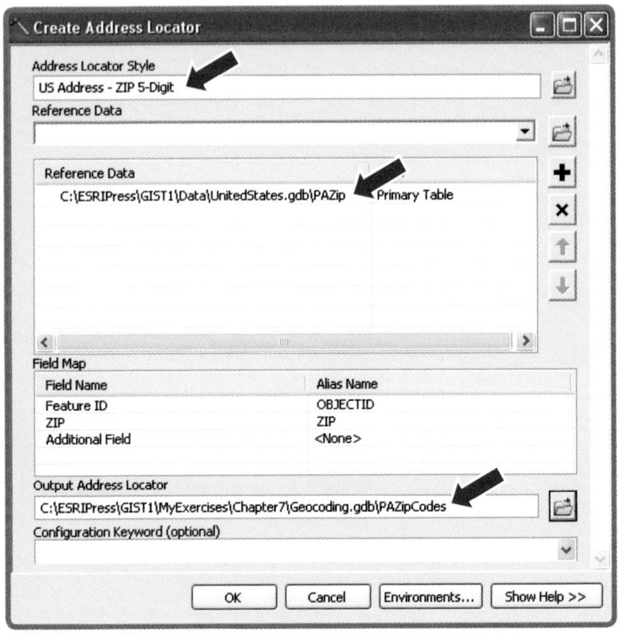

Geocode records by ZIP Code

1 In ArcMap, click Customize, Toolbars, Geocoding. The Geocoding taskbar appears.

2 On the Geocoding toolbar, click the Geocode Addresses button, click PAZipCodes to select it, click OK, check that the Attendees table is selected as the Address table, select ZIP_Code as the ZIPCode field, and change the name of the output to **\ESRIPress\GIST1\MyExercises\Chapter7\Geocoding.gdb\AttendeesZIP**.

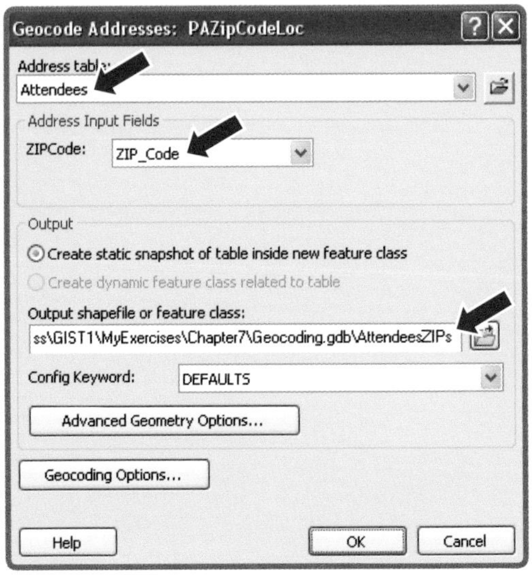

3 **Click OK.** ArcMap geocodes the addresses by ZIP Code with 86 percent of the records mapped—less than the 89 percent of addresses that are in Pennsylvania, as expected.

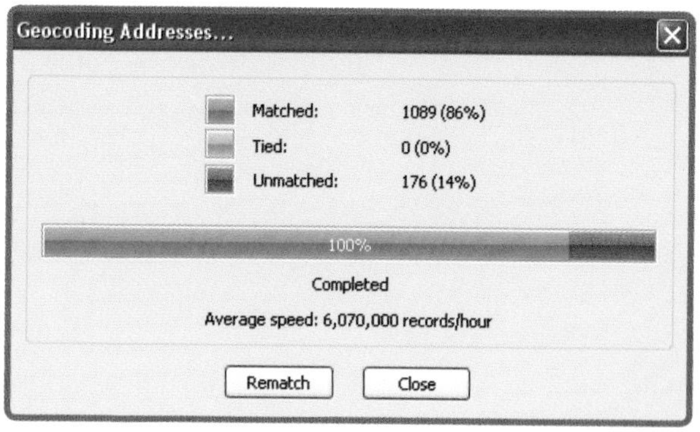

4 **Click Close.** ArcMap adds the geocoding results to the map with point markers at the centroids of ZIP Codes that have one or more attendees. As you might expect, attendees cluster around southwestern Pennsylvania near the location of the FLUX events, but attendees come from all over the state.

5 **Click the List By Drawing Order button in the TOC.**

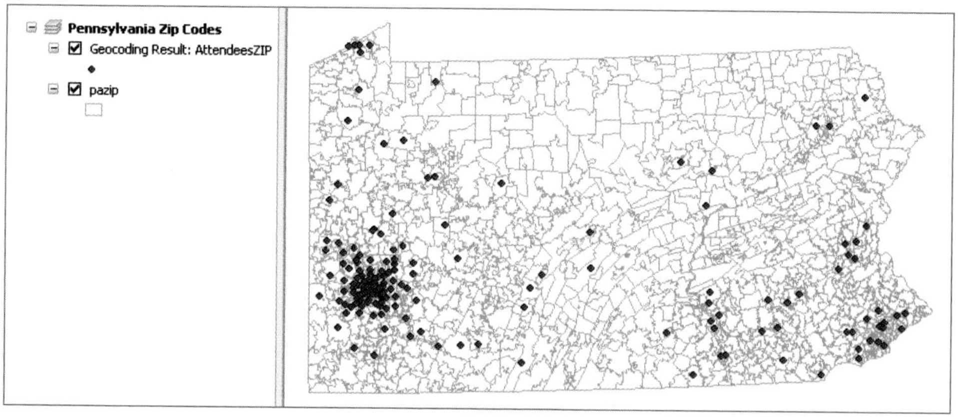

Count geocoded records by ZIP Code

You can get a better understanding of the geocoding output by next taking an extra step, aggregating geocoded points to obtain a count of attendees per ZIP Code area.

1 **Right-click GeocodingResult: AttendeesZIP in the TOC, and click Open Attribute Table.**

2 **Right-click the Match_Addr column header, click Summarize, change the output table to \ESRIPress\GIST1\MyExercises\Chapter7\Geocoding.gdb\CountAttendees, and click OK and Yes.**

3 **Close the geocoding results table.**

4 **Right-click CountAttendees in the TOC and click Open.**

5 **Right-click Count_Match_Addr column header and click Sort Descending.** In total there are 1,476 ZIP Codes in the state, but only 22 ZIP Codes with 10 or more attendees and 90 that have 2 or more out of the 174 matched.

6 **Close the table.**

YOUR TURN

Join CountAttendees to GeocodingResult:AttendeesZip using Join attributes to a table. Create a map displaying Count_Match_Addr using size-graduated point markers with five classes and quantiles. Zoom in to southwestern Pennsylvania.

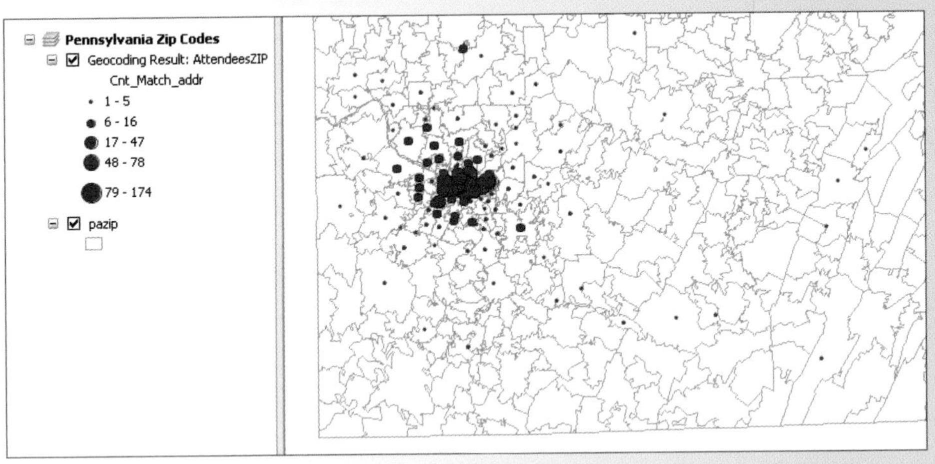

Fix and rematch ZIP Codes

1 **Click Geocoding Result: AttendeesZIP in the TOC, and click the Review/Rematch Addresses button on the Geocoding toolbar.** The Interactive Rematch window shows each unmatched record individually and allows you to manually edit the address values.

2 **Scroll down to the record with ObjectID=50, whose address value is 414 South Craig Street, and select that record.**

3 **Scroll horizontally (and adjust field widths by dragging their header boundaries) so you can see the Address, City, State, and ZIP_Code fields.** Notice that the ZIP_Code is missing for this record. That address's ZIP_Code is 15213.

4 **With the 414 SOUTH CRAIG STREET row selected, type 15213 in the ZIP_Code field and press the Tab key on your keyboard.** The Candidate panel shows one candidate with a perfect score of 100.

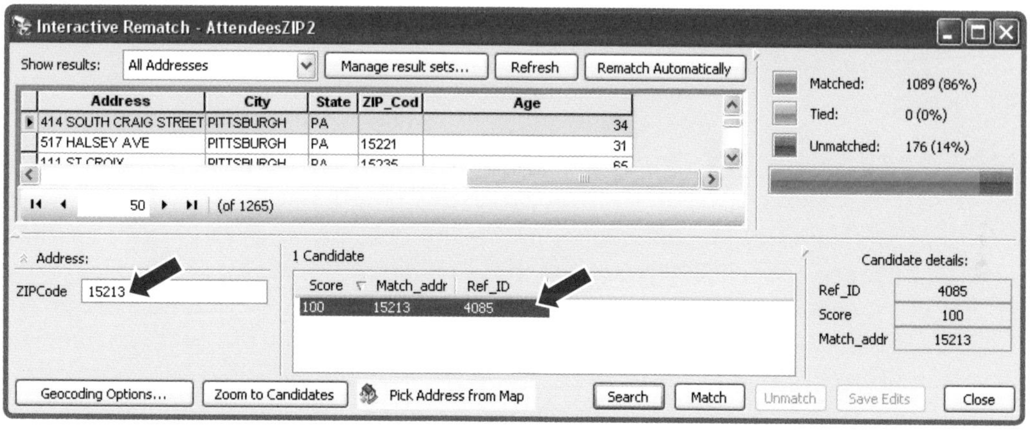

5 **Click the Match button.** The count of matched addresses goes up by one, from 1,089 to 1,090 as ArcMap is successful with that point. With some research, it is usually possible to make similar additions.

YOUR TURN

Use the U.S. Postal Service's ZIP Code lookup Web site, **`http://zip4.usps.com/zip4/welcome.jsp`**, to find the ZIP Code for the record with ObjectID=57, 11244 Azalea Dr, Pittsburgh, PA (it's 15235). Then use Interactive Rematch to match the address. Close the Interactive Rematch window, and save your map document when finished.

Tutorial 7-2

Geocode data by street address

In this exercise, you will again geocode the FLUX attendee records, but this time at the street level for Pittsburgh, Pennsylvania. Of course, the records you wish to map contain attributes for the street address, such as 123 Oak Street. In this case, you will also incorporate ZIP Code into the address locator, because some addresses may have the same house number and street name but be in different ZIP Code areas. This happens frequently in study areas that have two or more municipalities, such as in a county. In this case there are address records from other cities and states, so ZIP Code plays an important role to eliminate non-Pittsburgh addresses that have the same street address as in Pittsburgh.

Examine address data and street map

1 Open \ESRIPress\GIST1\Maps\Tutorial7-2.mxd.

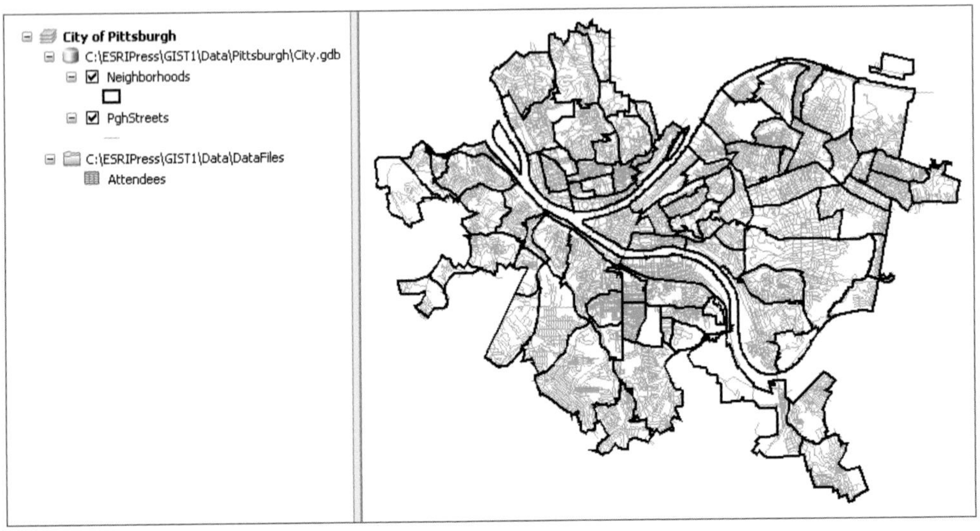

2 Click File and Save As, browse to \ESRIPress\GIST1\MyExercises\Chapter7\, and click Save.

3 Open the attribute tables for PghStreets and Attendees and review their contents, especially addresses. You will find that PghStreets has TIGER-style street address data, with starting and ending house numbers for each street segment. The table, Attendees, has street address in one field, Address, plus City, State, and ZIP_Code in their own fields. Only Address and ZIP_Code are necessary for geocoding, because with ZIP_Code you can look up city and state.

4 **Close the tables when you are finished reviewing them.**

Create an address locator for streets with a zone

1 **Click the Catalog icon in the right edge of the ArcMap window (or, if the icon is not there, click Windows, Catalog).**

2 **Expand the \ESRIPress\GIST1\MyExercises\Chapter7\ folder, right-click Geocoding.gdb, and click New and Address Locator.**

3 **In the Create Address Locator window, click the browse button for the AddressLocator Style, scroll down, click US Address – Dual Ranges, click OK, and ignore the warning icon and message.**

4 **Select PghStreets for Reference Data, click the browse button for Output Address Locator, browse to \ESRIPress\GIST1\MyExercises\Chapter7, double click Geocoding.gdb, type PghStreets for name, and click Save.** Notice that Catalog identifies all fields that it needs for the geocoding process in the lower panel of the Create Address Locator window. If it was unsuccessful in doing so, you would have to click in a cell on the right and select the needed field name from the table. Field names that start with an asterisk are required.

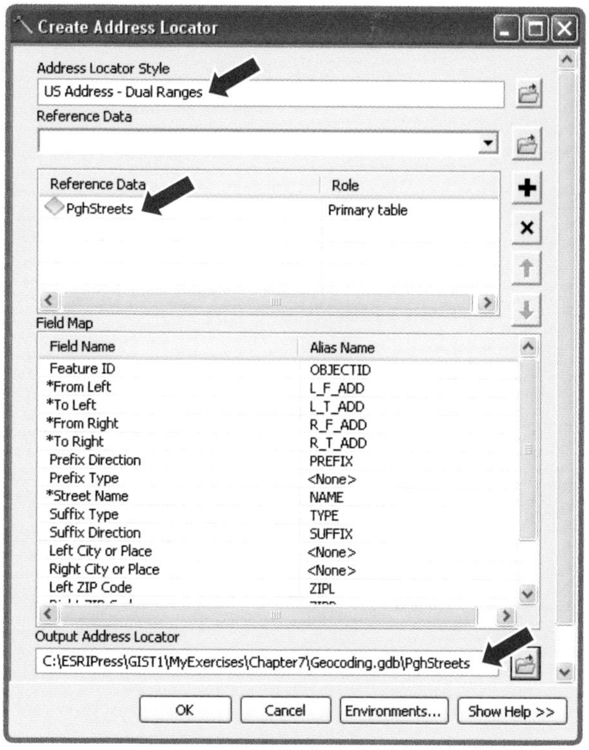

5 **Click OK and wait until Catalog informs you that the address locator is created.**

6 **Expand Geocoding.gdb in the Catalog tree and double-click PghStreets.** That brings up the Address Locator Properties sheet. First, notice that you can associate an alias table with the locator. It would contain the place-names such as PNC Ballpark and associated street addresses such as 115 Federal Street. With an alias table, ArcMap makes a pass through address data replacing place-names with their street addresses. Next, you see that the allowable connectors for street intersection addresses, such as Oak St & Pine Ave, are currently the "&", "@", and "|" characters, and the word "and". If your data has a different separator, you could type it here. Also, the point that ArcMap assigns to an address has a 20-foot offset on the correct side of the street (left or right). You can change the offset to another value if desired. Offsets are desirable, especially for aggregating geocoded data points up to counts by area, such as neighborhoods. Areas tend to use street centerlines as boundary lines, so an offset ensures that ArcMap will count each point once, in the correct polygon. If on the centerline, a point gets double-counted for polygons that share the street segment as a boundary line.

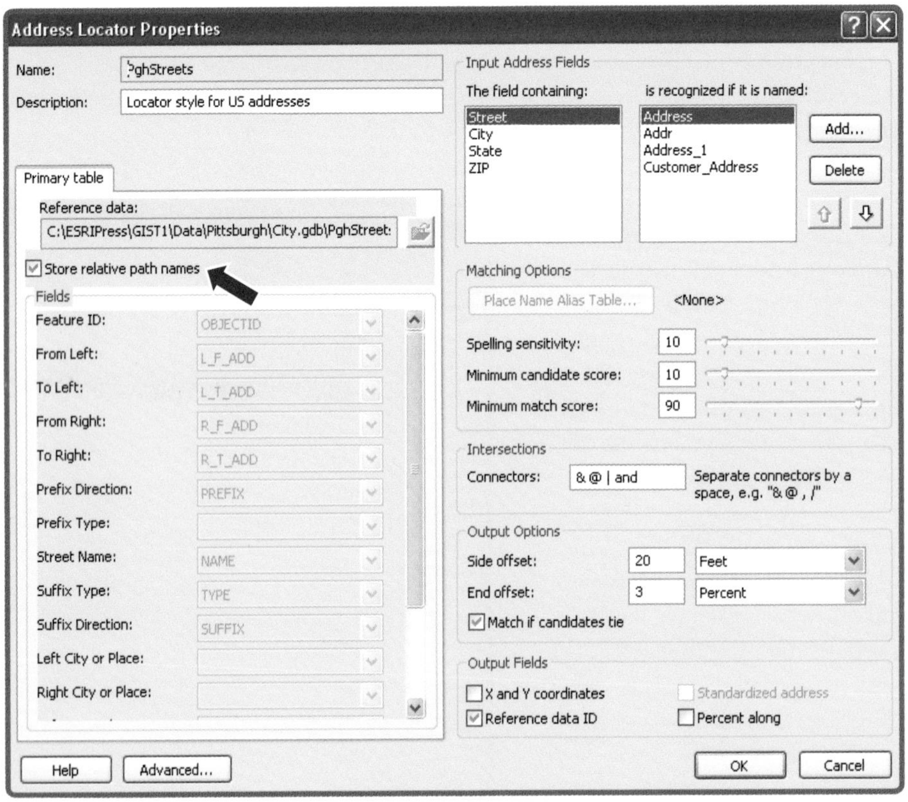

7 **Click the Store relative path names check box, click OK, close the Address Locator Properties window, and hide the Catalog window.**

Interactively locate addresses

Before geocoding the Attendees table to the streets layer, you will try out your locator with ArcMap's Find tool to locate individual addresses. The Find tool that you will use has the same methodology as geocoding to transform street address data into a point on the map. It matches the address you type in with similar data stored as attribute data in the street centerline map. It does the matching by finding good candidates and then computing a match score for each. Then for each identified problem or flaw of a candidate in matching the desired address, ArcMap subtracts a penalty from the match score. The candidate with the perfect score, 100, or highest score above a threshold value is chosen as the geocoded point.

1 On the Tools toolbar, click the Find button.

2 Click the Locations tab, select PghStreets, and type **3609 Penn Ave** in the Full Address field.

3 Click Find. The locator finds the address, briefly flashing it on the map.

4 Right-click the matched address of the lower panel to open a context menu.

5 Click the Add Labeled Point option.

6 Close the Find window. A point appears at the corresponding point on the map, along with a label for the street address.

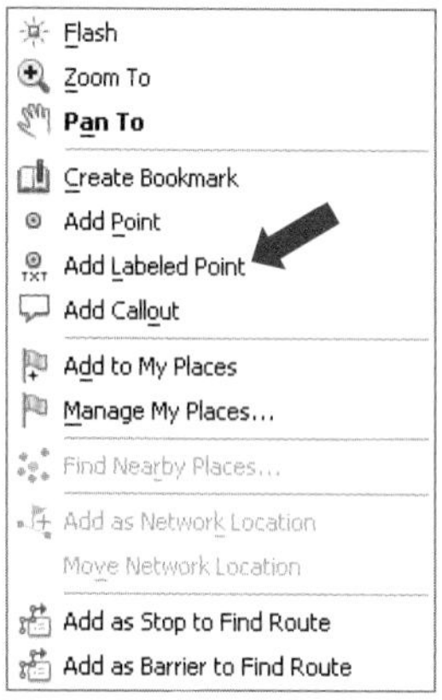

YOUR TURN

Use the Find tool to locate the following addresses. Add the best match in each case to the map as a labeled point.

- 1920 S 18th ST
- 255 Atwood ST
- 3527 Beechwood BLVD

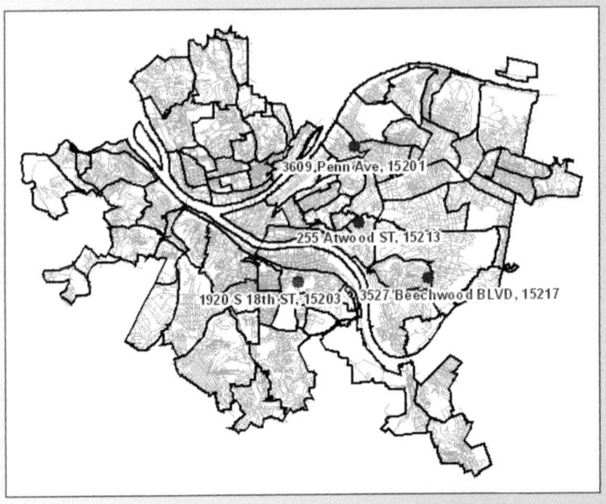

The finished map at right has some editing of the labels, which is optional for you. You can ungroup a label and point by clicking the label to select the graphic, right-clicking the graphic, and clicking Ungroup. Then you can format the font and point color by right-clicking an element, clicking Properties, and clicking Change Symbol.

When finished, delete the label and point graphics by dragging a rectangle around them to select them and press the Delete key on your keyboard.

Geocode address data to streets

We do not expect ArcMap to geocode a high percentage of records because the streets reference layer is only for Pittsburgh while many attendees live outside of Pittsburgh.

1 **If the Geocoding toolbar is not already open, click Customize, Toolbars, Geocoding.**

2 **On the Geocoding toolbar, click the Geocode Addresses button, select PghStreets, and click OK.**

3 **Type or make selections as shown in the image (but do not click OK).**

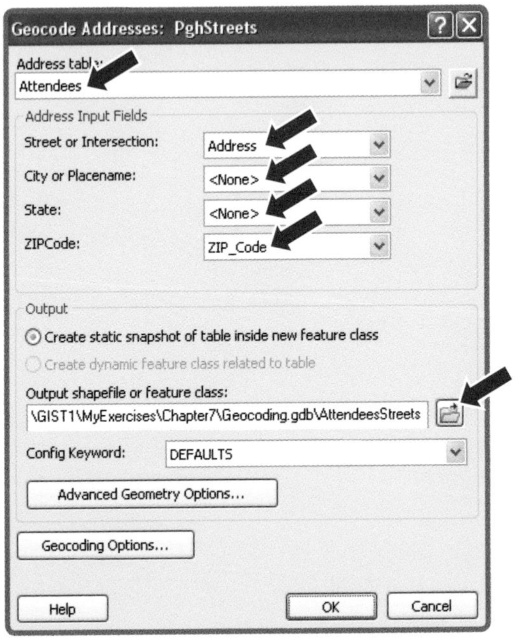

4 **Click Geocoding Options.** This is a user dialog box controlling the behavior of address matching and its outputs. You saw several of these options earlier. Here you can choose to add the X and Y coordinates of matched points to the output map layer's attribute table. You will leave all default settings as found here.

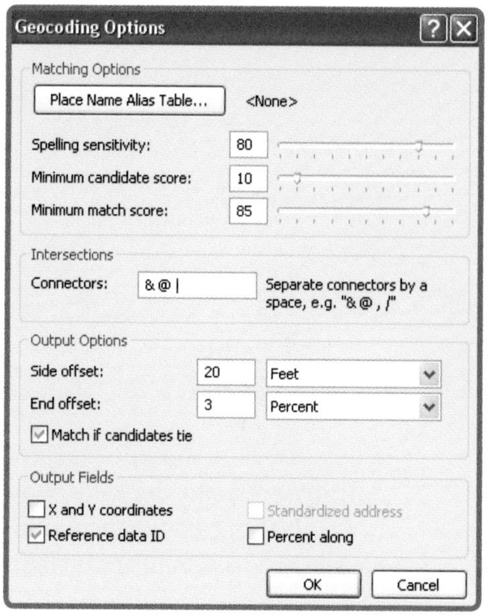

5 **Click Cancel, OK.** ArcMap only matches 43 percent of records, a relatively low number as expected, because there are many addresses outside of Pittsburgh as well as unmatched addresses in Pittsburgh.

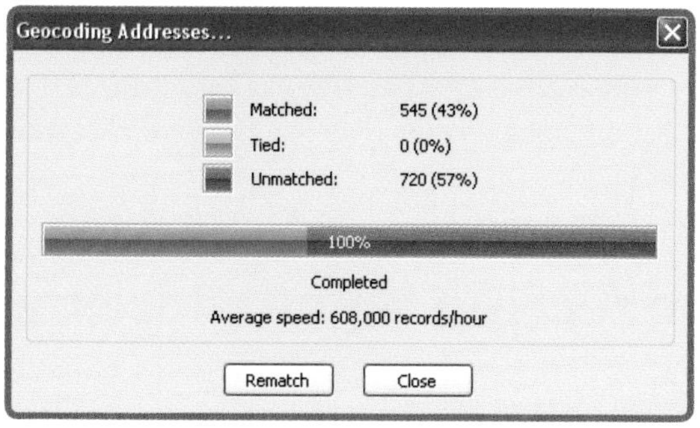

6 **Click Close.** The residences of attendees have distinctive spatial patterns across Pittsburgh neighborhoods, which could inform marketing campaigns to increase attendance. It looks like most attendees come from only 15 to 20 out of the 94 neighborhoods comprising Pittsburgh.

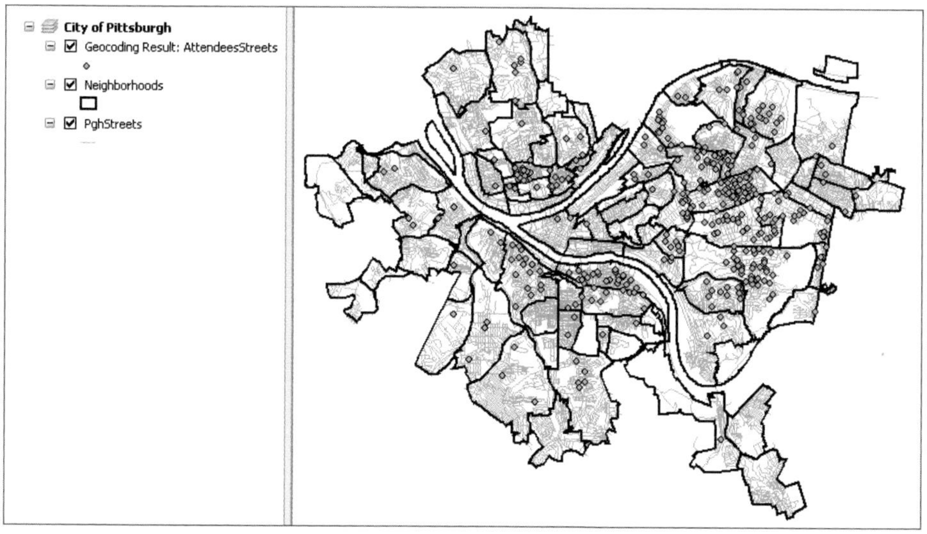

Tutorial 7-3

Correct source addresses using interactive rematch

While ArcMap did not match many of the addresses in the Attendees table to the PghStreets layer because they are outside of Pittsburgh, others did not match due to spelling errors or data omissions in the input table. Making corrections to the address data depends on the user's knowledge of local streets and addresses. In this tutorial, you will use ArcMap's interactive review process to correct and then match a few of the unmatched records.

Rematch interactively by correcting input addresses

1. Click the Geocoding Result: AttendeesStreets in the TOC. On the Geocoding toolbar, click the Review/Rematch Addresses button . You can increase the height of the panel with addresses by finding and dragging horizontal boundary lines.

2. Select Unmatched Addresses in the Show results field.

3. Horizontally scroll across the fields so you can see the Address, City, State, and Zip_Code fields, and then make those fields narrower by dragging vertical boundaries between fields so that you can see all of their values.

4. Right-click the Address column heading in the data panel and click Sort Ascending.

5. Scroll down to the record with address 1011 BRADISH STREET and select that record. ArcMap did not match this address because its ZIP Code is missing. The value is 15203.

6. Type **15203** in the ZIP Code field of the lower left panel, press the Tab key, click the resulting candidate record with Score 100, and click Match. ArcMap matches the records and the count of Matched addresses advanced by one from 545 to 546.

YOUR TURN

Rematch an additional record, 5879 SHADY-FORBES TERRACE. The problem here is that there should be a space between SHADY and FORBES, instead of a hyphen. Make the correction in the lower left panel of Interactive Rematch. Leave the Interactive Rematch window open when finished.

Rematch interactively by pointing on the map

Sometimes you will have unmatched records that you can find on the map using external information or expert knowledge, but your reference data (street map) simply will not have a corresponding address or street. Often too, TIGER street maps have missing house numbers or other data for street segments so that address matching is not possible. In such cases, ArcMap lets you point to the map to geocode.

1 **If the Interactive Rematch window is not open, click the Review/Rematch Addresses button and select Unmatched Addresses for the Show results field.**

2 **Sort the address data by ObjectID, scroll down to the record with ObjectID 826, and click that record to select it.** While this record only has a ZIP Code value, 15221, for address data, suppose that a comment field of the original survey data mentioned that the attendee lived on Canada Way. That street is only one block long in Pittsburgh, so you decide to point to the middle of that street to address match the point.

3 **On the Tools toolbar, click the Zoom In tool and zoom in to the eastern portion of Pittsburgh as seen in the image.** Highlighted is Canada Way. If you have difficulty locating that street, you can use any street segment in its vicinity for practice here.

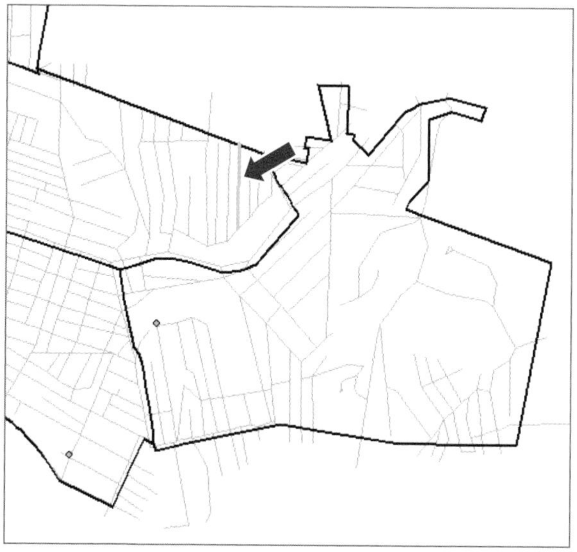

4 On the bottom of the Interactive Rematch window, click Pick Address from Map; move your cursor to the middle of Canada Way; right-click; click Pick Address; and scroll to the left in the address panel of the Interactive Match window. ArcMap adds the address as a point to the map where you clicked. It sets the record's Status to M for matched and sets the Match_type to PP for picked point (A is the code for address matched).

5 Close the Interactive Rematch window and click the Full Extent button on the Tools toolbar.

6 Save your map document as **\ESRIPress\GIST1\MyExercises\Chapter7\Tutorial7-3.mxd**.

Tutorial 7-4

Correct street reference layer addresses

In this tutorial, you will learn how to find and fix an incorrect address in a reference street layer used for geocoding. To do this, you will examine unmatched user addresses, identify candidate streets for revisions, and examine the attributes of the streets to look for misspellings or data omissions.

Open a map document

1 In ArcMap, open \ESRIPress\GIST1\Maps\Tutorial7-4.mxd. Tutorial7-4 contains a table of clients and a layer containing street centerlines in Pittsburgh's central business district.

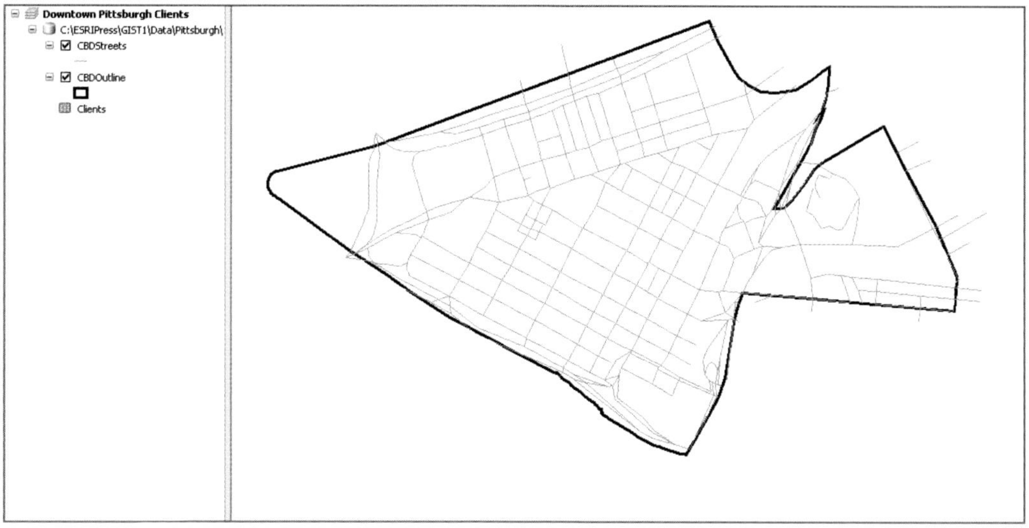

2 Click File and Save As, browse to \ESRIPress\GIST1\MyExercises\Chapter7\, and click Save.

3 Right-click Clients in the TOC and click Open. The data table has addresses for all 27 records, including ZIP Code. Notice that the last four records have place-names instead of street addresses. Later in the tutorial you will use an alias table to geocode those locations. The alias table includes street addresses for place-names.

OBJECTID *	LAST_NAME	FIRST_NAME	ADDRESS	ZIP
1	Roberts	Louisa	133 Seventh St	15222
2	Johnson	Tammy	615 Penn Ave	15222
3	Clark	Robert	309 Ross St	15222
4	Peterson	Jennifer	118 6th Street	15222
5	Young	Mike	490 Penn Ave	15222
6	Thompson	Samantha	711 Liberty Ave	15222
7	Reed	Rhonda	111 Hawksworth	15222
8	Baker	Sally	900 Smallman St	15222
9	Wilson	Laura	599 Smithfield St	15222
10	Jenkins	Amy	701 Grant Street	15222
11	Kelly	Tabitha	900 Lib Ave	15222
12	Riley	Jennifer	777 Illini Drive	15222
13	Smith	Emily	341 Stanwix St	15222
14	Williams	Polly	109 Washington Pl	15222
15	Davis	Kelly	1100 Liberty Ave	15222
16	Dobbins	Joshua	2 S. Market Place	15222
17	Perry	Catherine	651 Forbes Ave	15222
18	Nelson	Joseph	241 Forbes Ave	15222
19	Miller	Matthew	923 French St	15222
20	Lawson	Todd	295 Wood St	15222
21	Welch	Karen	301 5th Ave	15222
22	Sigler	Dan	119 9th Avenue	15222
23	Hampton	Frances	401 1st Ave	15222
24	Peters	Amanda	One PPG Place	15222
25	Franklin	John	One PPG Place	15222
26	Burns	Anthony	Two Gateway Center	15222

4 Close the table.

Create an address locator for CBD Streets

1 Click the Catalog icon in the right edge of the ArcMap window (or if the icon is not there, click Windows, Catalog).

2 Expand the \ESRIPress\GIST1\MyExercises\Chapter7\ folder, right-click Geocoding.gdb, and click New and Address Locator.

3 In the Create Address Locator window, click the browse button for the Address Locator Style, click US Address – Dual Ranges, click OK, and ignore the warning icon and message.

4 Select \ESRIPress\GIST1\Data\Pittsburgh\CentralBusinessDistrict.gdb\CBDStreets for the reference Data.

5 Click the browse button for Output Address Locator, browse to \ESRIPress\GIST1\MyExercises\Chapter7\, double-click Geocoding.gdb, type **PghCBDStreets** for name, and click Save.

6 Click OK.

7 Hide Catalog.

Geocode clients' addresses to CBD Streets

1 If the Geocoding toolbar is not open, click Customize, Toolbars, Geocoding.

2 On the Geocoding toolbar, click the Geocode Addresses button .

3 Select PghCBDStreets and click OK.

4 In the Geocode Addresses window, type **\ESRIPress\GIST1\MyExercises\Chapter7\Geocoding.gdb\CBDClients** for Output shapefile or feature class, click Save, and click OK. ArcMap matches 12 (44 percent) of the 27 records.

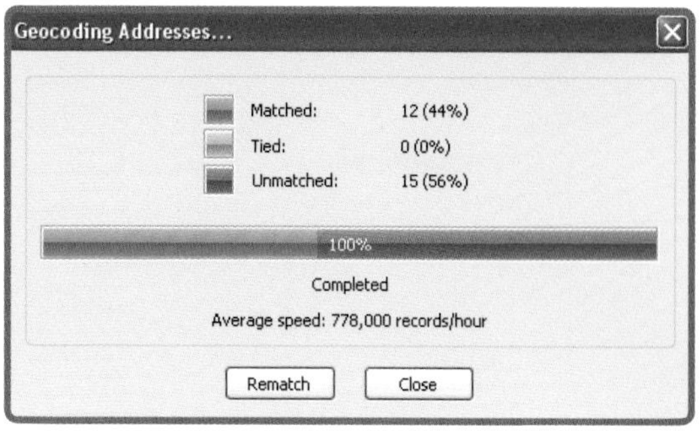

5 Click Close. The geocoded clients are widely scattered around the CBD.

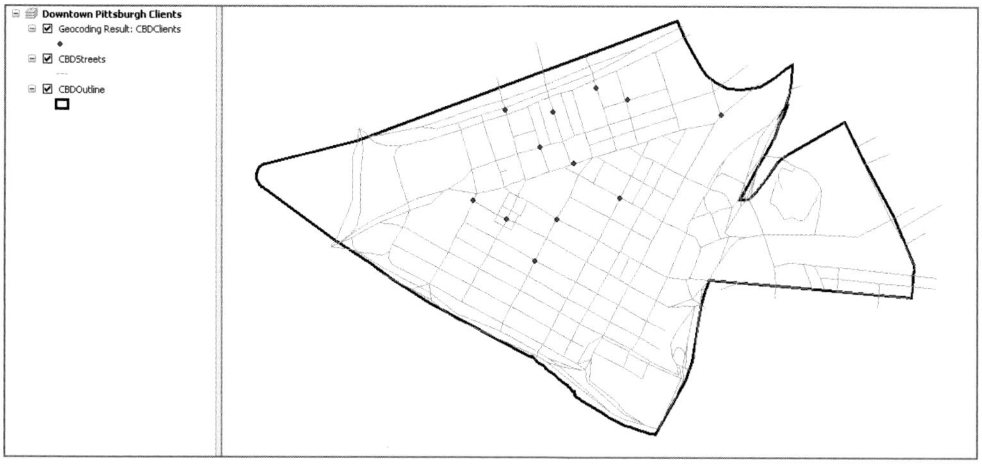

Identify a problem street segment record using Review/Rematch Addresses

1 Click Geocoding Result: CBD Clients in the TOC to select it, if it is not already selected.

2 On the Geocoding toolbar, click the Review/Rematch Addresses button. Select Unmatched Addresses for the Show results field.

3 Scroll to the right in the unmatched addresses, right-click the ADDRESS column, and click Sort Ascending.

4 Scroll down and select the record with ADDRESS 490 Penn Ave. There are many candidate street matches in the lower panel of the Interactive Rematch window. The closest match is 500 PENN AVE 15222. Select that record. ArcMap shows all of the candidates on the map, giving the selected candidate a yellow point marker. Street address numbers increase from left to right in the CBD, and you can see in the lower right of the Interactive Rematch

7-1
7-2
7-3
7-4
7-5
A7-1
A7-2

window that the best candidate's lowest number is 500. So the desired street segment must be to the immediate left of the yellow point marker.

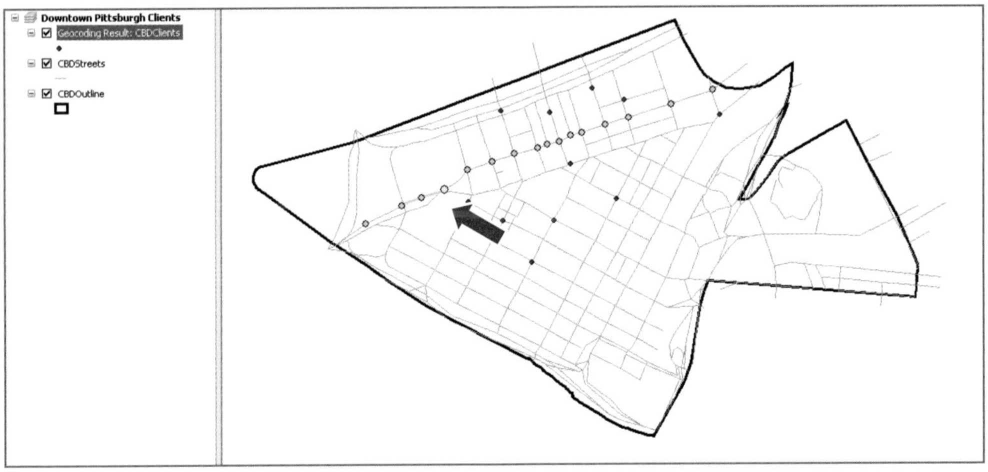

5 On the Tools toolbar, click the Select button and click the street segment indicated in the graphic above.

6 Close the Interactive Rematch window.

7 Right-click CBDStreets in the TOC, click Open Attribute Table, and click the Selected button at the bottom of the table.

8 Scroll to the right in the table to see that the selected street segment's record is missing the TIGER-style street numbers (from and to, left and right street numbers).

RECNUM	L_F_ADD	L_T_ADD	R_F_ADD	R_T_ADD	PREFIX	NAME	TYPE
24924586						PENN	AVE

Edit a street record

Suppose that you have obtained valid numbers for the street segment's missing attributes: 498 to 474 on the left side and 499 to 475 on the right side. You can use ArcMap's Editor toolbar to enter those values.

1 On the main menu, click Customize, Toolbars, Editor.

2 Click the list arrow on the Editor toolbar. Click Start Editing, CBDStreets, and OK.

3 In the CBDStreets table, type the following values: L_F_ADD=**498**, L_T_ADD=**474**, R_F_ADD=**499**, and R_T_ADD=**475**.

4 **Click the list arrow on the Editor, click Save Edits, and click Stop Editing.**

5 **Close the CBDStreets table.** Now, 490 Penn Ave., one of the unmatched client addresses, will geocode the next time you attempt to rematch the addresses. First, however, you have to rebuild the PghCBDStreets Locator so that it includes the edits you just made to the reference CBDStreets layer.

Rebuild a street locator

If you modify a location reference layer after you built a locator, you have to rebuild it.

1 **Unhide or open Catalog.**

2 **If necessary, expand the \ESRIPress\GIST1\MyExercises\Chapter7 folder and Geocoding.gdb.**

3 **Right-click PghCBDStreets and click Rebuild, OK.**

4 **Hide Catalog.**

Rematch interactively using edited street segment

1 **Click Geocoding Result: CBD Clients in the TOC to select it.**

2 **On the Geocoding toolbar, click the Review/Rematch Addresses button, select Unmatched Addresses for the Show results field, and select the 490 Penn Ave address.** Now there is a 100 score candidate for geocoding, for exactly the right address, 490 PENN AVE 15222, given your updated street segment.

3 **In the Interactive Rematch window, click the 100 score candidate record and click Match.** ArcMap matches the record, and the number of matched records increases from 12 to 13. In addition, any time you use CBDStreets in the future for geocoding, ArcMap will successfully geocode any input address records for the edited street segment.

4 **Close the Interactive Rematch window.**

5 **Save your map document.**

Tutorial 7-5

Use an alias table

Some places are commonly located by their landmark names instead of their street address. For example, the White House may be listed in a table as "White House" instead of 1600 Pennsylvania Avenue NW, Washington, D.C., 20500. In the following exercise, you will use an alias table to geocode records that are identified by their landmark name rather than their street address.

Add an alias table and rematch addresses

1 **Click the Add Data button and add the BldgNameAlias table from the \ESRIPress\GIST1\Data\Pittsburgh\CentralBusinessDistrict.gdb.**

2 **In the TOC, right-click BldgNameAlias and click Open.** The table contains the alias name and street address for three records. You can create such tables in Microsoft Excel, in NotePad with comma delimiters, or in other packages, and import them into a file geodatabase using Catalog.

OBJECTID *	BLDGNAME	ADDRESS
1	One PPG Place	200 4th Ave
2	Two PPG Place	200 4th Ave
3	Two Gateway Center	197 Stanwix St

3 **Close the alias table.**

4 **Click Geocoding Result: CBD Clients in the TOC to select it.**

5 **Click the Review/rematch button on the Geocoding toolbar. Select Unmatched Addresses for the Show results field.**

6 **Click Geocoding Options, Place Name Alias Table.**

7 **Select BldgNameAlias for the Alias table, BLDGNAME for the Alias field, and click OK twice.**

8 **Scroll down in the address records, click the record with ObjectID 24, and scroll to the right to see the address—One PPG Place.**

9 **Click the candidate with score 100 and click Match. Similarly, match the last three records.**

10 **Close the Interactive Rematch window, save your map document, and exit ArcMap.**

Assignment 7-1

Geocode household hazardous waste participants to ZIP Codes

Many county, city, and local environmental organizations receive inquiries from residents asking how they can dispose of household hazardous waste (HHW) materials that cannot be placed in regular trash or recycling collections. Homeowners continually search for environmentally responsible methods for disposing of common household products such as paint, solvents, automotive fluids, pesticides, insecticides, and cleaning chemicals.

The Pennsylvania Resources Council (PRC) (`www.prc.org`) is a nonprofit organization dedicated to protecting the environment. The PRC facilitates meetings, organizes collection events, spearheads fundraising and volunteer efforts, and develops education and outreach materials in response to the HHW problem.

At each event, the PRC collects residence data from participants. In this exercise, you will geocode participants by ZIP Code for a recent Allegheny County event.

Start with the following:

- \ESRIPress\GIST1\Data\UnitedStates.gdb\HHWZIPCodes—table of 5-digit ZIP Codes for a HHW Allegheny County event collected by the PRC

Note: PRC suppressed all attributes except ZIP Code to protect confidentiality.

- \ESRIPress\GIST1\Data\UnitedStates.gdb\PAZip—polygon layer of Pennsylvania ZIP Codes used for address matching
- \ESRIPress\GIST1\Data\UnitedStates.gdb\PACounties—polygon layer of Pennsylvania counties

Create a choropleth map of HHW participants by ZIP Code

Create a file geodatabase, **\ESRIPress\GIST1\MyAssignments\Chapter7\Assignment7-1 YourName.gdb**. Then create an address locator to use when geocoding HHW participants to ZIP Codes, **\ESRIPress\GIST1\MyAssignments\Chapter7\Assignment7-1.gdb\HHWZipLocator**. Store the geocoded ZIP Codes as **\ESRIPress\GIST1\MyAssignments\Chapter7\Assignment7-1 YourName.gdb\HHWZIPCodeResidences**. When aggregating points, count Match_addr in the geocoded attribute table.

In ArcMap, create a new map document called **\ESRIPress\GIST1\MyAssignments\Chapter7\ Assignment7-1YourName.mxd** that uses the UTM Zone 17N projection and includes a choropleth map in a layout showing the number of Household Hazardous Waste participants by ZIP Code in Pennsylvania. Add the PA County shapefile as a thick dark outline. Label counties with county names. Add a second copy of PAZip with outline and hollow fill to complete ZIP Codes throughout Pennsylvania where there were no HHW residences.

Export a layout with title, map, and legend to a file called **\ESRIPress\GIST1\MyAssignments\Chapter7\Assignment7-1YourName.jpg**. Include the layout image in a Word document, **\ESRIPress\GIST1\MyAssignments\Chapter7\Assignment7-1YourName.doc**, in which you describe residence patterns of HHW event attendees.

WHAT TO TURN IN

If your work is to be graded, turn in the following files:

File geodatabase: \ESRIPress\GIST1\MyAssignments\Chapter7\Assignment7-1YourName.gdb

ArcMap document: \ESRIPress\GIST1\MyAssignments\Chapter7\Assignment7-1YourName.mxd

Image file: \ESRIPress\GIST1\MyAssignments\Chapter7\Assignment7-1YourName.jpg

Word Document: \ESRIPress\GIST1\MyAssignments\Chapter7\Assignment7-1YourName.doc

If instructed to do so, instead of the above individual files, turn in a compressed file, **Assignment7-1YourName.zip**, with all files included. Do not include path information in the compressed file.

Assignment 7-2

Geocode immigrant-run businesses to Pittsburgh streets

As the 2000 Census shows, immigrants are largely becoming one of the most salient indicators of growth and wealth in a region. By looking at the immigrants who live in a city and analyzing where they decide to set up their businesses, city planners can investigate why certain neighborhoods are more immigrant-friendly than others, and in turn focus on the qualities that make a neighborhood open and diverse.

According to the 2000 Census, Pittsburgh, Pennsylvania, ranked 25th of all U.S. metropolitan areas in the number of immigrants who live there. GIS can geocode as points where these immigrants create businesses and then aggregate this to neighborhoods. The data used in this assignment is a sample of businesses that focuses on immigrant-run high-tech firms, restaurants, and grocery stores.

Start with the following:

- \ESRIPress\GIST1\Data\Pittsburgh\City.gdb\ImmigrantBusinesses—sample of immigrant-run businesses in Pittsburgh

Included in this sample are street address, city, state, ZIP Code, and type of business (Firm, Grocery, and Restaurant).

- \ESRIPress\GIST1\Data\Pittsburgh\City.gdb\PghStreets—TIGER layer of Pittsburgh street centerlines
- \ESRIPress\GIST1\Data\Pittsburgh\City.gdb\Neighborhoods—polygon layer of Pittsburgh neighborhoods

Create a point map of geocoded immigrant-owned businesses and choropleth map of businesses by neighborhood

Create a file geodatabase, **\ESRIPress\GIST1\MyAssignments\Chapter7\Assignment7-2YourName.gdb**. Then create an address locator to use when geocoding immigrant-run businesses to streets, **\ESRIPress\GIST1\MyAssignments\Chapter7\Assignment7-2YourName.gdb\BusinessStreetsLocator**. In the Geocode Addresses form, set Street or Intersection to ADDRESS, City or Intersection to <None>, State to <None>, and ZIPCode to ZIP. Save the geocoded businesses as **\ESRIPress\GIST1\MyAssignments\Chapter7\Assignment7-2YourName.gdb\ImmigrantBusinesses**.

In ArcMap, create a new map document called **\ESRIPress\GIST1\MyAssignments\Chapter7\Assignment7-2YourName.mxd** that includes a layout showing a point map of immigrant-run business locations in Pittsburgh. Symbolize the point map using type of business.

Export the map as a file called **\ESRIPress\GIST1\MyAssignments\Chapter7\Assignment7-2YourName.jpg**.

Further requirements

You should get about 30 percent that do not match. Reasons include wrong ZIP Code data, place-names instead of street addresses, and incorrect or misspelled street address.

- Use Internet sites such as `www.usps.com` or `maps.google.com` to rematch five unmatched addresses.
- Keep a log of steps you took to try and rematch addresses and turn this in with your assignment as **Assignment7-2YourName.doc** stored in **\ESRIPress\GIST1\MyAssignments\Chapter7**. For each address investigated, give the original address, its problem, source for additional information, and correction.

WHAT TO TURN IN

If your work is to be graded, turn in the following files:

File geodatabase: \ESRIPress\GIST1\MyAssignments\Chapter7\Assignment7-2YourName.gdb

ArcMap document: ESRIPress\GIST1\MyAssignments\Chapter7\Assignment7-2YourName.mxd

Image file: \ESRIPress\GIST1\MyAssignments\Chapter7\Assignment7-2YourName.jpg

Word document: \ESRIPress\GIST1\MyAssignments\Chapter7\Assignment7-2YourName.doc

If instructed to do so, instead of the above individual files, turn in a compressed file, **Assignment7-2YourName.zip**, with all files included. Do not include path information in the compressed file.

8

Geoprocessing

Generally you will need to extract or otherwise process study areas for GIS applications from available basemaps. In this chapter, you will learn how to extract a subset of spatial features from a map using attribute or spatial queries. You will also learn how to aggregate polygons to larger polygons and how to append two or more layers into a single layer. This sort of work using GIS processes is called "geoprocessing," and often it is necessary to string several such processes together to obtain a desired product. You will learn how to build, share, and document your multiprocess workflows by creating macros using ArcGIS's ModelBuilder as introduced in this chapter.

Learning objectives

- *Use data queries to extract features*
- *Clip features*
- *Dissolve features*
- *Merge features*
- *Intersect layers*
- *Union layers*
- *Automate geoprocessing with ModelBuilder*

Tutorial 8-1

Use data queries to extract features

New York City has an interesting geography. It is made up of five boroughs, each of which is also a county. The Bronx is also Bronx County, Brooklyn is Kings County, Manhattan is New York County, Queens is Queens County, and Staten Island is Richmond County. The map layer that you will open next has these areas denoted as boroughs. Other map layers that you will use have the alternative designation of counties.

Open a map document

1 On your desktop, click Start, All Programs, ArcGIS, ArcMap 10.

2 In the ArcMap – Getting Started window, click Browse for more. Browse to the drive and folder where you installed \ESRIPress\GIST1\Maps\, and double-click Tutorial 8-1.mxd. The Tutorial8-1.mxd file opens showing a map of the New York City metropolitan area including Manhattan, Brooklyn, the Bronx, Staten Island, and Queens.

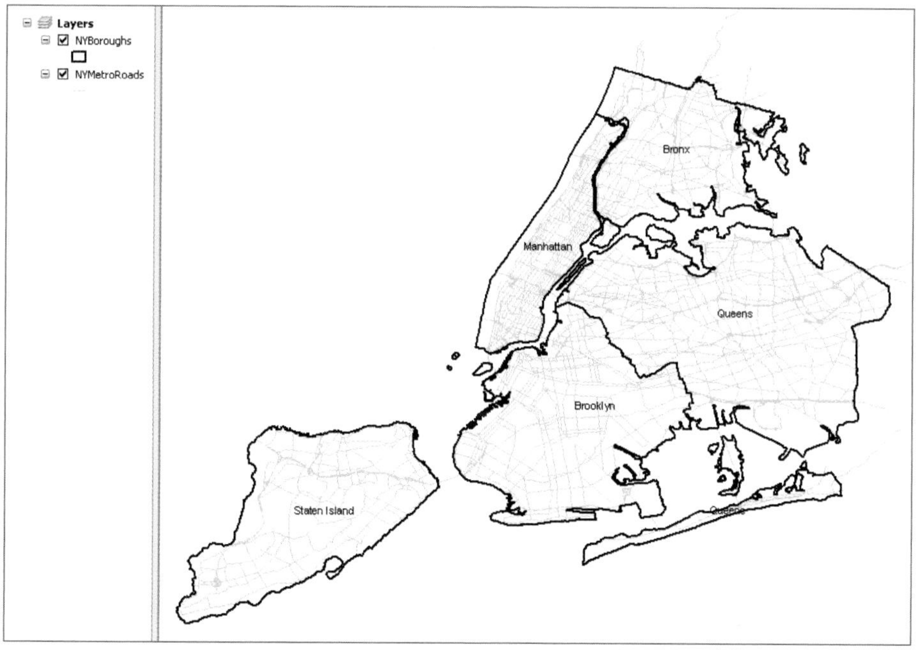

3 Click File and Save As, browse to \ESRIPress\GIST1\MyExercises\Chapter8\, and save the map document as **Tutorial8-1.mxd**.

Use the Select By Attributes dialog box

Here you will use ArcMap's Select By Attributes tool to create a study area for Manhattan extracted from the NYBoroughs layer.

1 On the main menu, click Selection, Select By Attributes.

2 From the Layer drop-down list, click NYBoroughs.

3 In the Fields box, double-click "NAME".

4 Click the = button.

5 Click the Get Unique Values button. Then, in the Unique Values box, double-click 'Manhattan'.

6 Click Apply and Close.

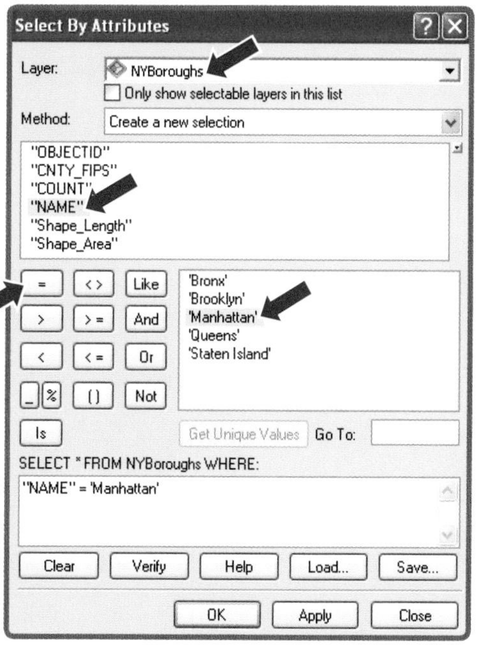

Show selected features and convert to feature class

1 In the TOC, right-click NYBoroughs. Click Selection and Zoom to Selected Features.

2 In the TOC, right-click the NYBoroughs layer. Click Data, Export Data.

3 Save the output features class as **\ESRIPress\GIST1\MyExercises\Chapter8\NewYork.gdb\Manhattan**.

4 Click OK, click Yes to add the layer to the map, and zoom to the full extent. Your map now contains a new feature class containing only the borough of Manhattan.

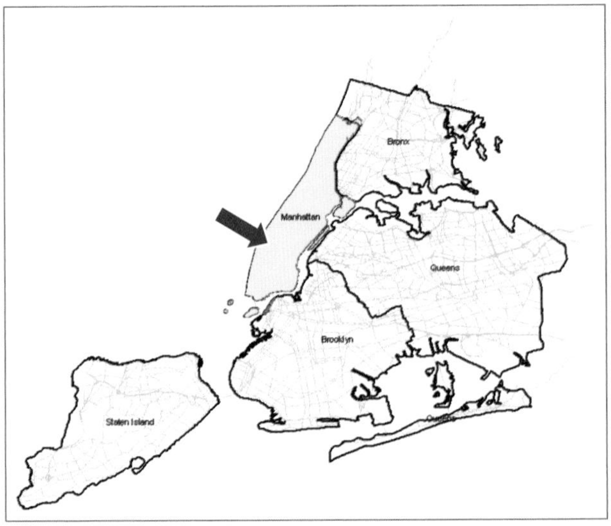

Use the Select Features tool

In the previous steps, you used an attribute query to select the feature you wanted to extract a layer. Sometimes, however, it is easier to manually select the feature(s) directly from the map display you want to extract instead of building a query expression in the Select By Attributes dialog box.

1 Make NYBoroughs the only selectable layer.

2 Click the Select Features button, and click inside the polygon feature for Brooklyn.

3 In the TOC, right-click the NYBoroughs layer. Click Data, Export Data.

4 Save the output feature class as **\ESRIPress\GIST1\MyExercises\Chapter8\NewYork.gdb\Brooklyn**, click OK, then click Yes to add the layer to the map. Your map now contains another new feature class, this one containing only the borough of Brooklyn.

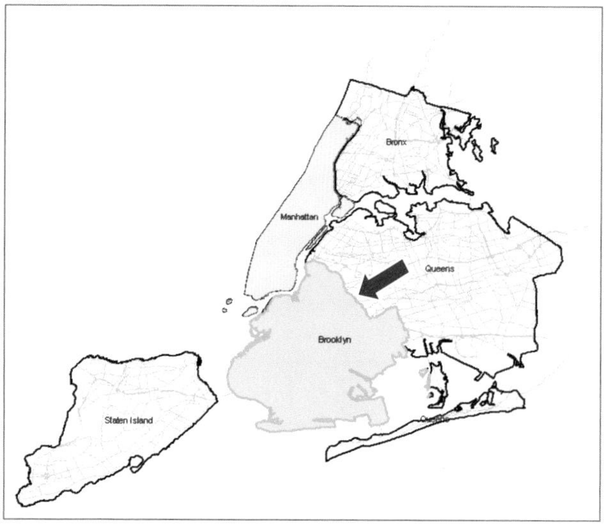

YOUR TURN

Use either the Select By Attributes dialog box or the Select Features tool to create study area feature classes for Queens, the Bronx, and Staten Island. When finished, clear all selections. Save your map document.

Tutorial 8-2

Clip features

Use the Select By Location dialog box

In the following steps, you will use the Select By Location dialog box to select the streets (or roads) in Manhattan only. After selecting the roads, you will create a new line feature class from them.

1 On the main menu, click Selection, Select by Location.

2 Make selections as shown in the image.

3 Click Apply, Close.

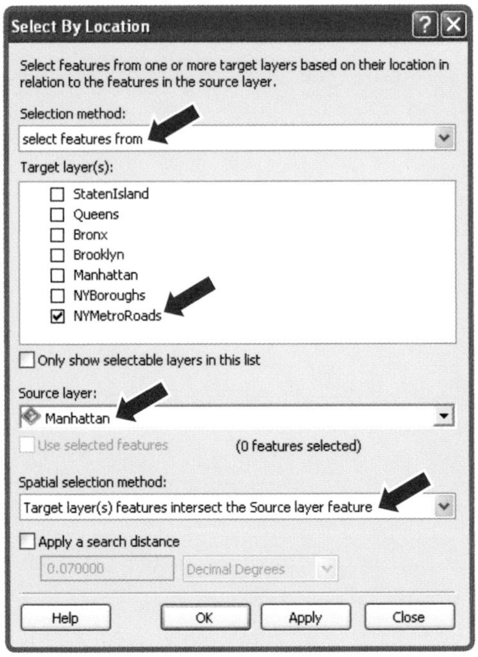

Examine selected features and convert to a feature class

1 In the TOC, right click NYMetroRoads. Click Selection and Zoom to Selected Features. The selected roads are only those within—or those that intersect—Manhattan borough.

2 In the TOC, right-click the NYMetroRoads layer. Click Data, Export Data.

3 Save the output feature class as **\ESRIPress\GIST1\MyExercises\Chapter8\NewYork.gdb\ManhattanRoads**.

4 Click OK, then click Yes to add the layer to the map.

5 Turn off the NYMetroRoads layer so the only roads visible in the map are those in the ManhattanRoads layer. Notice that some of the roads in the ManhattanRoads layer extend or "dangle" beyond the Manhattan borough outline.

Clip streets

Next, you will use the Clip geoprocessing tool to cut off the NYMetroRoads segments using the Manhattan feature class. Once this is done, the roads in the ManhattanRoads layer will have no dangling lines. Note that for geocoding tabular address data with these TIGER-style roads, you should use the roads version with dangles because ArcMap interpolates house numbers using the starting and ending house numbers of TIGER street segments. Roads that ArcMap clips will have the original starting and ending house numbers but shortened lengths. This will introduce location errors beyond those inherent in the approximate TIGER streets. Use the clipped roads for display purposes only in a study area.

8-1 8-2 8-3 8-4 8-5 8-6 8-7 A8-1 A8-2 A8-3

1 Click Geoprocessing, Clip.

2 In the Clip dialog box, click the Input Features drop-down list and choose NYMetroRoads.

3 Click the Clip Features drop-down list and choose Manhattan.

4 Save the Output Feature Class as **\ESRIPress\GIST1\MyExercises\Chapter8\NewYork.gdb\ClippedManhattanRoads**.

5 Click OK.

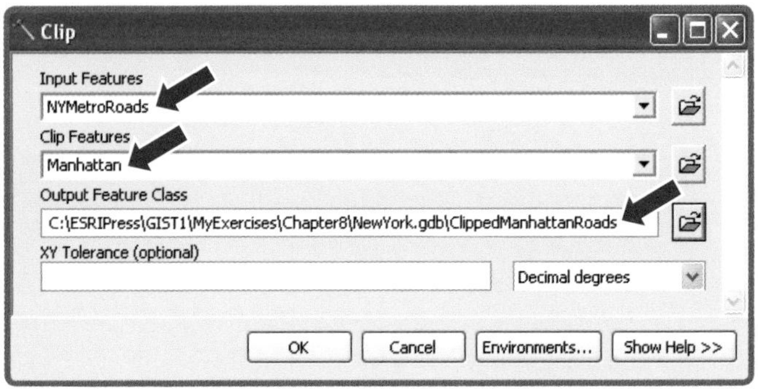

6 Turn on the ClippedManhattanRoads layer and turn off the ManhattanRoads layer. The streets in ClippedManhattanRoads layer do not cross the borough of Manhattan's boundary.

7 Save your map document as **\ESRIPress\GIST1\MyExercises\Chapter8\Tutorial8-2.mxd**.

YOUR TURN

Clip the NYMetroRoads layer to the Bronx borough. Save the Output Feature Class as **\ESRIPress\GIST1\MyExercises\Chapter8\NewYork.gdb\ClippedBronxRoads**. Add these to your map document and save it.

Tutorial 8-3

Dissolve features

You can create administrative or other types of boundaries by merging polygons in a feature class that share common attribute values. This type of a merge is called a dissolve, and in this tutorial you will use the Dissolve tool to dissolve ZIP Code polygons, based on their post office (PO) name, to create PO boundaries as the U.S. Postal Service defines them. Needed for this work is a so-called "crosswalk" that lists all ZIP Codes of the area and corresponding post office names, defining post offices by ZIP Code areas. In this case, the crosswalk already exists, built into the ZIP Code attribute table. In other cases, you will have to create a separate cross-walk table or add an attribute such as borough name in the input layer.

Open a map document

1 In ArcMap, open Tutorial8-3.mxd from the \ESRIPress\GIST1\Maps\ folder.
Tutorial8-3.mxd contains a map of the New York City Metro Area ZIP Codes, including Manhattan, Brooklyn, Staten Island, the Bronx, and Queens.

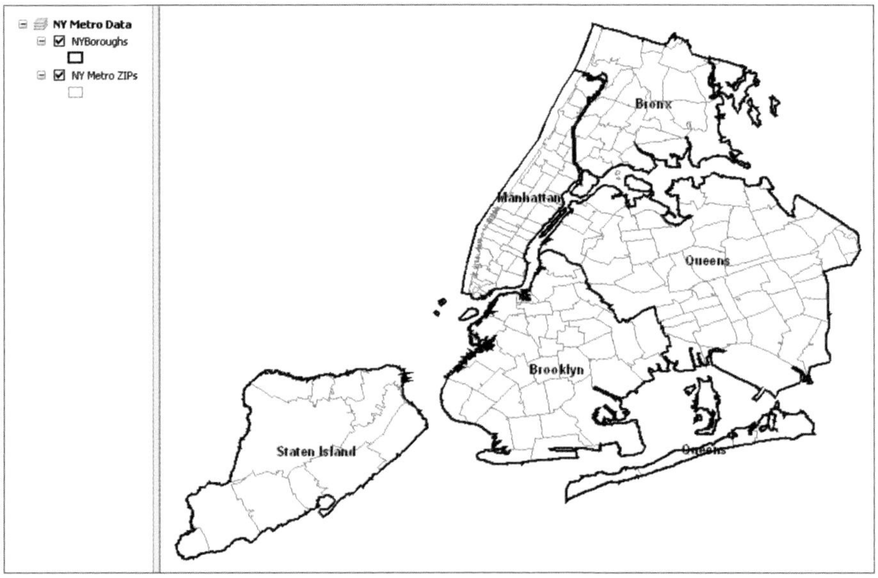

2 Click File and Save As, browse to \ESRIPress\GIST1\MyExercises\Chapter8\, and save the map document as **Tutorial8-3.mxd**.

Examine the crosswalk

1 In the TOC, right click NY Metro Zips. Click Open Attribute Table. The ZIP and PO_NAME attributes provide the crosswalk data. The PO_NAME attribute provides data on which ZIP Codes to aggregate into POs, which have different boundaries than the boroughs.

Table

NY Metro ZIPs

OBJECTID_1 *	Shape *	ObjectID	ZIP	PO_NAME	STATE	AREA	SUMBLKPO	POP2003	Shape_Length	Shape_Area	Borough
70	Polygon	87621660	10036	NEW YORK	NY	0.40841	17205	18240	0.055216	0.000113	Manhattan
71	Polygon	87621661	10038	NEW YORK	NY	0.27295	15092	14987	0.051109	0.000075	Manhattan
72	Polygon	87621662	10280	NEW YORK	NY	0.14531	6537	7473	0.03882	0.00004	Manhattan
73	Polygon	87687168	10004	NEW YORK	NY	0.44583	1363	903	0.096208	0.000123	Manhattan
103	Polygon	87818240	00083	CENTRAL PARK	NY	1.37335	18	-99	0.099532	0.00038	Manhattan
104	Polygon	87818241	10021	NEW YORK	NY	0.9538	103118	102543	0.069627	0.000264	Manhattan
105	Polygon	87818242	10022	NEW YORK	NY	0.46542	31790	32079	0.054812	0.000129	Manhattan
106	Polygon	87818243	10025	NEW YORK	NY	0.69699	95103	94773	0.0628	0.000193	Manhattan
107	Polygon	87818244	10028	NEW YORK	NY	0.35308	48162	47996	0.05916	0.000098	Manhattan
108	Polygon	87818245	10029	NEW YORK	NY	0.82243	76826	76591	0.06349	0.000227	Manhattan
109	Polygon	87818246	10035	NEW YORK	NY	1.39075	33381	33351	0.126671	0.000384	Manhattan
110	Polygon	87818247	10044	NEW YORK	NY	0.23066	9520	9407	0.065011	0.000064	Manhattan
111	Polygon	87818248	10128	NEW YORK	NY	0.3923	55430	55115	0.04718	0.000108	Manhattan
102	Polygon	87752721	11385	RIDGEWOOD	NY	4.47001	99443	100888	0.225835	0.001234	Queens
112	Polygon	87818249	11101	LONG ISLAND CIT	NY	2.93407	27161	28376	0.181332	0.00081	Queens
113	Polygon	87818250	11102	ASTORIA	NY	0.68435	34378	35898	0.067566	0.000189	Queens
114	Polygon	87818251	11103	ASTORIA	NY	0.72587	44580	44109	0.069908	0.000201	Queens
115	Polygon	87818252	11104	SUNNYSIDE	NY	0.42534	26294	26975	0.064259	0.000117	Queens
116	Polygon	87818253	11105	ASTORIA	NY	1.7505	40703	40899	0.117921	0.000484	Queens
117	Polygon	87818254	11106	ASTORIA	NY	0.85696	41031	43040	0.075663	0.000237	Queens

1 (0 out of 181 Selected)

NY Metro ZIPs

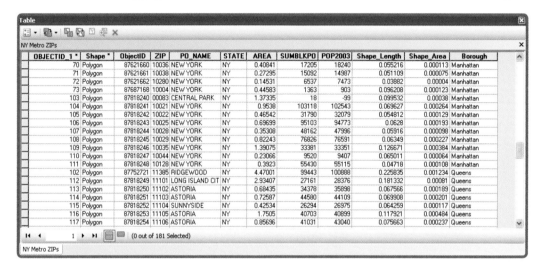

2 Close the table.

Dissolve ZIP Codes

1 On the main menu, click Geoprocessing, Dissolve.

2 In the Dissolve dialog box, click the Input Features drop-down list and choose NY Metro Zips.

3 Save the Output Feature Class as **\ESRIPress\GIST1\MyExercises\Chapter8\NewYork.gdb\DissolvedNYPOs**.

4 Click PO_NAME as the Dissolve field.

5 Click the Statistics Field drop-down list and choose POP2003.

6 Click the Statistic Type drop-down and choose SUM. This is an optional setting. When the dissolve runs, ArcGIS will sum the values in the POP2003 field for each group of polygons with the same PO_NAME value. In other words, it will aggregate the population up to the new polygon feature of POs.

7 **Verify your selections with the graphic at the right.**

8 Click OK.

9 **Symbolize the new layer with no fill, a Mars Red outline, and an outline width of 2.0.** Once the dissolve process completes, ArcMap adds the DissolvedNYPOs feature class to the map. The following map shows the input ZIP Codes and dissolved POs.

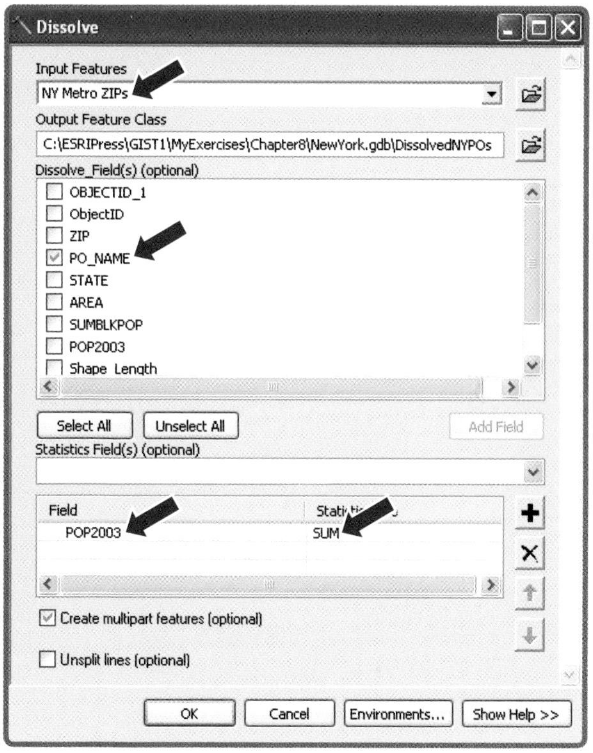

10 **Use the Identify button to view the attribute information for the Jamaica PO.**

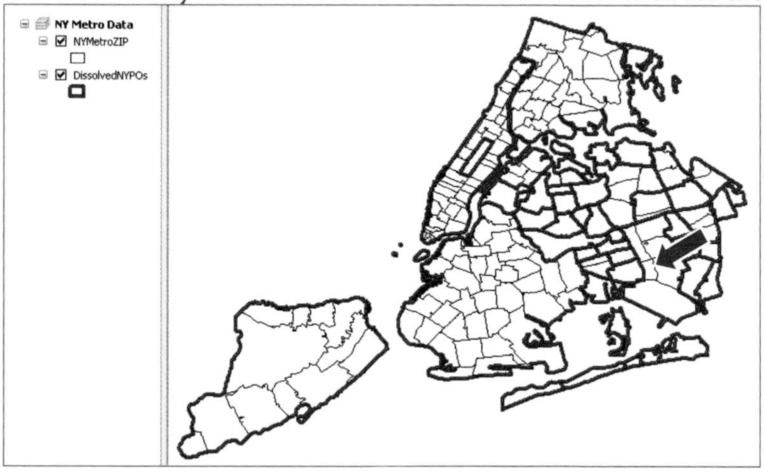

8-1
8-2
8-3
8-4
8-5
8-6
8-7
A8-1
A8-2
A8-3

In addition to the name of the PO, you will see a population value in the SUM_POP2003 field, which the dissolve process derived from the POP2003 values of the ZIP Code layer.

11 Save your map.

Tutorial 8-4

Merge features

Sometimes it is necessary to merge two or more separate but adjacent layers into a single layer. For example, you may want to build a water layer for an environmental study that includes several adjacent counties. Using the Merge tool, you could merge county water layers into a single layer, then use the layer for further analysis. Here you will merge water layers for New York City's counties.

Open a map document

1 In ArcMap, open Tutorial8-4.mxd from your \ESRIPress\GIST1\Maps\ folder.
Tutorial8-4.mxd contains a map of the New York City area counties. Each county's water layer in the map exists in a separate layer.

2 Click File and Save As, browse to \ESRIPress\GIST1\MyExercises\Chapter8\, and save the map document as **Tutorial8-4.mxd**.

Merge several feature layers into one feature layer

1 On the main menu, click Geoprocessing, Merge.

2 In the Merge dialog box, click the Input Datasets drop-down list and choose all five layers for the New York area water polygons.

3 Save the Output Feature Class as **\ESRIPress\GIST1\MyExercises\Chapter8\NewYork.gdb\NewYorkWater**.

4 Verify that the Merge settings match the graphic at the right.

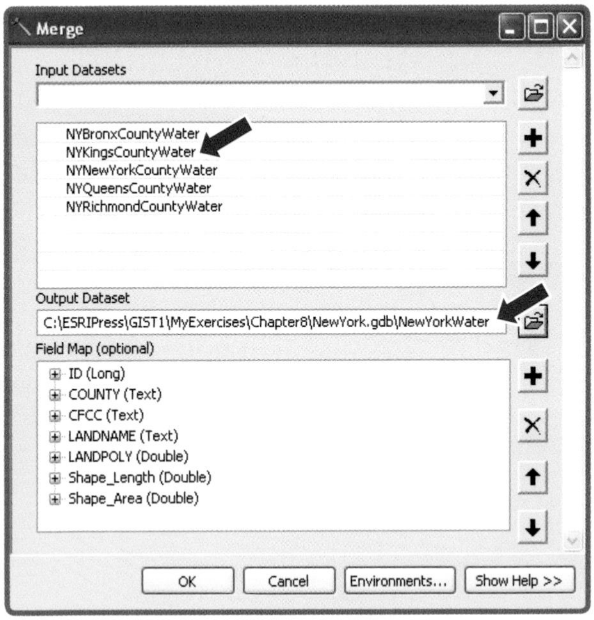

5 Click OK. The NewYorkWater layer now contains water boundaries for all five counties around New York City.

6 Save your map document.

Tutorial 8-5

Intersect layers

The Intersect tool creates a new feature class from all the features of two input, overlaying feature classes. For example, an emergency preparedness official might like to know the name of the water boundary that each road crosses over (or intersects). ArcGIS can provide such information using the Intersect tool. Intersect excludes any parts of the two or more input layers that do not overlay each other.

Open a map document

1 **In ArcMap, open Tutorial8-5.mxd from your \ESRIPress\GIST1\Maps\ folder.** Tutorial 8-5.mxd contains a map of the merged water boundaries from the previous tutorial and NYMetroRoads. Notice that many roads in the New York area cross water boundaries.

2 Click File and Save As, browse to \ESRIPress\GIST1\MyExercises\Chapter8\, and save the map document as **Tutorial8-5.mxd**.

Open tables

1 From the TOC, right-click the NYMetroRoads layer and click Open Attribute Table. There is no data about the water boundaries in this file.

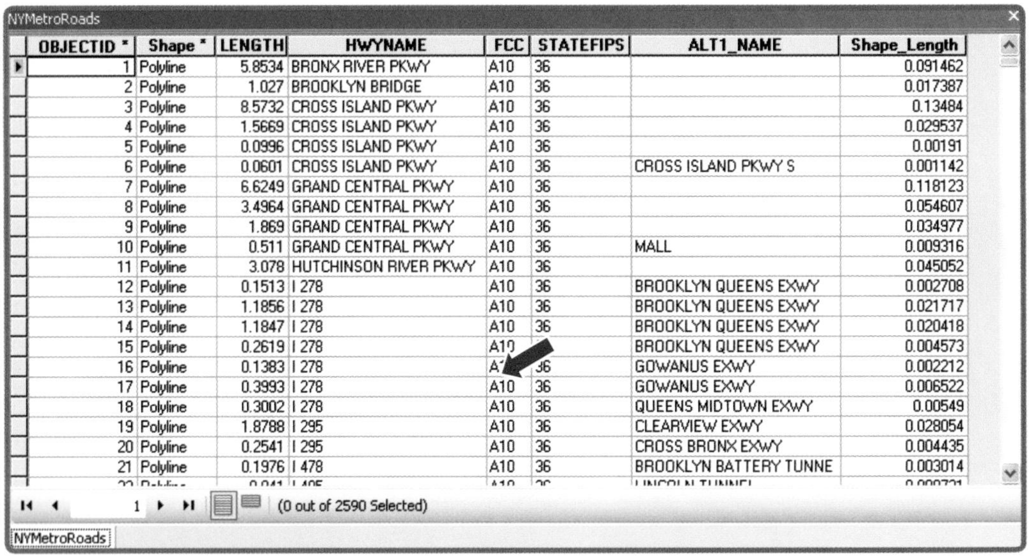

NYMetroRoads

OBJECTID *	Shape *	LENGTH	HWYNAME	FCC	STATEFIPS	ALT1_NAME	Shape_Length
1	Polyline	5.8534	BRONX RIVER PKWY	A10	36		0.091462
2	Polyline	1.027	BROOKLYN BRIDGE	A10	36		0.017387
3	Polyline	8.5732	CROSS ISLAND PKWY	A10	36		0.13484
4	Polyline	1.5669	CROSS ISLAND PKWY	A10	36		0.029537
5	Polyline	0.0996	CROSS ISLAND PKWY	A10	36		0.00191
6	Polyline	0.0601	CROSS ISLAND PKWY	A10	36	CROSS ISLAND PKWY S	0.001142
7	Polyline	6.6249	GRAND CENTRAL PKWY	A10	36		0.118123
8	Polyline	3.4964	GRAND CENTRAL PKWY	A10	36		0.054607
9	Polyline	1.869	GRAND CENTRAL PKWY	A10	36		0.034977
10	Polyline	0.511	GRAND CENTRAL PKWY	A10	36	MALL	0.009316
11	Polyline	3.078	HUTCHINSON RIVER PKWY	A10	36		0.045052
12	Polyline	0.1513	I 278	A10	36	BROOKLYN QUEENS EXWY	0.002708
13	Polyline	1.1856	I 278	A10	36	BROOKLYN QUEENS EXWY	0.021717
14	Polyline	1.1847	I 278	A10	36	BROOKLYN QUEENS EXWY	0.020418
15	Polyline	0.2619	I 278	A10	[illegible]	BROOKLYN QUEENS EXWY	0.004573
16	Polyline	0.1383	I 278	[illegible]	36	GOWANUS EXWY	0.002212
17	Polyline	0.3993	I 278	A10	36	GOWANUS EXWY	0.006522
18	Polyline	0.3002	I 278	A10	36	QUEENS MIDTOWN EXWY	0.00549
19	Polyline	1.8788	I 295	A10	36	CLEARVIEW EXWY	0.028054
20	Polyline	0.2541	I 295	A10	36	CROSS BRONX EXWY	0.004435
21	Polyline	0.1976	I 478	A10	36	BROOKLYN BATTERY TUNNE	0.003014

(0 out of 2590 Selected)

NYMetroRoads

2 Open the NewYorkWater table. Examine the attributes of this table.

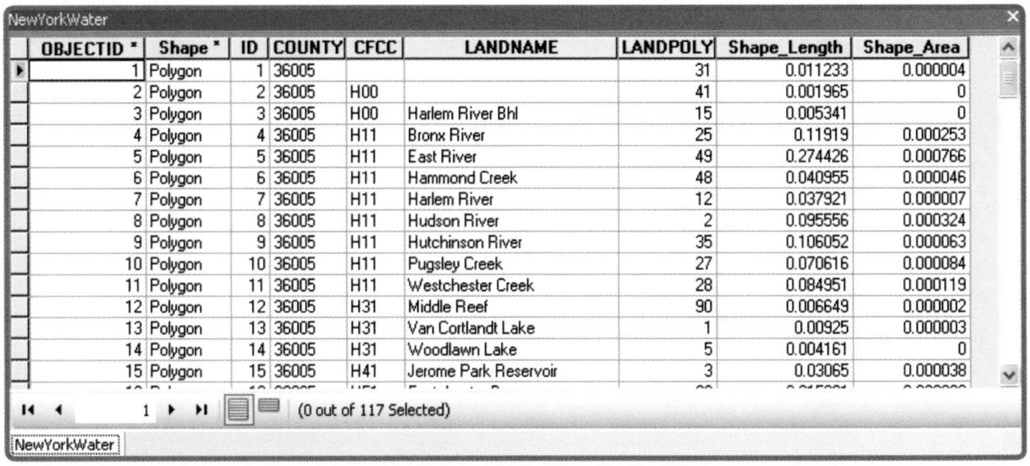

NewYorkWater

OBJECTID *	Shape *	ID	COUNTY	CFCC	LANDNAME	LANDPOLY	Shape_Length	Shape_Area
1	Polygon	1	36005			31	0.011233	0.000004
2	Polygon	2	36005	H00		41	0.001965	0
3	Polygon	3	36005	H00	Harlem River Bhl	15	0.005341	0
4	Polygon	4	36005	H11	Bronx River	25	0.11919	0.000253
5	Polygon	5	36005	H11	East River	49	0.274426	0.000766
6	Polygon	6	36005	H11	Hammond Creek	48	0.040955	0.000046
7	Polygon	7	36005	H11	Harlem River	12	0.037921	0.000007
8	Polygon	8	36005	H11	Hudson River	2	0.095556	0.000324
9	Polygon	9	36005	H11	Hutchinson River	35	0.106052	0.000063
10	Polygon	10	36005	H11	Pugsley Creek	27	0.070616	0.000084
11	Polygon	11	36005	H11	Westchester Creek	28	0.084951	0.000119
12	Polygon	12	36005	H31	Middle Reef	90	0.006649	0.000002
13	Polygon	13	36005	H31	Van Cortlandt Lake	1	0.00925	0.000003
14	Polygon	14	36005	H31	Woodlawn Lake	5	0.004161	0
15	Polygon	15	36005	H41	Jerome Park Reservoir	3	0.03065	0.000038

(0 out of 117 Selected)

NewYorkWater

3 Close both tables.

Intersect features layers

1 On the main menu, click Geoprocessing, Intersect.

2 From the Input Features drop-down list, choose NYMetroRoads and NewYorkWater one at a time.

3 Save the Output Feature Class as **\ESRIPress\GIST1\MyExercises\Chapter8\NewYork.gdb\RoadsWaterIntersection**.

4 From the Output Type drop-down list, choose LINE.

5 Verify that the Intersect settings match the graphic at the right.

6 Click OK.

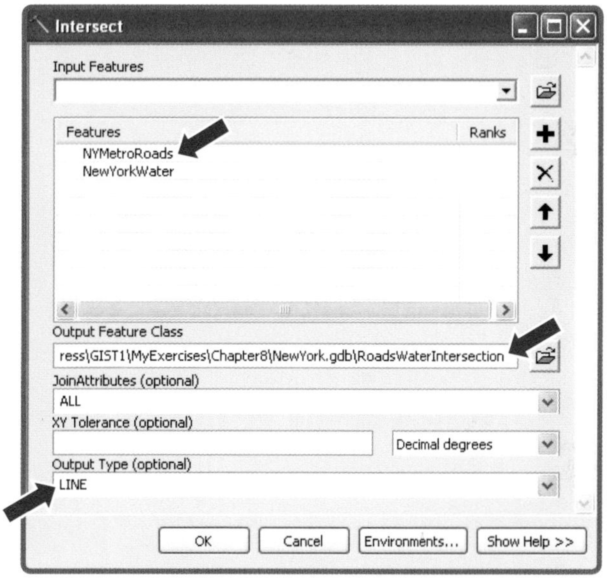

7 Turn the NYMetroRoads layer off and symbolize the RoadsWaterIntersection layer as a thick black line. The output added to your map will be roads that intersect the water polygons.

Examine intersection table

1 **From the TOC, right-click the RoadsWaterIntersection layer and click Open Attribute Table.** Each road now has data about the body of water that it intersects.

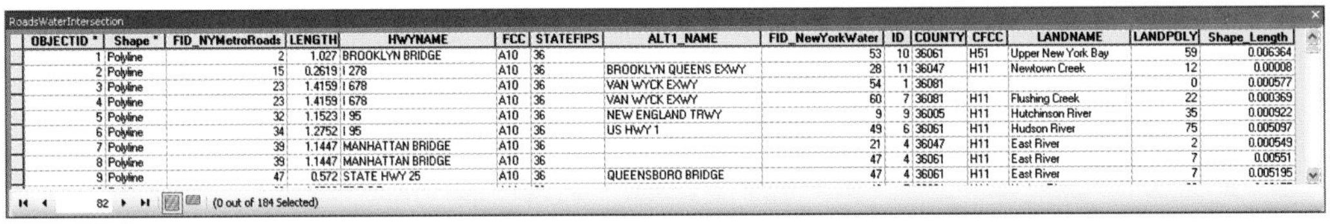

RoadsWaterIntersection

OBJECTID *	Shape *	FID_NYMetroRoads	LENGTH	HWYNAME	FCC	STATEFIPS	ALT1_NAME	FID_NewYorkWater	ID	COUNTY	CFCC	LANDNAME	LANDPOLY	Shape_Length
1	Polyline	2	1.027	BROOKLYN BRIDGE	A10	36		53	10	36061	H51	Upper New York Bay	59	0.006364
2	Polyline	15	0.2619	I 278	A10	36	BROOKLYN QUEENS EXWY	28	11	36047	H11	Newtown Creek	12	0.00008
3	Polyline	23	1.4159	I 678	A10	36	VAN WYCK EXWY	54	1	36081			0	0.000577
4	Polyline	23	1.4159	I 678	A10	36	VAN WYCK EXWY	60	7	36081	H11	Flushing Creek	22	0.000369
5	Polyline	32	1.1523	I 95	A10	36	NEW ENGLAND TRWY	9	9	36005	H11	Hutchinson River	35	0.000922
6	Polyline	34	1.2752	I 95	A10	36	US HWY 1	49	6	36061	H11	Hudson River	75	0.005097
7	Polyline	39	1.1447	MANHATTAN BRIDGE	A10	36		21	4	36047	H11	East River	2	0.000549
8	Polyline	39	1.1447	MANHATTAN BRIDGE	A10	36		47	4	36061	H11	East River	7	0.00551
9	Polyline	47	0.572	STATE HWY 25	A10	36	QUEENSBORO BRIDGE	47	4	36061	H11	East River	7	0.005195

82 (0 out of 184 Selected)

2 **Close the table and save your map document.**

Tutorial 8-6

Union layers

The Union tool combines the geometry and attributes of two input polygon layers to generate a new output polygon layer. In this example, you will use the Union tool to combine ZIP Codes in the borough of Manhattan with a census tract layer for the same borough. The output of the union will be a new feature layer of smaller polygons, each with combined boundaries and attributes of both census tracts and ZIP Codes. Union keeps all features of the input layers, even if they do not overlap.

Open a map document

1 In ArcMap, open Tutorial8-6.mxd from your \ESRIPress\GIST1\Maps\ folder. Tutorial8-6.mxd contains both ZIP Codes and census tracts for Manhattan.

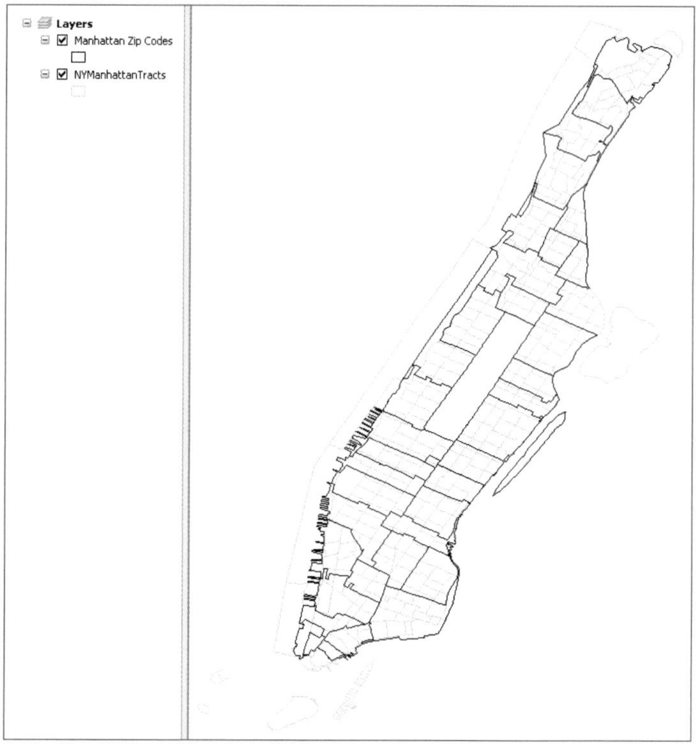

2 Click File and Save As, browse to \ESRIPress\GIST1\MyExercises\Chapter8\, and save the map document as **Tutorial8-6.mxd**.

Union feature classes

1 On the main menu, click Geoprocessing, Union.

2 Type or make selections as shown in the image.

3 Click OK. The output added to your map contains many small polygons with both census and ZIP Code data attached to each polygon. Note that the population for each new small polygon is incorrect. The Union tool merely joins the layers together and does not apportion the data across the smaller polygons. You will learn how to apportion data in chapter 9 of this book.

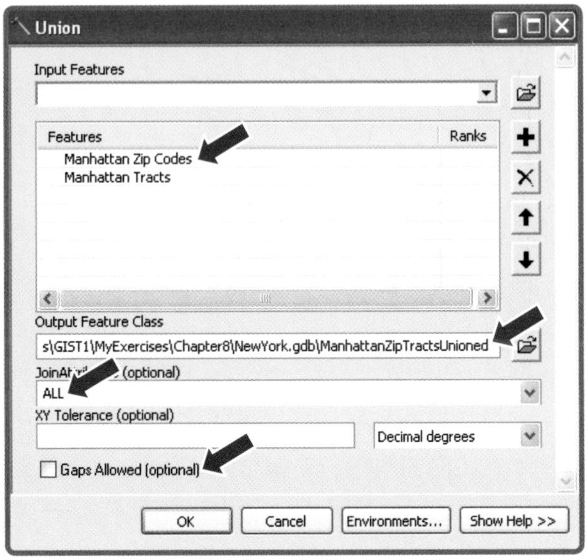

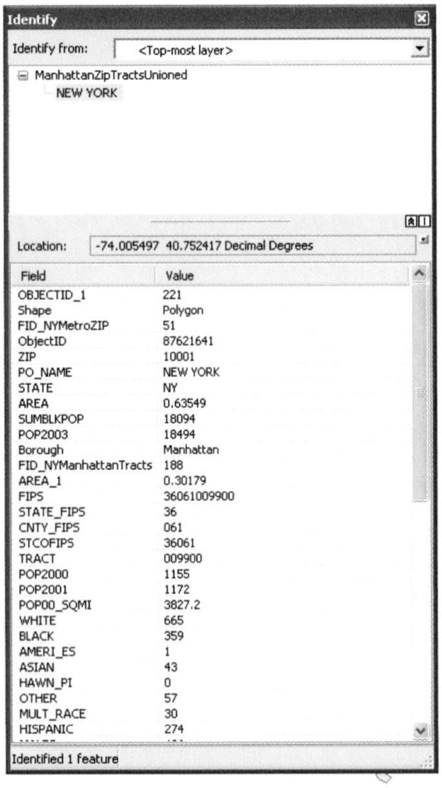

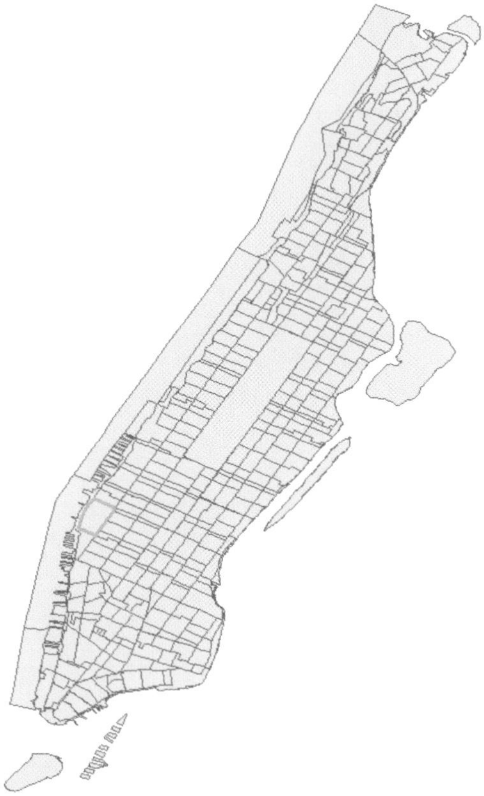

4 Save your map document.

Tutorial 8-7

Automate geoprocessing with ModelBuilder

Spatial data processing often requires several steps and geoprocessing tools to produce desired results. ModelBuilder is an application in ArcGIS for creating macros—custom programs that document and automate geoprocessing workflows. After you build a model, you can run it once, or save it and run it again using different input parameters. In this exercise, you will build a model with several steps for dissolving census tracts to make neighborhoods for a selected city within a county. Before putting you to work using ModelBuilder, it is helpful to examine the inputs and outputs, and then the finished model that you will build.

The starting map document, shown at right, has all municipalities (cities) and census tracts in Allegheny County, Pennsylvania, as downloaded from the U.S. Census Bureau's TIGER basemaps. In this tutorial, you will create neighborhoods for Pittsburgh in the center of the map.

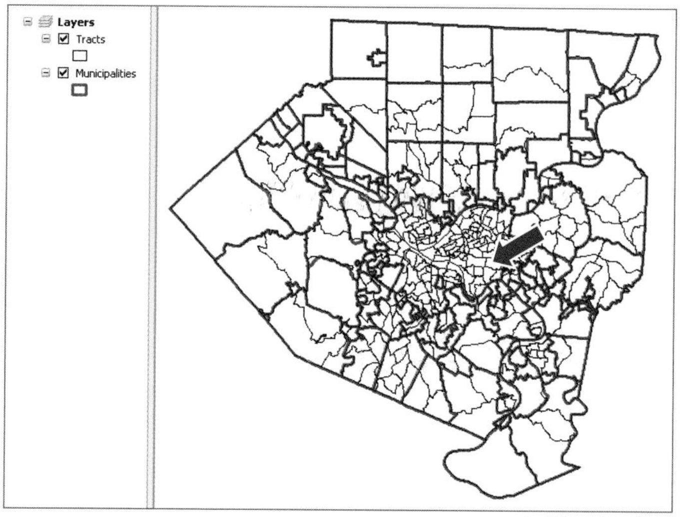

The user must supply a crosswalk table that lists the tracts that define neighborhoods in the city. In this case, each Pittsburgh neighborhood is made up of one or more tracts as seen in the partial crosswalk table listing at right.

OID	STFID	HOOD
0	42003220400	Allegheny Center
1	42003220100	Allegheny West
2	42003180300	Allentown
3	42003160300	Arlington
4	42003160400	Arlington Heights
5	42003202300	Banksville
6	42003050900	Bedford Dwellings
7	42003191600	Beechview
8	42003192000	Beechview
9	42003180900	Beltzhoover
10	42003080200	Bloomfield
11	42003080400	Bloomfield
12	42003080600	Bloomfield
13	42003080900	Bloomfield
14	42003090300	Bloomfield

The output is the dissolved set of neighborhoods below. You can see the tracts that ModelBuilder dissolved for each neighborhood as interior black lines for the red neighborhoods.

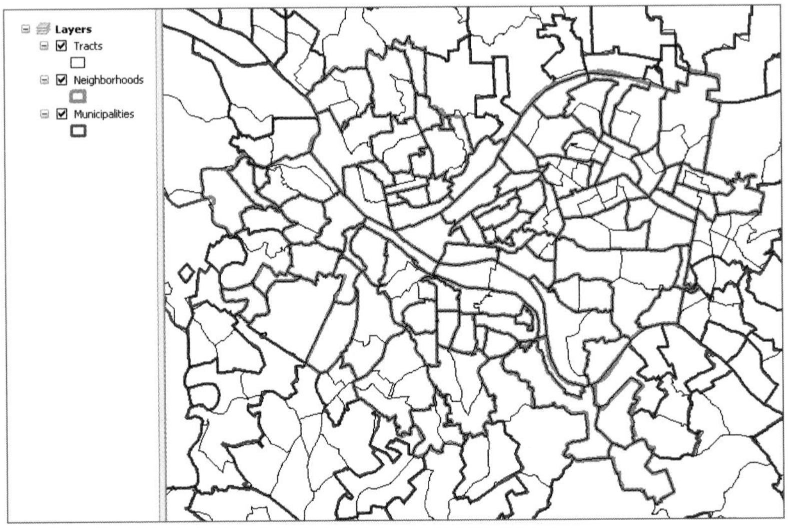

When you run the model, a form opens asking you to supply parameters—all of the elements in the model that the user needs to change for the particular run. With the model, it is possible to create dissolved polygons for any subset of a polygon basemap. See the following for a description of this model's parameters. This is the user interface for the model that has documentation and parameters that the user can change.

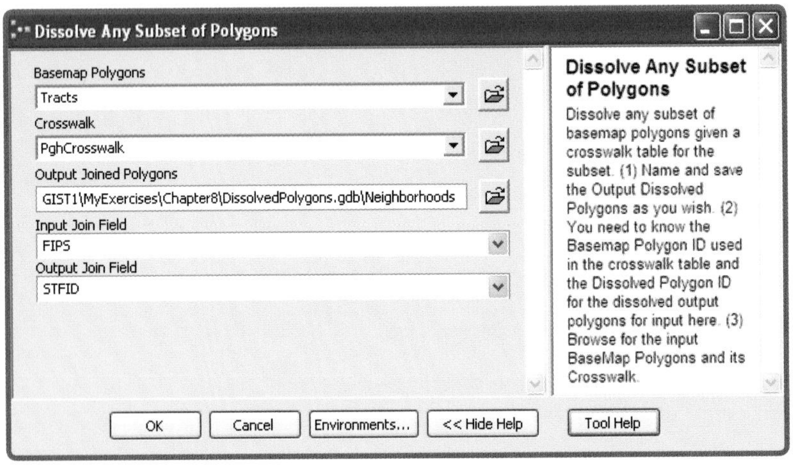

Finally, on the next page is the work flow model diagram that you will build. Earlier in this chapter you ran geoprocessing steps interactively from ArcMap's main menu. For models, however, you access the same functionality using toolboxes and tools. Each tool becomes a process (the yellow boxes seen in the model on the next page) with blue inputs and green outputs in a model diagram.

ModelBuilder shows input and output elements with black arrow lines that go to and from processes. Each element with a "P" near its upper right is a parameter. Input Join Field and Output Join Field are variables that store input parameter values for further processing (an explanation is given later in this tutorial).

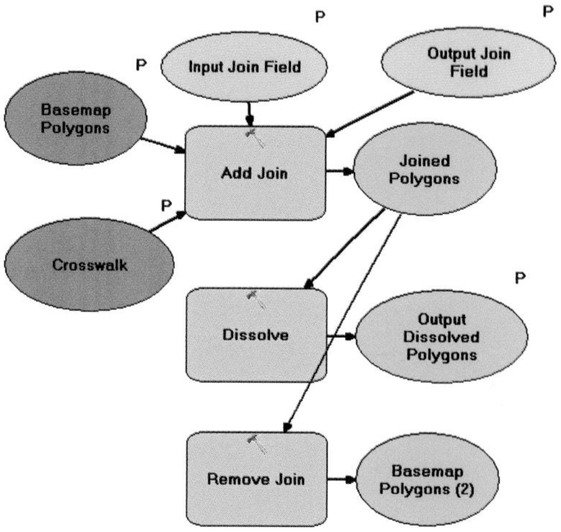

The first step of the model is to join the crosswalk table to the basemap polygons. The user can supply any two consistent inputs to Add Join that have matching polygon IDs. The output, Joined Polygons, only has polygons included in the crosswalk table. In the case that you will run, the crosswalk is for Pittsburgh tracts, so only Pittsburgh tracts are output from the larger county basemap.

Next, the Dissolve process uses the crosswalk data to carry out dissolving, resulting in Output Dissolved Polygons. Lastly, the model removes the join so that you can rerun the model with the same or different initial inputs. Otherwise, an error would result—that the join already exists.

Open a map document

1 **In ArcMap, open Tutorial8-7.mxd from your \ESRIPress\GIST1\Maps folder.** The map of Allegheny County opens displaying TIGER file census tract and municipality polygons. Municipalities is just for reference, while Tracts is an input for dissolving. The other input, the crosswalk table PghCrosswalk, is also available in the map document.

2 **In the TOC, click the List by Drawing Order button.**

3 **Click File and Save As, browse to \ESRIPress\GIST1\MyExercises\Chapter8\, and click Save.**

Set geoprocessing options

1 **On the main menu, click Geoprocessing, Geoprocessing Options.**

2 **If not already selected, make sure that the following option is checked: Overwrite the outputs of geoprocessing operations.** With this option on, you can rerun the model repeatedly without having to delete model outputs first, which saves time when debugging and getting your model to work properly.

3 **Click OK.**

Create a new model

1 **On the main menu, click Windows, Catalog.**

2 **Expand Home – Chapter 8 in the folder/file tree.**

3 **Right-click Home – Chapter 8, click New and Toolbox, and rename the new toolbox Chapter8.tbx.**

4 **Right-click Chapter8.tbx and click New, Model.** ArcGIS opens the Model window that you will use to create your model.

Join the crosswalk table to the layer to dissolve

Next, you will browse through system tools to find the Dissolve tool. When you pursue model building on your own, you will need to systematically browse through all of the tools available to get ideas and see what is possible. When you find a tool and want to learn about it, right-click it and click Help.

1 **On the main menu, click Windows, Search.**

2 **Click Tools from the Search window.** The result is a listing and links to ArcGIS's classification of tools available for use directly or as elements in models.

3 **Click the Data Management tools link.** Here you see the first page of many pages with data management tools.

4 **Scroll down and click Joins.**

5 **Drag the Add Join tool to your model and drop it there.**

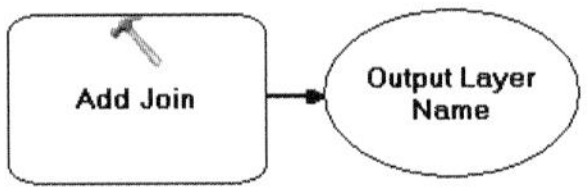

6 **Double-click the Add Join process in your model and make selections using the drop-down list in each field as shown in the image.** Be sure to clear the Keep All (optional) check box. With this option off, the only features kept in the output are those in the crosswalk table, which will be Pittsburgh census tracts. This saves a Clip tool step.

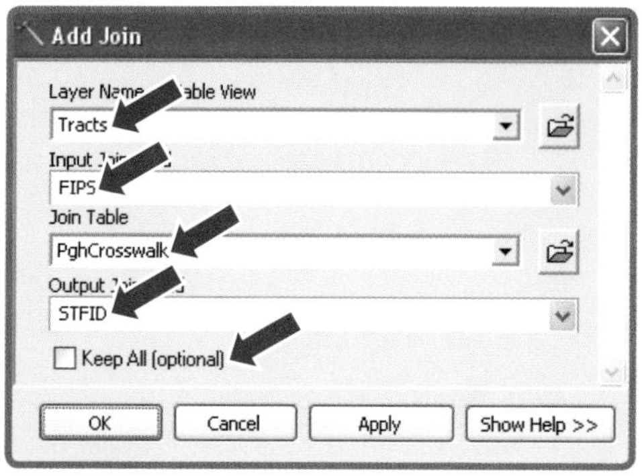

7 **Click OK and resize and reposition model elements as shown in the image.**

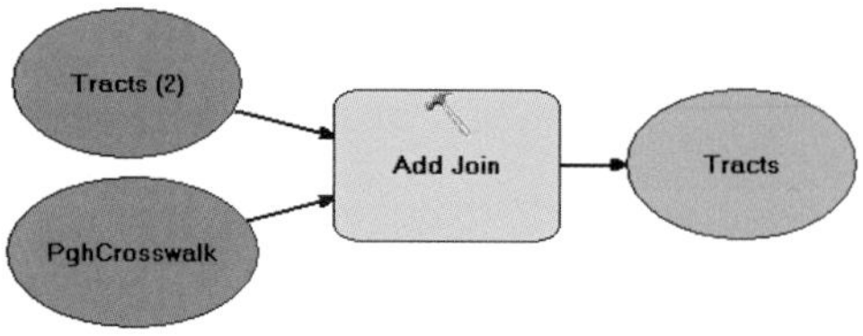

8 **Click the model's Save button.**

Run the partial model

ArcGIS appends table names to attribute names in the joined data. You will need the exact spelling of the PghCrosswalk attribute that will identify the dissolved neighborhoods, so next you will run the Add Join process. Then the desired attribute will be available in a drop-down list of attributes in subsequent processes.

1 **Right-click Add Join in your model, and then click Run.** As the process runs, a report window opens on the task and its status.

2 **Click Close.** ArcGIS adds shadows to the process and its output to indicate that they have executed. Note that if you made an error and have to rerun the model, first you have to click Model on the Model window's main menu and then click Validate Entire Model. This resets all processes to the unrun state.

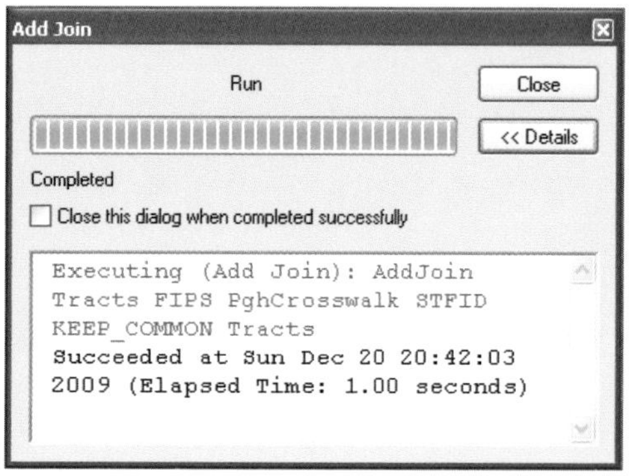

Dissolve tracts

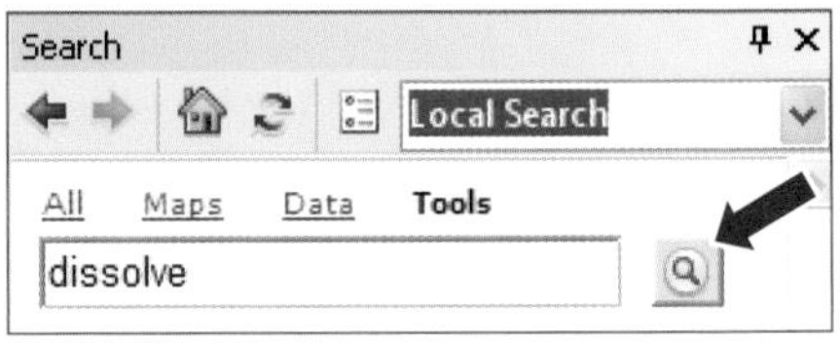

1 Type **dissolve** in the search text box and click the Search button .

2 Drag the Dissolve tool and drop it below the Join process in your model.

3 Click the Connect button on the model window's Standard toolbar, click Tracts output from the Add Join process in the model, click the Dissolve process, and click Input Features in the resulting context menu.

4 Click the model's Select button .

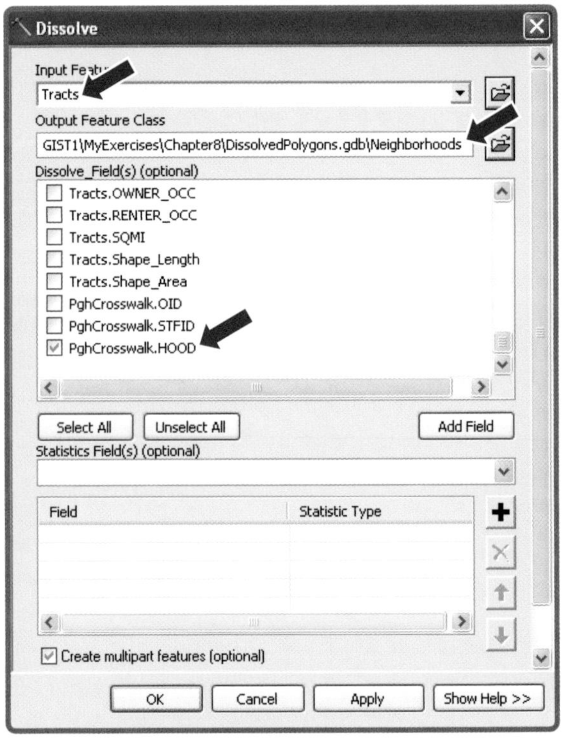

5 Double-click the Dissolve process in your model and make selections using the drop-down list in each remaining field as shown in the image at the right (but do not click OK).

6 Select Tracts.POP2000 in the Statistics Field(s) and ignore the resulting warning.

7 Click in the Statistics Type cell to the right of Tracts.POP2000, click the resulting drop-down arrow, and select SUM.

8 Repeat steps 6 and 7 for two additional attributes, Tracts.WHITE and Tracts.BLACK, using SUM for both. Click OK.

9 Right-click the Neighborhoods output of the Dissolve process, click Add To Display, and save your model.

10 Right-click the Dissolve process, click Run, and close the resulting window when the model has finished running.

Note: If several of your model processes and outputs lose their color fill, meaning that an input is missing, double-click the Add Join process and add Tracts as the Layer Name or Table View. Then delete the original input that will now be disconnected.

YOUR TURN

The basic model is almost complete. The last step is to have the model remove the join in the Tracts output of the Add Join process so that the user can run the model again without manually doing so as you just did in step 10. Otherwise, the Add Join process would fail because a join already exists. Search for and add the Remove Join tool to the model as the last process. Use the output of Add Join as its input. The Remove Join tool automatically identifies PghCrosswalk as the join to remove (check this by opening the Remove Join process). Do not add the output of Remove Join to the display.

Run the Remove Join process. Symbolize Output Joined Polygons (the neighborhoods) with hollow fill and red size 3 outline, then move Tracts to the top of the TOC and compare Tracts and the new Neighborhoods. You should see the output display that is at the start of this tutorial on ModelBuilder.

Reset the model so that you can run it again by clicking Model, Validate Entire Model. Run the entire model by clicking the model Run button ▶. Again, symbolize Output Joined Polygons with hollow fill and red size 3 outline, move Tracts to the top of the TOC, and compare Tracts and the new Neighborhoods. You should see the output display that is at the start of this tutorial on ModelBuilder.

Generalize element labels

Your model is capable of being a general tool for dissolving any polygons. As a first step to making the model general, you will change several element labels.

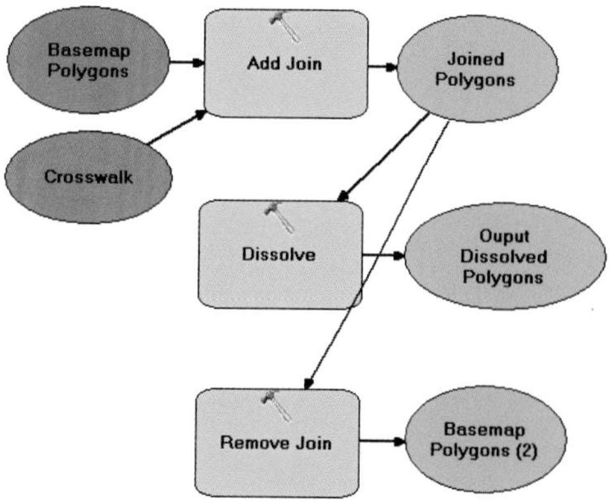

1 Right-click the Tracts (2) element, click Rename, type **Basemap Polygons**, and click OK.

2 Similarly, change labels of other elements as shown in the image.

3 Click the model's save button.

8-1 8-2 8-3 8-4 8-5 8-6 8-7 A8-1 A8-2 A8-3

Add model parameters

Currently, your model is "hardwired" with inputs and outputs fixed in processes. Next, you will make several elements parameters that users can change without modifying the model itself. Instead, users will type or make selections in a form when opening the model.

1 Right-click Basemap Polygons and click Model Parameter. A "P" appears above and to the right of the element indicating the ArcGIS will ask the user to browse for an input map layer.

2 Similarly, make Crosswalk and Output Dissolved Polygons model parameters.

Add variables to model

To be general, the Add Join process needs to get two of its inputs from the user, the Input Join Field and Output Join Field. You can make these inputs parameters, but first you need to create variables to store them in the model.

1 Right-click the Add Join process. Click Make Variable, From Parameter, and Input Join Field. ArcGIS creates the variable for you.

2 Click anywhere in the white area of the model to deselect elements. Move the new variable above the top left of the Add Join process and make its element a bit wider so that its entire label displays.

3 Make Input Join Field a model parameter.

4 Repeat steps 1–3, except make the variable for the Output Join Field of the Add Join process. Move the new element above and to the right of the Add Join process.

Add labels for documentation

Labels can help document the model. You will add a model title and some notes about the variables.

1 If necessary, select all model elements and make some room at the top for a label.

2 Right-click the white area at the top, click Create Label, double-click the resulting Label, and type **Model to Dissolve a Subset of Basemap Polygons**.

3 Right-click the new label, click Display Properties, click in the cell to the right of Font, click the resulting builder button, change the font to Bold size 14, and click OK.

4 Right-click the Input Join Field element, click Create Label, click anywhere in the model white area, click the new label and move it above and to the left of the element, double-click it, and type **"You must examine the Basemap Polygons attribute table and Crosswalk table and note their field names that share the same polygon IDs or names for use as parameter values on opening the model."**

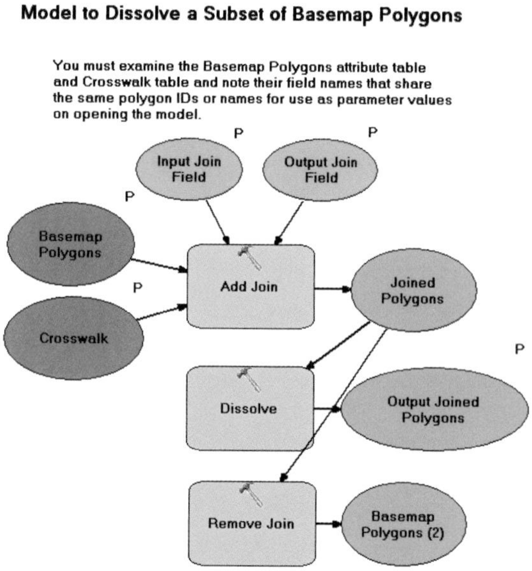

5 Right-click the new label, click Display Properties, click the cell to the right of Text Justification, click Left instead of Center, and close the properties window.

6 Break this label up into several lines by placing your cursor after words to create new lines, press the Shift key, and press Enter.

7 Click the model's save button.

Add model name and description for documentation

You can add help documentation to the form that will open on running the model in model properties.

1 Click Model on the model's main menu, and click Model Properties.

2 Type **DissolvePolygons** for Name and **Dissolve Any Subset of Polygons** for Label.

3 Type as much as you want of the following help message in the Description text box:

Dissolve any subset of basemap polygons given a crosswalk table for the subset.

(1) Name and save the Output Dissolved Polygons as you wish.

(2) You need to know the Basemap Polygon ID used in the crosswalk table and the Dissolved Polygon ID for the dissolved output polygons for input here.

(3) Browse for the input BaseMap Polygons and its Crosswalk.

4 Check the Store relative path names check box.

5 Click OK.

6 Click the model's save button and close the model window.

Open and run the finished model

You can add help documentation to the form that will open on running the model in model properties.

1 **In Catalog, right-click the Dissolve Any Subset of Polygons model, click Open, and close the warning at the top of the resulting form.**

2 **Click the Show Help button at the bottom of the form.** You do not need to change any of these input parameters.

3 **Click OK to run the model.** The model runs, adding the dissolved neighborhoods to your map display.

4 **Save your map document and close ArcMap.**

Assignment 8-1

Build a study region for Colorado counties

In this assignment, you will create a study area for two rapidly growing counties in Colorado: Denver and Jefferson counties. You will create new feature classes for an urban area study using polygon layers downloaded from the U.S. Census Web site. Because we want to study two counties, you will need to join some of the layers together and clip two layers from the country and state levels to the smaller study area.

Start with the following:

- \ESRIPress\GIST1\Data\UnitedStates.gdb\COCounties—polygon feature class of Colorado counties
- \ESRIPress\GIST1\Data\UnitedStates.gdb\COStreets—TIGER/Line shapefile of Jefferson County streets
- \ESRIPress\GIST1\Data\UnitedStates.gdb\COStreets2—TIGER/Line shapefile of Denver County streets
- \ESRIPress\GIST1\Data\UnitedStates.gdb\COUrban—urban area feature class for Jefferson County
- \ESRIPress\GIST1\Data\UnitedStates.gdb\COUrban2—urban area feature class for Denver County
- \ESRIPress\GIST1\Data\UnitedStates.gdb\USCities_dtl—point feature class of detailed cities

Create a study area map of Colorado urban areas

In Catalog, create a new file geodatabase called **\ESRIPress\GIST1\MyAssignments\Chapter8\Assignment8-1YourName.gdb**. This is where you will store the study area feature classes you are about to create.

Create a new map document called **\ESRIPress\GIST1\MyAssignments\Chapter8\Assignment8-1YourName.mxd** with all of the above feature classes added. Perform the geoprocessing operations necessary to create the following study area features in your file geodatabase:

- Jefferson and Denver counties boundaries, StudyAreaCounties
- Graduated point layer showing populations of detailed cities for Jefferson and Denver counties only, StudyAreaCities
- One layer showing urban areas for both counties, StudyUrbanArea
- One streets layer for the new urban area study, StudyAreaUrbanStreets

Create an 8.5-by-11-inch map layout displaying your new datasets zoomed to the study urban area. Export the layout as **\ESRIPress\GIST1\MyAssignments\Chapter8\Assignment8-1YourName.pdf**.

STUDY QUESTIONS

Perform the necessary queries and spatial analysis to answer the following questions. Save your answer in a document called **\ESRIPress\GIST1\MyAssignments\Chapter8\YourNameAssignment8.doc.**

1. How many cities are within the study urban area?
2. What is the total population of these cities?
3. What cities are within one mile of Wadsworth Street in the study urban area? (**Hint:** FE_NAME=Wadsworth)

WHAT TO TURN IN

If your work is to be graded, turn in the following files:

File geodatabase: \ESRIPress\GIST1\MyAssignments\Chapter8\Assignment8-1YourName.gdb

ArcMap document: \ESRIPress\GIST1\MyAssignments\Chapter8\Assignment8-1YourName.mxd

Exported map: \ESRIPress\GIST1\MyAssignments\Chapter8\Assignment8-1YourName.pdf

Word document: \ESRIPress\GIST1\MyAssignments\Chapter8\Assignment8-1YourName.doc

If instructed to do so, instead of the above individual files, turn in a compressed file, **Assignment8-1YourName.zip**, with all files included. Do not include path information in the compressed file.

Assignment 8-2

Dissolve property parcels to create a zoning map

In this assignment, you will dissolve a parcel map to create a zoning map that highlights a proposed commercial development in what is now a residential area. A commercial company wants to apply for a zoning variance so that it can use the land in residential parcels with PARCEL_ID values 623, 633, 641, 651, and 660 for a commercial purpose. Change the zoning code of these properties to X and highlight them on your map with a red color fill. The Zoning Department wants the map for a public hearing on the proposal and will use it in a PowerPoint presentation.

Start with the following:

- \ESRIPress\GIST1\Data\Pittsburgh\EastLiberty\EastLib—coverage for the East Liberty neighborhood boundary
- \ESRIPress\GIST1\Data\Pittsburgh\EastLiberty\Parcel—coverage for land parcels in the East Liberty neighborhood of Pittsburgh

 The parcel coverage includes the following attribute fields:

 ZON_CODE—an attribute with zoning code values

 TAX_AREA, TAX_BLDG, and TAX_LAND_A—have annual property tax components

- \ESRIPress\GIST1\Data\Pittsburgh\EastLiberty\Curbs—coverage that has street curbs and annotation with street names

Notice in the attributes of the parcels that groups of the zoning codes start with the same letter:

A – development
C – commercial
M – industrial
R – residential
S – special

The digit or character following the first letter further classifies land uses. For example, R4 is a residential dwelling with four units.

Prepare map layers

In Catalog, create a new file geodatabase called **\ESRIPress\GIST1\MyAssignments\Chapter8\Assignment8-2YourName.gdb**. This is where you will store the study area feature classes you are about to process. Also, create a new map document called **Assignment8-2YourName.mxd** stored in the same folder with all of the above feature classes added.

Perform the geoprocessing operations necessary to convert EastLib, Parcel, and Curbs to feature classes in \ESRIPress\GIST1\MyAssignments\Chapter8\Assignment8-2YourName.gdb\.

You will create an aggregate-level zoning code by adding a new field to the parcels attribute table that has just the first character of the full zoning code. Call the new field **Zone** with Text data type and length 1. Use the Field Calculator on the new field. Click the String Option button, click the Left( ) function, type to yield Left(**[ZON_CODE],1**), and click OK. The left function extracts the number of characters entered—1 in this case—starting on the left of the input field, ZON_CODE. Edit the new field to change the Zone values of the parcels in the zoning variance proposal to X.

Dissolve the parcel's shapefile using the Dissolve, using your new field (Zone) as the dissolve field, and adding SUM statistics for the three tax fields in parcels. Click in the Statistics Type cells to select SUM. Save the output feature class as **\ESRIPress\GIST1\MyAssignments\Chapter8\Assignment8-2YourName.gdb\Zoning**.

Map document

Add the new Zoning feature class to your map document, as well as the curbs arcs and annotation, and East Liberty outline. Use the unique values option of the categories method of classification for symbolizing the Zone field. Use colors for the various Zone values, including green for residential and bright red for the parcels of the proposal. Create output **\ESRIPress\GIST1\MyAssignments\Chapter8\Assignment8-2YourName.jpg** from an 11-by-8.5-inch landscape layout, zoomed in to the upper left quarter of the neighborhood, and including a legend.

WHAT TO TURN IN

Note: Do not submit any of the interim files that are not in your final map document (e.g., original counties, streets, or urban areas).

If your work is to be graded, turn in the following files:

File geodatabase: \ESRIPress\GIST1\MyAssignments\Chapter8\Assignment8-2YourName.gdb

ArcMap document: \ESRIPress\GIST1\MyAssignments\Chapter8\Assignment8-2YourName.mxd

Image file: \ESRIPress\GIST1\MyAssignments\Chapter8\Assignment8-2YourName.jpg

If instructed to do so, instead of the above individual files, turn in a compressed file, **Assignment8-2YourName.zip**, with all files included. Do not include path information in the compressed file.

Assignment 8-3

Build a model to create a fishnet map layer for a study area

Below is the fishnet map layer that you will create with several steps saved in a ModelBuilder model. It consists of uniform, square grid cells saved in a polygon map layer, and as an option includes an additional layer of centroid points for each cell. Both the cells and centroids are useful for spatial analysis, for example, for displaying counts of fires and other point data by grid cell. The centroids allow you to display a second attribute using size-graduated point markers on a color ramp, while a choropleth map displays the first variable. You can create cell-level data from point data, such as residential fire incidents or crimes locations, using spatial overlay as done in chapter 9. An advantage of data analysis with uniform grid cells is that that configuration holds shape and area of each spatial unit constants, leaving all observed variation due to the variable or variables under investigation.

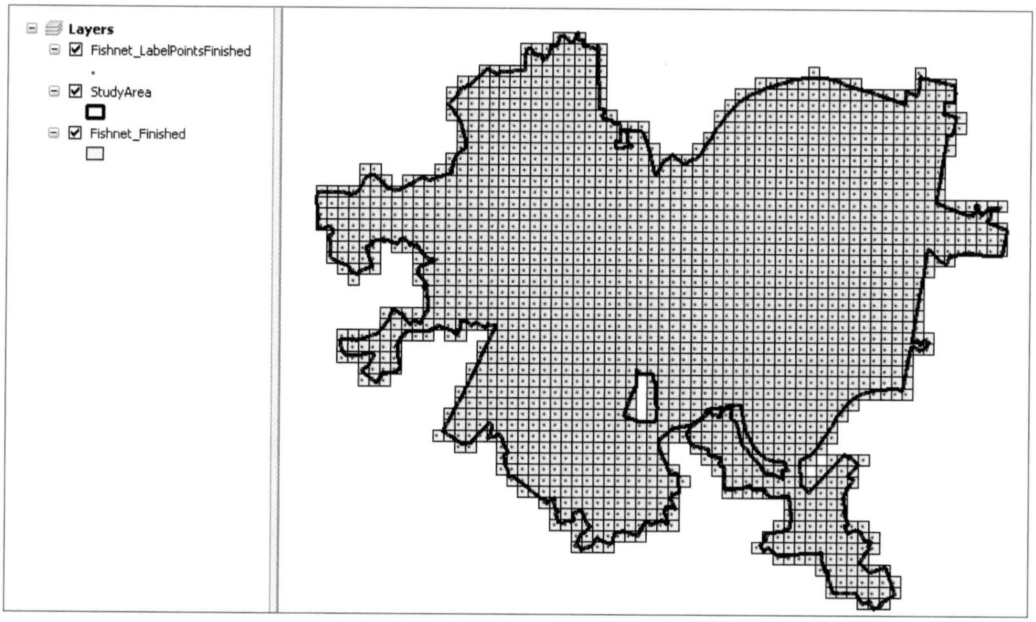

Start with the following:

- \ESRIPress\GIST1\Data\Pittsburgh\City.gdb\Pittsburgh—feature class for Pittsburgh boundary

Create a file geodatabase called **\ESRIPress\GIST1\MyAssignments\Chapter8\Assignment8-3YourName.gdb** and write all outputs to it. Likewise, create a map document, **Assignment 8-3YourName.mxd**; toolbox, **Assignment83YourName.tbx** (without a hyphen between the 8 and 3); and model, **FishnetStudyArea** all stored in the same **\ESRIPress\GIST1\MyAssignments\Chapter8** folder.

Requirements and hints

The input map layer is any polygon layer that defines the study area. It could be a boundary, such as for Pittsburgh as seen on the previous page; census tracts; or any other set of polygons. The Fishnet tool that you will use needs several parameters to construct the grid cell map layer. In particular, it needs the cell size (while the tool has inputs for both width and height, you will almost always want square grid cells so that width will equal height), the number of rows and columns in the extent, and the extent coordinates of the output. The Fishnet tool has provision to import the needed map extent coordinates from the input study area polygons, so that part is easy.

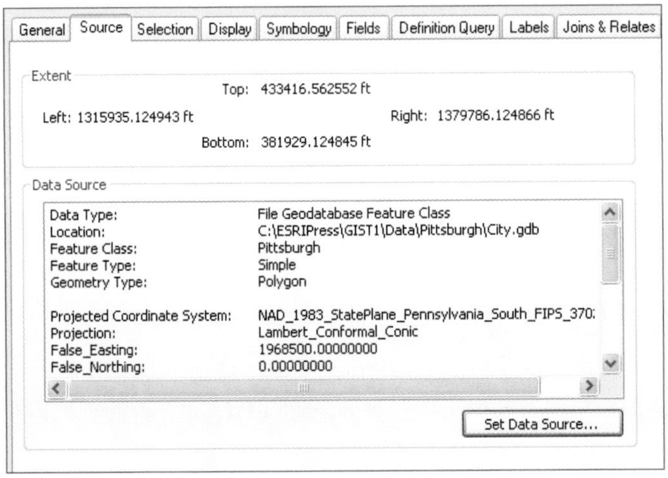

To calculate the number of rows and columns of the fishnet, you must have projected coordinates for the input layer (the Pittsburgh layer has projection state plane in feet, so this case meets that requirement) and you must look up its map extent coordinates in its Source properties as seen in the image as Top, Bottom, Left, and Right.

So, if you want cells that are squares 1,000 feet by 1,000 feet, you calculate:

Number Rows = (433,416–381,929) / 1,000 + 1 = 52 (rounded down)

Number Columns = (1,379,786–1,315,935) / 1,000 + 1 = 65 (rounded up)

To the right is the model you will create. Note that you will use the Cell Size parameter for both the height and width of the desired square cells.

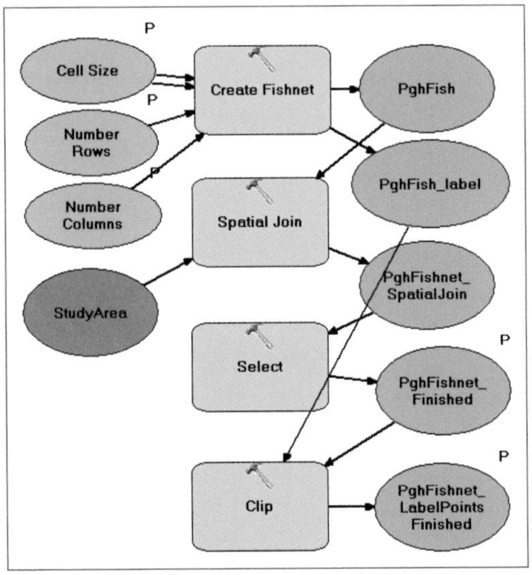

Notes on the Create Fishnet process follow:

Create the three variables. Select Double for the data type of Cell Size and Long for the other two variables' data type (which is an integer).

For the Create Fishnet process, name the study area polygon layer (Pittsburgh) **StudyArea** and select it for the Template Extent input.

For the Fishnet's Cell Size Width and Height inputs, use the drop-down lists and select Cell Size (the name of your model variable). For the Number of Rows and Number of Columns inputs, use the drop-down lists and select your corresponding variables.

For the Geometry, choose POLYGON.

The Fishnet process creates a rectangular map layer, but what is needed is a grid cell map limited to the shape of the study area. It is best to leave grid cells whole squares, with some overhang, rather than clipping them to the study area's boundary. The model needs two steps to accomplish this. First you have to assign study area attributes to grid cells with the Spatial Join process. Then all grid cells in the interior or crossing the boundary of the study area will have non-null attributes while remaining cells will have null values. Then you can use the Select process with a criterion such as

NAME > ''

where there are two single quotes after the greater than sign, signifying a null value. The criterion selects all cells having non-null values.

Finally, you can clip the label points using the finished fishnet. Making the final outputs (Fishnet_Finished and Fishnet_LabelPointsFinished) parameters allows the user to rename and save them wherever desired.

WHAT TO TURN IN

If your work is to be graded, turn in the following files:

File geodatabase: \ESRIPress\GIST1\MyAssignments\Chapter8\Assignment8-3YourName.gdb

ArcMap document: \ESRIPress\GIST1\MyAssignments\Chapter8\Assignment8-3YourName.mxd

Toolbox: \ESRIPress\GIST1\MyAssignments\Chapter8\Assignment83YourName.tbx

If instructed to do so, instead of the above individual files, turn in a compressed file, **Assignment8-3YourName.zip**, with all files included. Do not include path information in the compressed file.

Part 3
Learning advanced GIS applications

9 Spatial analysis

Spatial analyses are advanced applications of GIS such as determining relationships between locations, identifying locations that meet criteria, using models for estimation, and so forth. For instance, it is possible to place buffers around features to retrieve nearby features for proximity analysis. An example is to retrieve crime locations near properties at high-risk of criminal activity such as taverns and bars. A classical spatial analysis consists of a site selection or suitability analysis for facilities, especially when this involves several selection criteria such as being in a business area, on a major street, and centrally located. Finally, it is possible to carry out complex spatial processing for approximation purposes, for example, to estimate demographic attributes for administrative areas that do not follow census tracts. In this case, census tracts may be subdivided among two or more administrative areas, so an approximation is needed to split up or apportion tract data to the administrative areas.

Learning objectives

- *Buffer points for proximity analysis*
- *Conduct a site suitability analysis*
- *Apportion data for noncoterminous polygons*

Tutorial 9-1

Buffer points for proximity analysis

Some land uses attract crime, such as taverns and bars. So, it's a good idea for police to monitor crimes in the vicinity of bars, and this is possible with GIS using circular buffers.

Set up for analysis

1 **Start ArcMap and open \ESRIPress\GIST1\Maps\Tutorial9-1.mxd.** Tutorial9-1.mxd contains a map of the Lake Precinct of the Rochester, New York, Police Department. Shown are aggravated assault crime offense points, bars, police car beats (with one patrol car assigned to each beat), and street centerlines.

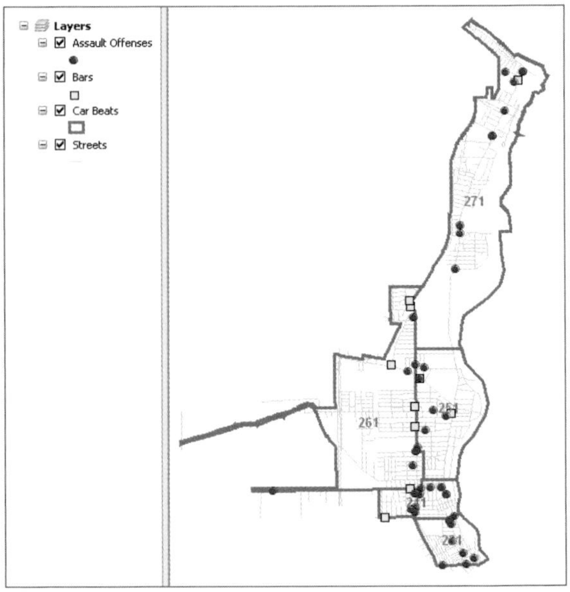

2 **Click File and Save As, browse to \ESRIPress\GIST1\MyExercises\Chapter9\, and click Save.**

Buffer bars

1 **On the main menu, click Windows, Search.**

2 **In the Search window, click the following links: Tools > Analysis tools > Analysis > Proximity > Buffer.**

3 Type or make selections as shown in the image at right.

4 Click OK, wait for ArcMap to finish processing, and hide the Search window.

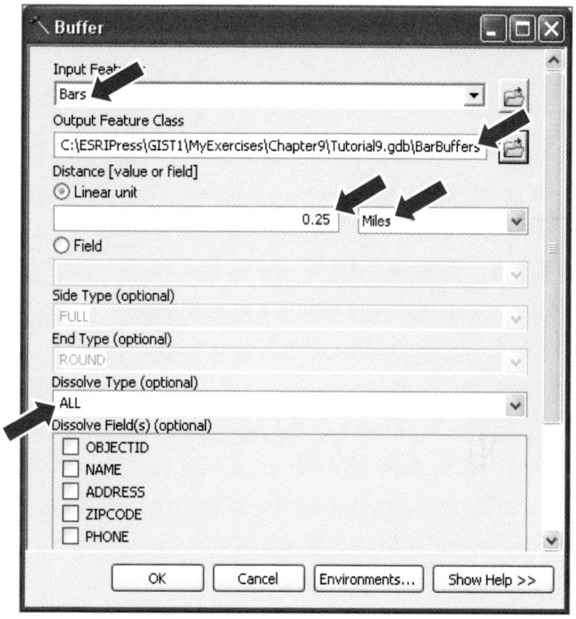

5 Move BarBuffers to the top of the TOC and change its symbology to a hollow fill.

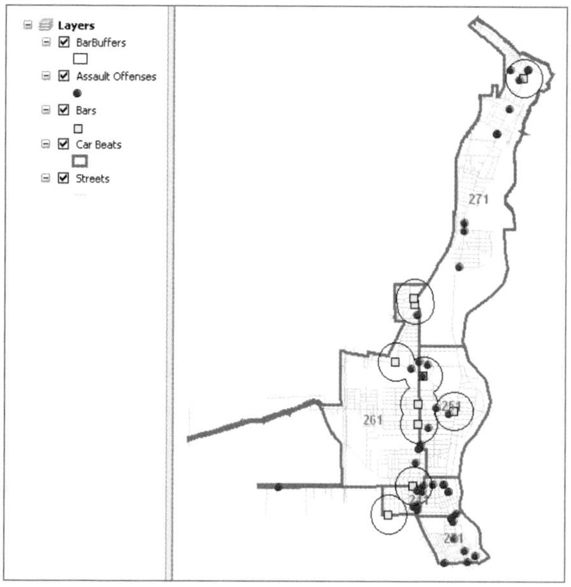

Extract assault offenses in bar buffers

1 In ArcMap, click Selection, Select by Location.

2 Make selections as shown in the image at the right.

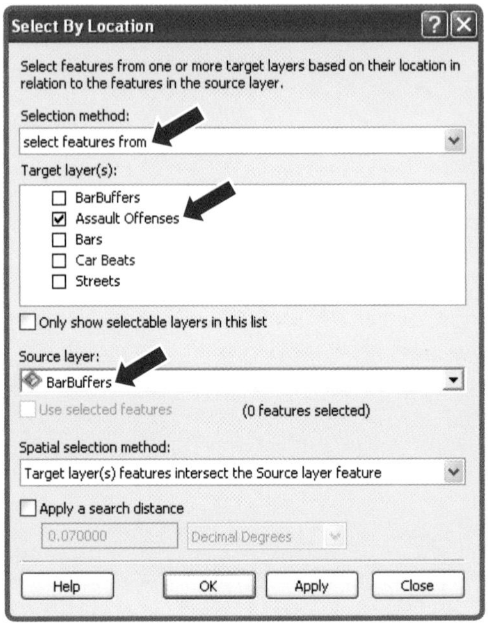

3 Click Apply and Close.

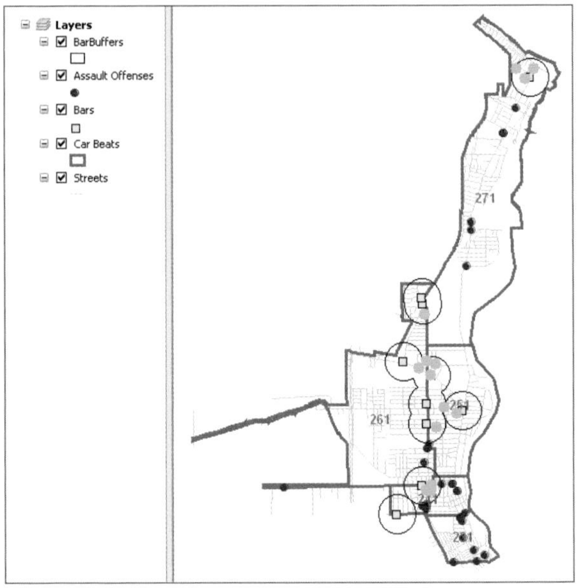

4 In the TOC, right-click the Assault Offenses layer. Click Data, Export Data.

5 In the Export Data dialog box, make sure that Selected features is selected, make the output feature class **\ESRIPress\GIST1\MyExercises\Chapter9\Tutorial9.gdb\AssaultsInBarBuffers**, click OK, then click Yes.

6 Symbolize the new layer using a bright red Circle 1 marker, size 8.

7 **Remove Assault Offenses from the TOC.** This is a map that a task force would want for making the case to enforce laws at bars or close bars down. For example, three of the bars have three or four assaults in their vicinity.

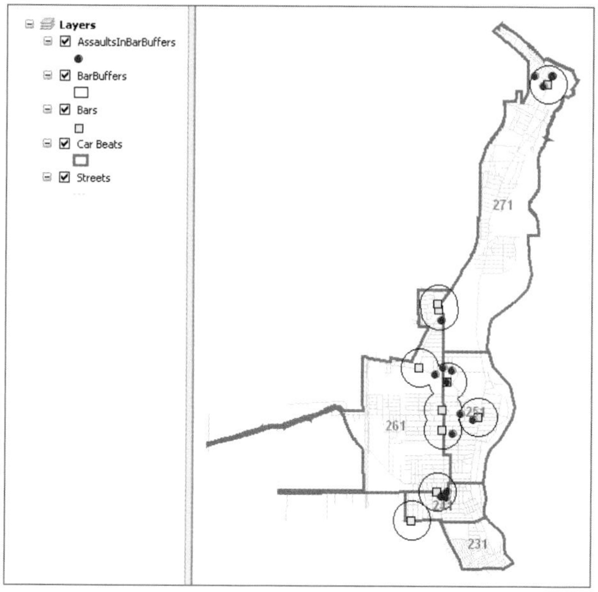

8 Save your map document.

Tutorial 9-2

Conduct a site suitability analysis

Suitability analysis for facility location is a classic GIS application. In this exercise, you will perform suitability analysis for the purpose of locating potential areas for new police satellite stations in each car beat of Lake Precinct. Criteria for locating these stations are that the site must be centrally located in each car beat (within a 0.33-mile radius buffer of car beat centroids), in retail/commercial areas (within 0.10 mile of a least one retail business), and within 0.05 mile of major streets. Typically, analysis consists of several steps that include attribute- and location-based queries, buffers, spatial joins, and other geoprocessing steps.

Open a map document

1 Open **\ESRIPress\GIST1\Maps\Tutorial9-2.mxd.** Tutorial9-2.mxd contains a car beat map of the Lake Precinct of the Rochester, New York, Police Department. It also has police car beats, retail business points, and street centerlines.

2 **Click File and Save As, browse to \ESRIPress\GIST1\MyExercises\Chapter9\, and click Save.**

Add X and Y columns to car beats

1 **In the TOC, right-click the Car Beats layer and click Open Attribute Table.**

2 **In the Attributes of Car Beats table, click the Table Options button and Add Field.**

3 In the Add Field dialog box, name the new field **X**. Choose Double from the Type drop-down list, and click OK.

4 Repeat steps 2 and 3, but name the field **Y**.

5 In the Attributes of Car Beats table, right-click the X column heading. Click Calculate Geometry, and click Yes to calculate outside of an edit session.

6 In the Calculate Geometry dialog box, select X Coordinate Centroid for Property. Click OK.

7 Repeat steps 5 and 6 for the Y column, but select Y Coordinate of Centroid. Each record in the Attributes of Car Beats table now contains an x- and y-coordinate value. Each x,y pair represents the centroid of a police car beat. The display with these coordinates has some fields turned off and not shown.

OBJECTID	Shape *	BEAT	X	Y
1	Polygon	261	1397139.549464	1165093.524371
2	Polygon	271	1404519.587559	1179549.785764
3	Polygon	251	1402264.268132	1165279.502132
4	Polygon	241	1400522.350349	1158989.873773
5	Polygon	231	1402631.8189	1156203.703772

Map car beat centroids

1 Click the Table Options button ▾, and click Export.

2 In the Export Data dialog box, click the Browse button for the Output table field. Change the Save as type to File and Personal Geodatabase tables, browse to \ESRIPress\GIST1\MyExercises\Chapter9\Tutorial9.gdb, change the Name to **CarBeatCentroids**, click Save, then click OK and No.

3 Close the Attributes of Car Beats table.

4 Unhide Catalog and expand the folder/file tree to \ESRIPress\GIST1\MyExercises\Chapter9\Tutorial9.gdb.

5 Right-click CarBeatCentroids in Tutorial9.gdb. Click Create Feature Class and From XY Table.

6 In the Create Feature Class from XY Table dialog box, click the Coordinate System of Input Coordinates button.

7 In the Spatial Reference Properties dialog box, click Import. Browse to \ESRIPress\GIST1\Data\RochesterNY\LakePrecinct.gdb. Click lakecarbeats, Add, and OK.

8 Click the browse button for Output in the Create Feature Class From XY Table window, browse to Tutorial9.gdb, and click Save and OK.

9 Right-click Tutorial9.gdb in Catalog and click Refresh.

YOUR TURN

Add XYCarBeatCentroids from Tutorial9.gdb to your map document and symbolize with a Circle 2 marker, Mars Red, size 10. Label the centroids with Beat instead of the original Car Beat polygons.

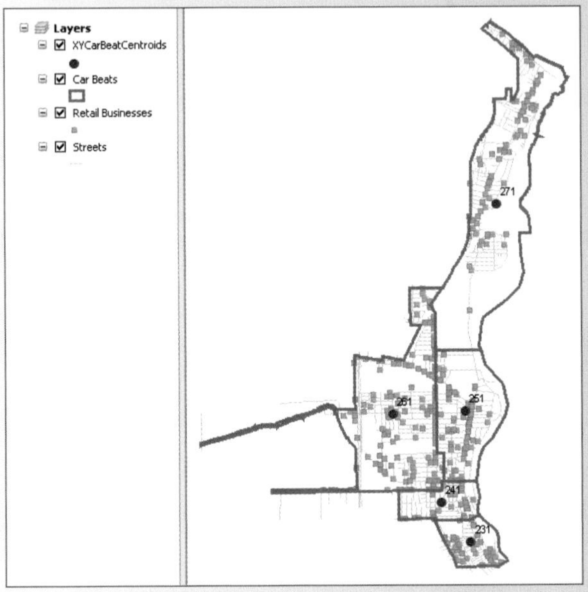

Buffer car beat centroids

1 On the main menu, click Windows, Search, Tools, Analysis tools, and Proximity.

2 Double-click the Buffer tool.

3 Type or make selections as shown in the image.

4 Click OK.

5 Hide the Search window and close the Buffer window.

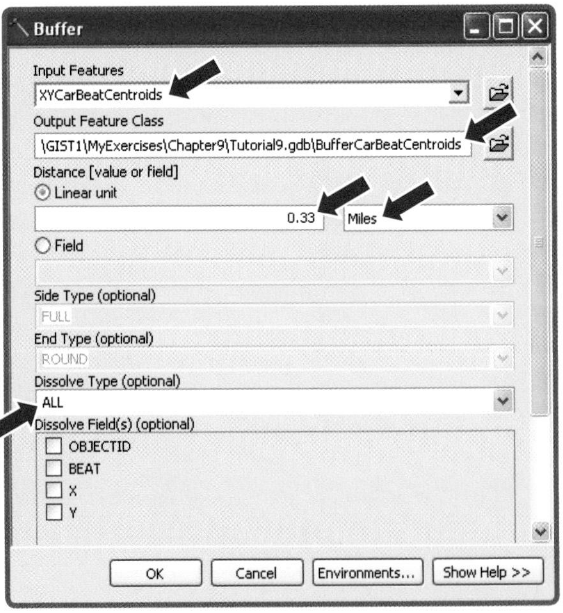

6 **Symbolize the buffers polygons with a hollow fill, set the Outline Color to Mars Red, then set the Outline Width to 1.** Next, you need to find areas within the car beat buffers that meet the remaining criteria.

Buffer retail businesses

1 **Unhide the Search window.**

2 **Double-click the Buffer tool and type or make selections as follows:**

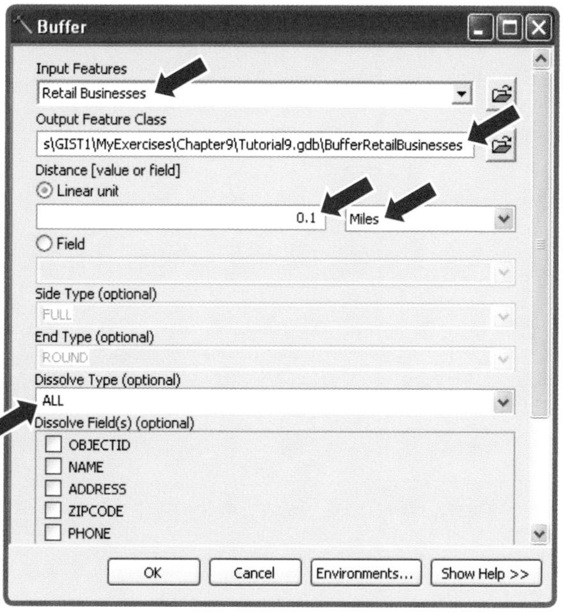

3 Click OK and close the Buffer window.

4 Symbolize the new buffer layer with a hollow fill, set the Outline Color to Ultra Blue, and the Outline Width to 1.

5 In the TOC, turn off the XYCarBeatCentroids and Retail Businesses layers. The intersection of the two buffers nearly satisfies the suitability criteria, but you still need to buffer the streets. Then you can take an intersection of all buffers to find suitable areas.

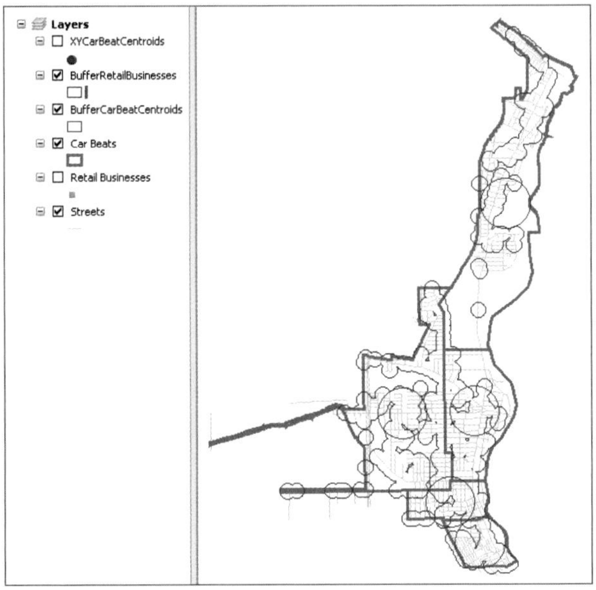

Select major streets

TIGER street records have an FCC (feature classification code) that classifies streets by type. Major and commercial streets have FCC values of A40 and A41. You will select only those streets and then buffer them.

1 In ArcMap, click Selection, Select By Attributes.

2 Type or make selections as shown in the image at right (be sure to click the Get Unique Values button for FCC values).

3 Click OK. Major and commercial streets turn the selection color.

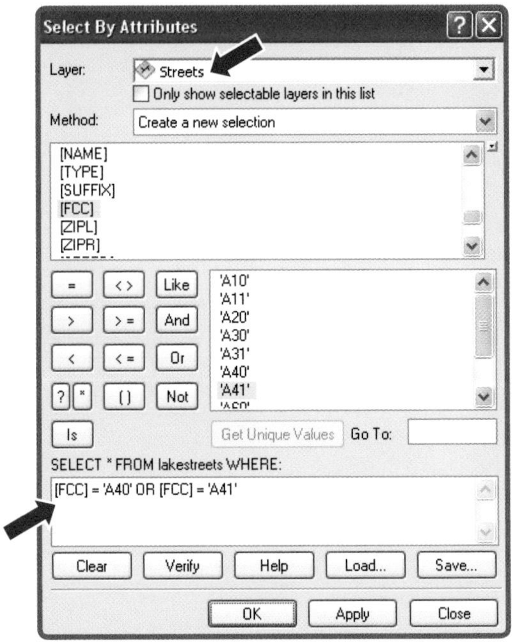

Buffer major streets

1 If necessary, unhide the Search window.

2 Double-click the Buffer tool and type or make selections as shown in the image at the right.

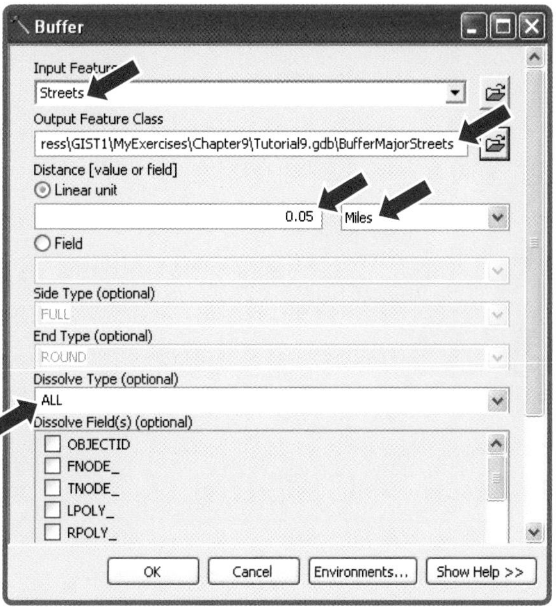

3 Click OK and close the Buffer window.

4 Symbolize the new buffer layer with a hollow fill, Medium Apple Outline Color, and Outline Width 1.

5 Click Selection and Clear Selected Features. In the TOC, turn off the Streets layer. It is getting difficult to identify areas that meet all three proximity criteria, but an intersection of all three buffers will show those areas directly.

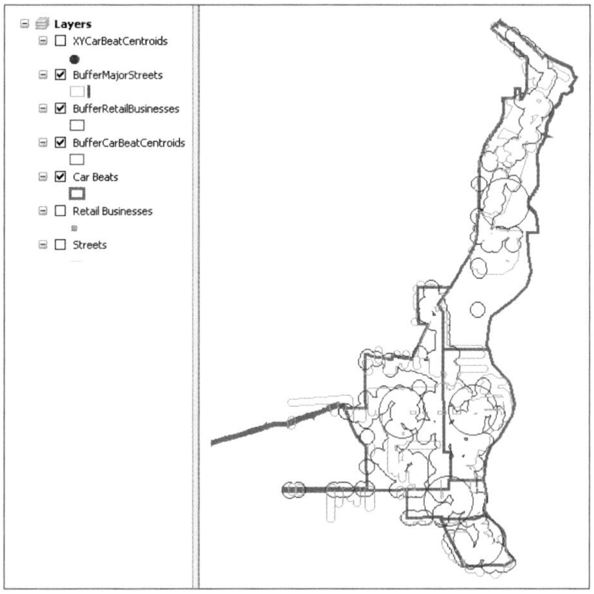

6 Save the map document.

Intersect buffers

1 Unhide the Search window, type **Intersect** in its search text box, press the search button, and click the Intersect tool.

2 Type or make selections as follows:

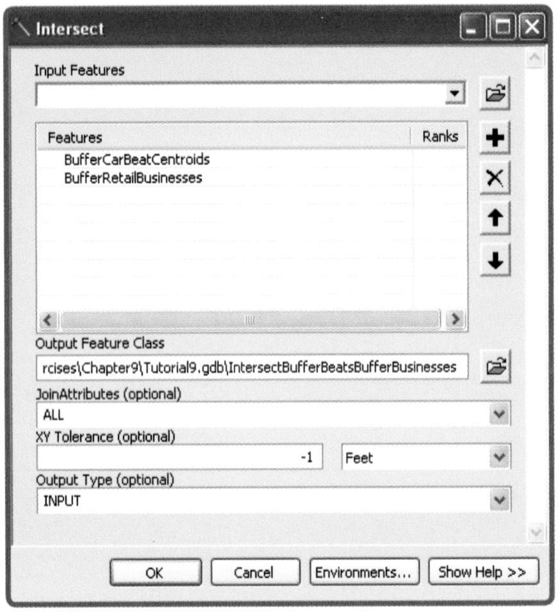

3 Click OK.

YOUR TURN

Intersect the intersection you just created with BufferMajorStreets to produce BufferSuitability. Symbolize the end product as you wish. Turn layers on/off as shown in the image. Notice that some of the suitable area for car beat 241 is in car beat 261. Obviously, you would not choose a site in the wrong car beat. Otherwise, the results are ready for use for finding sites for the satellite police stations.

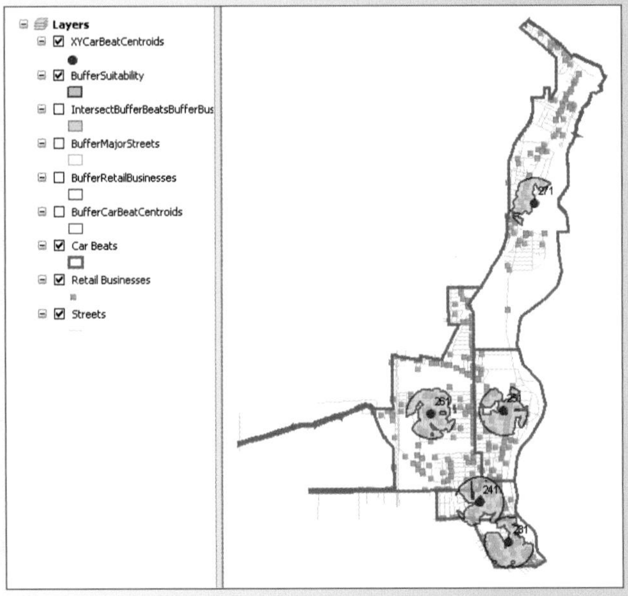

Tutorial 9-3

Apportion data for noncoterminous polygons

If you have point data on individual events, persons, or things, using a spatial join you can always aggregate data up to counts or sums for any polygon map layer with the same extent as the point layer. Sometimes, however, you will not have point data, but only aggregate data for given polygons. A good example is census data, which is tabulated for polygon layers from states down to blocks. Detailed census data from the SF 3 file is only available down to the block group level. Nevertheless, your need may be for other polygon boundaries—for example, administrative areas such as police car beats that are not coterminous with census boundaries. The Rochester, New York, Police Department designed its car beats to meet police needs, which do not always coincide with census-taking purposes. Consequently, car beat boundaries do not always follow census tracts. If you want census data by car beat polygons, you have to apportion (make approximate splits of) each tract's data to two or more car beats. The end result will have approximation errors and some apportionment methods will have fewer errors than others.

Open a map document

1 **In ArcMap, open Tutorial9-3.mxd from the \ESRIPress\GIST1\Maps\ folder.** Tutorial9-3 contains a map of car beats and census tracts in the Lake Precinct of the Rochester Police Department. You can find several cases where car beats contain only portions of tracts; for example, between car beats 251 and 261.

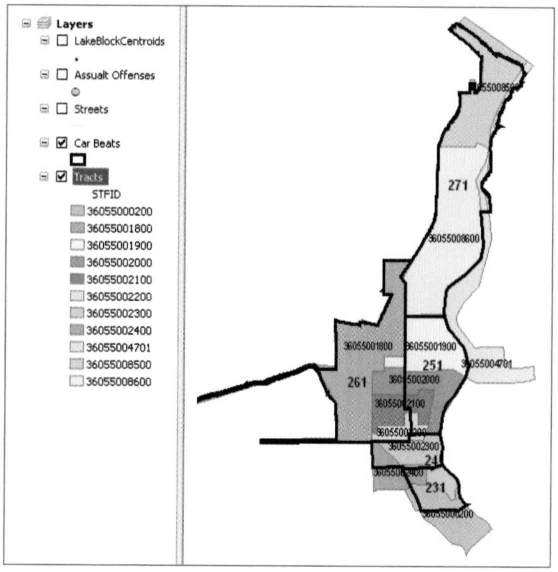

2 **Open the Tracts attribute table.** See that this table has attributes POP25_, which is the population 25 or older, and NOHISCH, which is the population 25 and older with less than a high school education. NOHISCH is the census attribute you need to have at the car beat level, and so you will apportion it from tracts to car beats. As explained below, NOHISCH is not available at the block level, but if it were you would not need to apportion it.

3 **Close the table and open the LakeBlockCentroids table.** See that this table has attributes for population in intervals from under 5 to 65 and over. You will use these attributes as the basis for apportionment as explained below.

4 **Close the table. Click File and Save As, browse to \ESRIPress\GIST1\MyExercises \Chapter9\, and click Save.**

The math of apportionment

Suppose that you want to apportion census data to administrative areas. If you only need short-form census data (various tabulations of population, families, households by sex, race, and age, as well as housing units and occupancy), then you have a simple task because the Census Bureau provides this data at the block level. A safe assumption is that any administrative areas that you would want to build would be aggregates of blocks, so a simple dissolve of census block centroid points (with needed census attributes in its table) based on a crosswalk table to administrative areas would complete the task.

If, however, you need long-form census data (place of work, transportation to work, employment status, educational attainment, school enrollment, disability, armed forces status, income, poverty status, characteristics of housing, etc.) for administrative areas, then you need apportionment. The smallest areas for long-form census data are block groups or tracts and their polygons are too large to dissolve. They would not match the desired administrative areas. So instead you need to apportion or split up block group or tract polygons that cross administrative polygons.

Let's take a look at one example. On the next page is a close-up of tract 360550002100, which is split between car beats 261 and 251. Tract 360550002100 has 205 people aged 25 or older with less than a high school education. For short, let's call this the "undereducated population." How can we divide those 205 undereducated persons between car beats 261 and 251?

One approach would be to assume that the target population is uniformly distributed across the tract. Then you could split undereducated population up by the fraction of the area of the tract in each car beat. What if, however, the tract has a cemetery, park, or other unoccupied areas? Then the apportionment could have sizable errors.

A better approach is to use a block-level, short-form census attribute as the basis of apportionment with the assumption that the long-form attribute of interest is uniformly distributed across the short-form population. This at least accounts for unoccupied areas.

One limitation of the block-level data is that the break points for age categories do not match those of the educational attainment data (persons 25 or older). The best that can be done with the block data is to tabulate persons aged 22 or older. Nevertheless, the resulting data should be close enough for approximation.

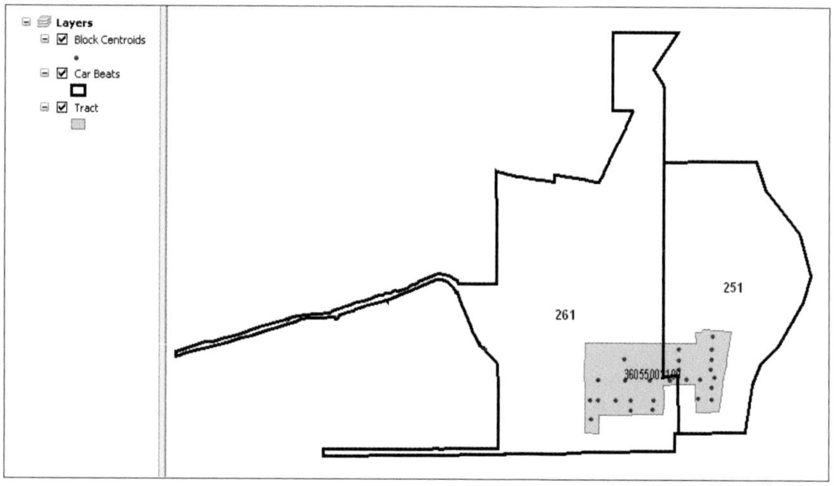

Let's work through the example at hand.

- Of the 26 blocks making up the tract, the 13 that lie in car beat 261 have 1,177 people aged 22 or older. The other 13 blocks in car beat 251 have 1,089 such people for a total of 2,266 for the tract.

- Apportionment assumes that the fraction of undereducated people aged 25 or older is the same as that for the general population aged 22 or older. This fraction, called the weight, is 1,177 ÷ 2,266 = 0.519. For the other car beat, the weight is 1,089 ÷ 2,266 = 0.481.

- Thus, we estimate the contribution of tract 36055002100 to car beat 261's undereducated population to be (1,177 ÷ 2,266) × 205 = 106. For car beat 251, it is (1,089 ÷ 2,266) × 205 = 99.

- Eventually, by apportioning all tracts, we can sum up the total undereducated population for car beats 261 and 251.

In this example, apportionment by area would have worked nearly as well as apportionment by block centroid population because the populations are evenly distributed across the tract. There is no way of assessing approximation errors in either case. You would need the Census Bureau's block tabulation of the undereducated population, which is not available to the public.

Preview of apportionment steps

The following is a summary of apportionment steps, starting from the beginning. These steps are for reading only and not for use on your computer. The preliminary steps are already familiar to you and so are finished and included in Tutorial9-3.mxd to save you some time and allow you to focus on apportionment itself.

Completed preliminary steps are as follows:

1. Download census block and tract polygons from the Census or ESRI Web sites for the county containing the administrative area polygons. Download the short-form census data for blocks that are the basis of apportionment, in this case the population of age 22 and greater. Download the long-form census attribute(s) at the tract level that you wish to apportion to the administrative area; in this case, the population aged 25 or greater with less than high school education.

2. Create a new tract layer that intersects the administrative boundaries. If a tract is only partially inside the administrative area, you must include the entire tract for apportionment to work correctly. An example tract is the southernmost tract in Tutorial9-3.mxd.

3. Create a new centroid point layer for blocks, clip the centroids with the new tract layer, and join census short-form data to the clipped block centroids. This is the layer that is the basis for apportionment.

4. Sum the short-form census attributes in age categories to create Age22Plus in the Clipped block centroids table. This step is unique to this problem. Also, this table has a new TractID attribute which concatenates FIPSSTCO & TRACT2000 to create an ID matching the Tracts map layer.

5. In the attribute table for block centroids, sum the field for persons aged 22 or older by TractID to create a new table, SumAge22Plus. This table provides the denominator for the weight used in apportionment.

Next are the five steps of apportionment:

1. Spatially join the tract and car beats layers to create new polygons that each have a tract ID and car beat number.

2. Spatially join the joined layer of tracts and car beats with the block centroids to assign all the tract attributes (including the attribute of interest: undereducated population) and car beat attributes to each block centroid.

3. Join SumAge22Plus to block centroids to make the apportionment weight denominator, total population aged 22 or older by tract, available to each block centroid.

4. For each block centroid, create new fields to store apportionment weight and apportioned undereducated population values, then calculate these values.

5. Sum the apportionment weights by tract as a check for accuracy (they should sum to 1.0 for each tract). Then sum the undereducated population per car beat, storing the results in new tables.

With apportionment completed, the last task is to join the table containing undereducated population by car beat to the car beats layer, then symbolize the data for map display.

Intersect tracts and car beats

1 **If necessary, click Windows, Search to open the Search window.**

2 **Type Intersect in its search text box, press the search button, and click the Intersect tool.**

3 **Type or make selections as shown in the image.**

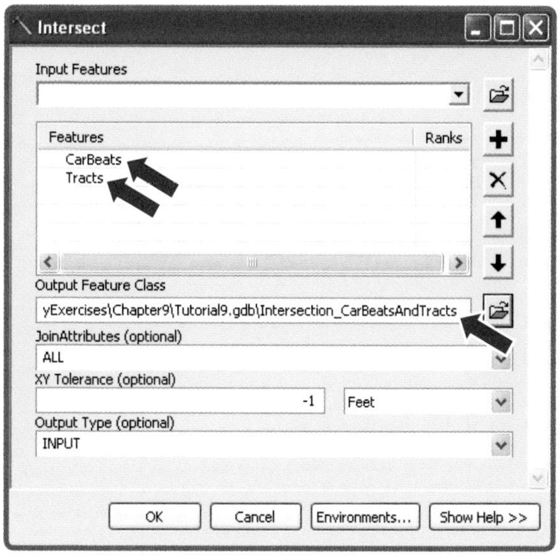

4 Click OK. The resulting layer contains polygons representing the areas where the car beats and census tract polygons overlap.

5 **Hide the Search window.**

Spatially join the intersection of car beats and tracts with block centroids

Next, you will spatially overlay the intersection layer of tracts and car beats on the block centroids to assign the tract and car beats attributes to the census blocks.

1 **In the TOC, right-click the LakeBlockCentroids layer, click Joins and Relates, and click Join.**

2 **Type or make selections as shown in the image.**

3 Click OK.

4 **Open the resulting point layer's attribute table.** See that each block now has the TractID in which it lies and Beat attribute with car beat number.

5 Close the table.

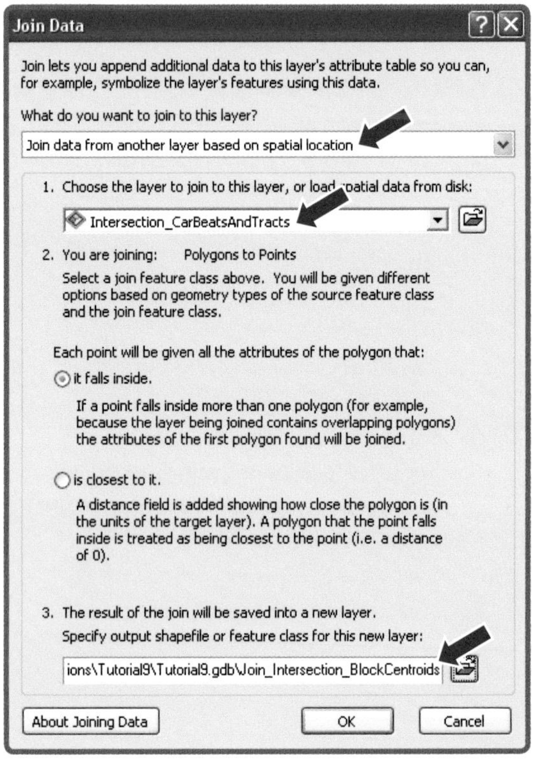

Join summary attributes to the spatial join output

Next, you will join the SumAge22Plus table to the block centroids to make the apportionment weight denominator—total population age 22 or greater by tract—available to each block centroid.

1 Right-click Join_Intersection_BlockCentroids, click Joins and Relates, click Join, and type or make selections as shown in the image at the right.

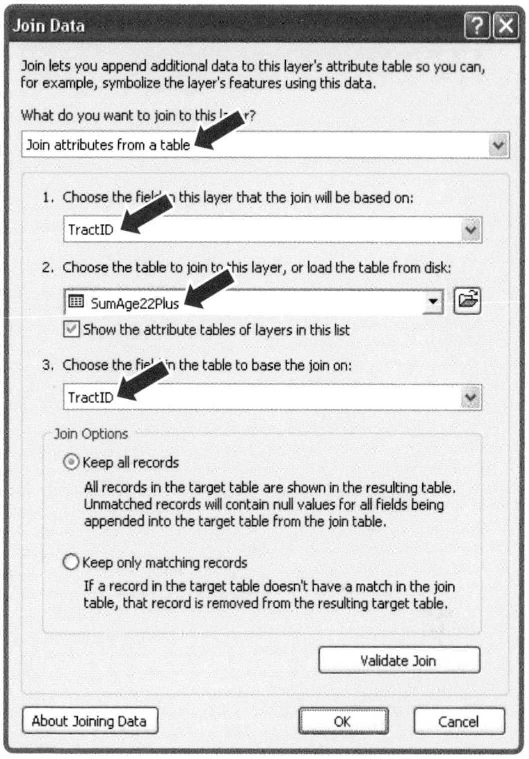

2 Click OK. Next, as a precaution, you will export the resulting joined map layer to a new layer in which all joins will become permanent fields. This makes the final calculations more stable.

3 Right-click Join_Intersection_BlockCentroids, click Data, click Export Data, save the Output feature class as **\ESRIPress\GIST1\MyExercises\Chapter9\Tutorial9.gdb\BlockCentroidsFinal**, click OK, and click Yes.

Compute apportionment weights

For each block centroid in the new table, now you will create and calculate new attributes for the apportionment weight and apportioned undereducated population.

1 Open the attribute table of BlockCentroidsFinal.

2 Click the Table Options button ▾ and Add Field.

3 Name the new field **Weight**, set its Type to Float, then click OK.

4 Repeat steps 2 and 3 except name the attribute **UnderEdu**.

5 In the Attributes of the BlockCentroidsFinal table, scroll to the right, right-click the Weight column heading, and click Field Calculator.

6 **In Fields box, double-click the AGE22Plus field, click the / button, and double-click Sum_ Age22Plus.** The resulting expression is [Age22Plus] / [Sum_Age22Plus].

7 Click OK.

Compute apportionment values

1 **In the BlockCentroidsFinal table, right-click the UnderEdu column heading. Click Field Calculator.**

2 **Clear your previous expression, scroll near the bottom of the Fields list, double-click the Weight field, click the * button, and double-click the NOHISCH field.**

3 Click OK. The first six rows of resulting values follow:

Weight	UnderEdu
0	0
0.006391	1.738241
0.006902	1.877301
0.001789	0.486708
0.000511	0.139059
0.01227	3.337423

4 Leave the table open.

Sum weights by tract

As a check, sum the apportionment weights by tract; they should add up to 1.0 for each tract.

1 **Right-click the TractID column heading, and click Summarize.**

2 **Type or make selections as shown in the image at the right.**

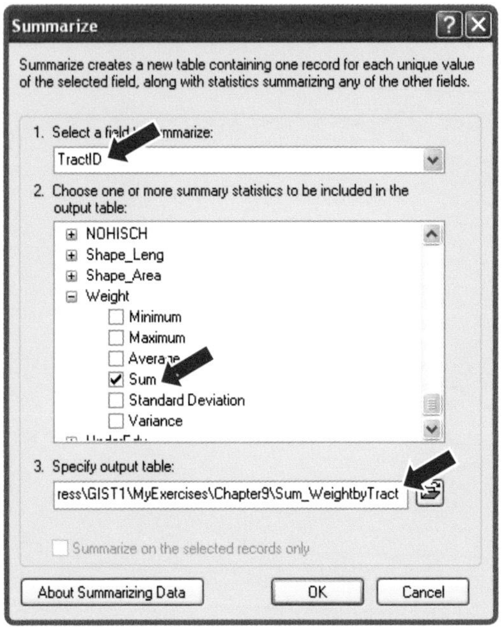

3 Click OK, Yes.

4 **In the TOC, right-click the Sum_WeightByTract table and click Open.** Each tract that is totally within car beats will have weights summing to 1. Those partially within car beats sum to less than 1. Check out your results by comparing tracts on the map with tabled values.

TractID	Count_TractID	Sum_Weight
36055000200	103	0.73674
36055001800	95	0.993354
36055001900	41	1
36055002000	38	1
36055002100	26	1
36055002200	32	1
36055002300	42	1
36055002400	29	0.387164
36055004701	37	0
36055008500	58	0.968186
36055008600	57	1
36055009200	1	0

5 **Close the Attributes of Sum_WeightByTract table.**

Sum undereducated population by car beat

This is the final step of apportionment.

1 **Open the BlockCentroidsFinal table, right-click the BEAT column heading, and click Summarize.**

2 **Type or make the selections as shown in the image at the right.**

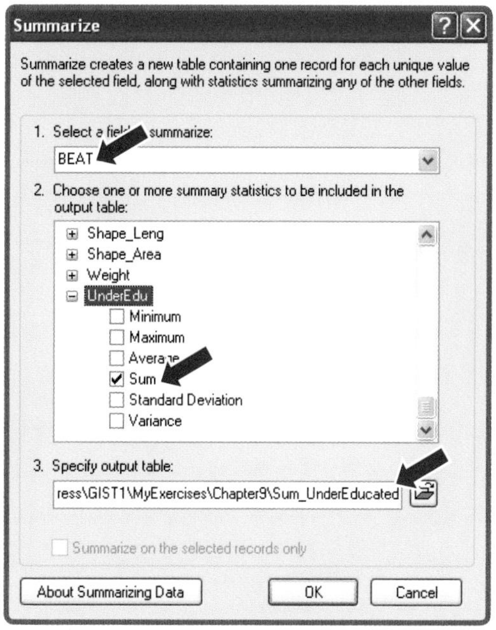

3 **Click OK, Yes.**

4 **Right-click the Sum_UnderEducated table in the TOC and click Open.** The extra row with no beat value is of no consequence, because it will not join to the car beats table in the next steps.

OBJECTID_12 *	BEAT	Count_BEAT	Sum_UnderEdu
1		0	0
2	231	57	392.602994
3	241	50	452.003292
4	251	76	460.155723
5	261	141	644.936157
6	271	116	473.283043

5 **Close all open tables.**

Join Sum_underEducated to the car beat layer

In the following steps, you will join the table for undereducated population by car beat to the car beats layer and use it to symbolize the map.

1 Right-click Car Beats in the TOC, click Joins and Relates, and click Join.

2 Type or make the selections as shown in the image at the right.

3 Click OK, then Yes.

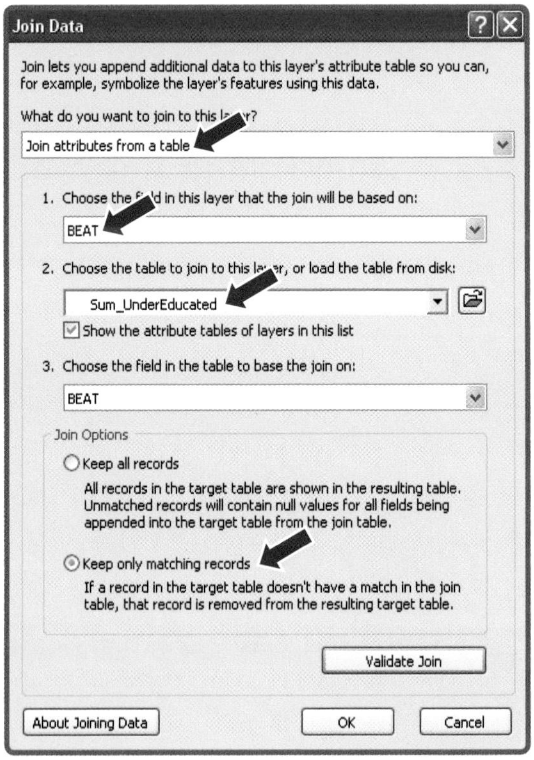

Map undereducated population by car beat

1 In the TOC, turn all layers off except Assault Offenses, Streets, and Car Beats.

2 Right-click Car Beats, click Properties, and click the Symbology tab.

3 In the Show box, click Quantities, Graduated colors.

4 From the Value drop-down list, choose Sum_UnderEdu, then click Classify.

5 In the Classification dialog box, select Quantile for the Method and click OK twice. There is some variation in the undereducated population in Lake Precinct car beats, but not a great deal. You can see that assault offenses tend to be in car beats or on the boundary of car beats with high undereducated populations.

6 Save your map document and close ArcMap.

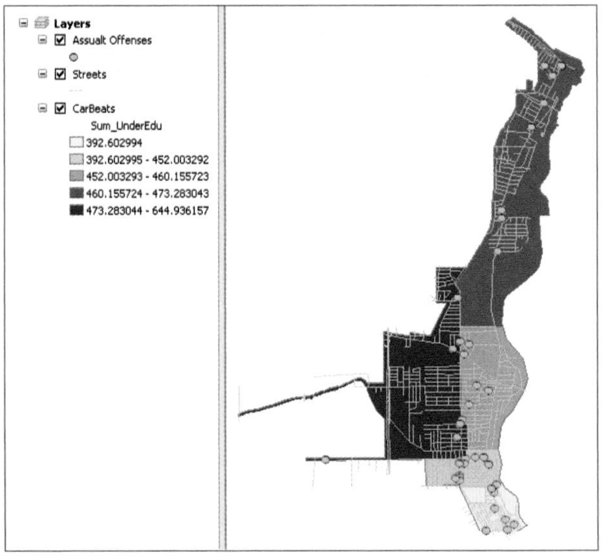

Assignment 9-1

Analyze population in California cities at risk for earthquakes

When natural disasters such as earthquakes occur, officials need to move quickly to find resources to aid affected people. In this assignment, you will use GIS to create buffers around major earthquakes that have occurred in California and analyze how many people live in urban areas near these events.

Start with the following:

- \ESRIPress\GIST1\Data\UnitedStates.gdb\CACounties—polygon boundary of California counties
- \ESRIPress\GIST1\Data\UnitedStates.gdb\USCities_dtl—point locations for cities in the United States
- \ESRIPress\GIST1\Data\DataFiles\Earthquakes.dbf—table of earthquakes in California with latitude and longitude attributes in North American Datum 1983 coordinates

Create a map showing California earthquakes and population

Create a map document called **\ESRIPress\GIST1\MyAssignments\Chapter9\Assignment9-1YourName.mxd** with a layout showing a 20-mile buffer around earthquakes whose magnitude is greater than 7. Use a UTM projection appropriate for California in your data frame. Include a label with the total population within buffers. See Hints. Export a layout with your map to **\ESRIPress\GIST1\MyAssignments\Chapter9\Assignment9-1YourName.jpg**.

Create a new file geodatabase, **\ESRIPress\GIST1\MyAssignments\Chapter9\Assignment9-1YourName.gdb,** and add any new files that you create in this assignment to it. Start by creating a new feature class called CACities that includes only California cities.

Hints

It takes a couple of steps to get the desired buffers for this assignment. The map needs each separate buffer area to have a label displaying the total urban population in that area. If you were to use the ALL dissolve type when buffering earthquakes, a single polygon would result for all buffers, even though they are separate areas. Some 20-mile buffers overlap. Each set of overlapping buffers needs to be dissolved to form a single buffer polygon. Many 20-mile buffers do not overlap. These need to be separate polygons. The approach to building the needed buffers uses a dissolve field that you can create in the following steps:

- Use Catalog to create a feature class from the XY table, Earthquakes.dbf, saved as **\ESRIPress\GIST1\MyAssignments\Chapter9\Assignment9-1YourName.gdb\CAEarthquakes**. Display only earthquakes with magnitude (attribute MAG). The XY coordinates in Earthquakes.dbf have the same coordinate system as CACounties, so you can import that system for earthquakes from CACounties. After creating this new feature class, add it to your map document. Create a definition query so that only earthquakes with magnitude greater than 7 are included.

- Buffer earthquakes with magnitude greater than 7 using a 20-mile radius, the NONE dissolve type, and saved in the feature class CAEarthquakes_Buffer. Several circular buffers overlap.
- Open the attribute table of CAEarthquakes_Buffer and add a new field called **BufferGroup** with the Short Integer data type.
- Work from north to south. Use the Select Features tool to select the five overlapping buffers along the north coast of California. Then, using the Field Calculator or Editor toolbar, set the members of the selected buffers to the value 1 in the BufferGroup field. Repeat this step for the other set of overlapping buffers but use the value 2 for BufferGroup. Finally, give each remaining buffer a unique value for BufferGroup: 3, 4, ... 10.
- Use the Dissolve tool to dissolve Buffer_None using BufferGroup as the dissolve field. Call the new feature class **Buffer _Earthquakes**, and save it in your file geodatabase. The end result is separate polygon buffers for each nonoverlapping buffer plus two more polygons for the two sets of overlapping buffers.
- Spatially join CACities with Buffer_Earthquakes_Buffer to create **Join_CitiesBuffer**. Start the join by right-clicking Buffer_Earthquakes and be sure to use SUM so that city attributes are summed by buffer polygon for labeling your map. Use SUM_POP_98 to label the buffers.

STUDY QUESTIONS

Save the answers to your questions in a Microsoft Word document called **\ESRIPress\GIST1\MyAssignments\Chapter9\Assignment9-1YourName.doc.**

1. Which major earthquake (magnitude greater than 7) has the most cities within 20 miles?
2. According to the cities table, how many people are within 20 miles of that earthquake?
3. What California cities with population over 350,000 have not yet been hit by an earthquake whose magnitude is over 7?

WHAT TO TURN IN

If your work is to be graded, turn in the following files:

File geodatabase: \ESRIPress\GIST1\MyAssignments\Chapter9\Assignment9-1YourName.gdb

ArcMap document: \ESRIPress\GIST1\MyAssignments\Chapter9\Assignment9-1YourName.mxd

Image file: \ESRIPress\GIST1\MyAssignments\Chapter 9\Assignment9-1YourName.jpg

Word document: \ESRIPress\GIST1\MyAssignments\Chapter 9\Assignment9-1YourName.doc

If instructed to do so, instead of the above individual files, turn in a compressed file, **Assignment9-1YourName.zip**, with all files included. Do not include path information in the compressed file.

Assignment 9-2

Analyze urban walking distances

Walkable catchments, sometimes referred to as "ped sheds," are areas within short walking distances of urban attractions or amenities. Promoting economic development and increased use of downtown areas is a major priority in many cities. Making maps of ped sheds is one way to promote downtown environments.

Study area background

In this assignment, you will study the "walkability" of an urban area of Pittsburgh, Pennsylvania. National Geographic Magazine featured ZIP Code 15222, in the heart of Pittsburgh, as one of the most interesting areas in the country (see **http://ngm.nationalgeographic.com/ngm/0308/feature6/**). Areas making up the 15222 ZIP Code include the Strip District (so-named because it is a narrow strip of land), Pittsburgh's Cultural District with many theaters and galleries, and the Central Business District (CBD). For this area, you will create a buffer for short walking distances from major streets with parking and a selection of restaurants. It turns out that this part of Pittsburgh is very "walkable".

Start with the following:

- \ESRIPress\GIST1\Data\Pittsburgh\15222.gdb\Streets—line layer ofstreet centerlines within the 15222 ZIP Code
- \ESRIPress\GIST1\Data\Pittsburgh\15222.gdb\Curbs—line layer of pavement curbs within the 15222 ZIP Code
- \ESRIPress\GIST1\Data\Pittsburgh\15222.gdb\Restaurants—point layer of selection of restaurants within the 15222 ZIP Code
- \ESRIPress\GIST1\Data\AlleghenyCounty.gdb\Rivers—line layer of rivers and water bodies in Allegheny County

These layers all have state plane projection in feet, so the map layer that they constitute will also have the same coordinates (recall that the layers take on the projection of the map layer that you add first).

Create a map showing "walkable" catchment areas for neighborhood and grocery store site.

Create a new map document, **\ESRIPress\GIST1\MyAssignments\Chapter9\Assignment9-2YourName.mxd** that includes a layout with the above layers. Create a new file geodatabase, **\ESRIPress\GIST1\MyAssignments\Chapter9\Assignment9-2YourName.gdb** and add any new files that you create in this assignment to it.

- The Penn Ave and Smallman Street corridor includes Pittsburgh's Cultural and Strip Districts along with many parking garages. Create a two-minute walking buffer to these two streets with major attractions. Use a buffer radius the length that an adult can comfortably walk in two minutes at three miles per hour. Use the All dissolve type for this and the next buffer.

- Create a buffer with the same radius for restaurants.
- The ped shed is the combined areas of both buffers.

Create a layout and export it to a JPEG file called **\ESRIPress\GIST1\MyAssignments\Chapter9\Assignment9-2YourName.jpg**.

WHAT TO TURN IN

If your work is to be graded, turn in the following files:

File geodatabase: \ESRIPress\GIST1\MyAssignments\Chapter9\Assignment9-2YourName.gdb

ArcMap document: \ESRIPress\GIST1\MyAssignments\Chapter9\Assignment9-2YourName.mxd

Image file: \ESRIPress\GIST1\MyAssignments\Chapter9\Assignment9-2YourName.jpg

If instructed to do so, instead of the above individual files, turn in a compressed file, **Assignment9-2YourName.zip**, with all files included. Do not include path information in the compressed file.

Assignment 9-3

Apportion census block group data to administrative areas

In tutorial 9-3 you apportioned census tract data to Lake Precinct car beats of the Rochester, New York, Police Department. A better approach is to use block groups instead of tracts as the census data source because block groups are smaller than tracts and thus lead to smaller approximation errors. Block groups are the smallest census geography for which the public can obtain long-form census data. In this problem you will use block group data and will apportion income data by households. Income is a good crime indicator, especially low income areas.

Start with the following:

- \ESRIPress\GIST1\Data\RochesterNY\LakePrecinct.gdb\LakeBlockCentroids—point layer of block centroids for Lake Precinct (Use Households as the basis for apportionment.)
- \ESRIPress\GIST1\Data\RochesterNY\LakePrecinct.gdb\lakecarbeats—polygon layer of police administrative areas, car beats
- \ESRIPress\GIST1\Data\RochesterNY\LakePrecinct.gdb\LakeBlockGroups—polygon layer of Census block groups for Lake Precinct (BKGPIDFP00 is the key for block groups and the field to use to join data.)
- \ESRIPress\GIST1\Data\RochesterNY\LakePrecinct.gdb\SumHouseholds—table containing the SumHOUSEHOLDS attribute that serves as the apportionment weight denominator (analogous to Sum_Age22Plus of table SumAge22Plus in tutorial 9-3)
- \ESRIPress\GIST1\Data\RochesterNY\HouseholdIncome.xls—Census long-form data, including the following:

  BlockGroupID—ID for joining with LakeBlockGroups

  Households—Number of households per block group

  AggrIncome—Aggregate household income ($) per block group

Apportion data

Create a new map document, **\ESRIPress\GIST1\MyAssignments\Chapter9\Assignment 9-3YourName.mxd** that includes the map layers and data table above. Create a new file geodatabase, **\ESRIPress\GIST1\MyAssignments\Chapter9\ Assignment9-3YourName.gdb** and add to it any new files that you create in this assignment.

The objective is to estimate average aggregate income per household in Lake Precinct car beats. Because the final output is a ratio (aggregate income divided by number of households) you have to be careful to apportion the numerator and denominator separately to car beats before dividing as a last step. The numerator must be apportioned because it is only available at the block group level, but the denominator, number of households by car beat, is available by direct aggregation from a spatial join of car beats and block centroids because number of households is available at the block level.

Hints

- To aggregate HOUSEHOLDS in LakeBlockCentroids to the car beat level, use a spatial join (join data from another layer based on spatial location) of LakeBlockCentroids and lakecarbeats to create **SpatialJoin_CarBeatsAndBlocCentroids**. Then summarize the BEAT field in the new layer using Sum for HOUSEHOLDS to create the **HouseholdsCarBeat** table. The result has final values ready for the last step of calculating average household income per car beat.
- To apportion aggregate income to the car beat polygons, start by joining the data table in HouseholdIncome.xls to LakeBlockGroups, using BlockGroupID in the former and BKGPIDFP00 in the latter for matching keys. This is just the common step of adding census data downloaded from the Census Web site to matching polygons for GIS processing.
- Then you can follow the steps of tutorial 9-3 to apportion each block group's aggregate income data to car beats. In the steps you have to substitute block groups for tracts and the SumHousehold table for the SumAge22plus table. Note that in this case Weight = [HOUSEHOLDS] / [Sum_HOUSEHOLDS] and apportioned income is AggrInc = [Weight] × [AggregateIncome_AggrIncome].
- After you perform the last step of apportionment, summarizing BEAT with a sum of AggrInc, you have a few more steps to complete the assignment. Join HouseholdsCarbeat and Sum_AggrInc tables to lakecarbeats. Add a new field to lakecarbeats called **HouseholdIncome** and calculate it to be [Sum_AggrInc.Sum_AggrInc] / [HouseholdsCarbeat.Sum_HOUSEHOLDS].

Make a layout with a choropleth map of the final apportioned ratio per car beat, using quantiles with 5 classes.

WHAT TO TURN IN

If your work is to be graded, turn in the following files:

File geodatabase: \ESRIPress\GIST1\MyAssignments\Chapter9 \Assignment9-3YourName.gdb

ArcMap document: \ESRIPress\GIST1\MyAssignments\Chapter9 \Assignment9-3YourName.mxd

If instructed to do so, instead of the above individual files, turn in a compressed file, **Assignment9-3YourName.zip**, with all files included. Do not include path information in the compressed file.

10

ArcGIS 3D Analyst

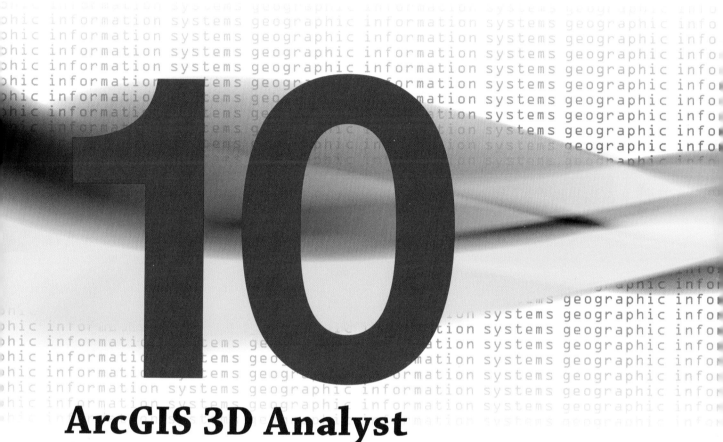

This chapter is an introduction to ArcGIS 3D Analyst, an extension to ArcGIS Desktop that enables 3D display and processing of maps. 3D viewing can provide insights that would not be readily apparent from a 2D map of the same data. For example, instead of inferring the presence of a valley from 2D contours, in 3D you actually can see the valley and perceive the difference in height between the valley floor and a ridge. This chapter uses topography, curb, and building data from the city of Pittsburgh's Mount Washington and Central Business District neighborhoods to show you how to display and analyze data in 3D. It also introduces ArcGlobe, a Web service from ESRI based on 3D that includes rich basemaps.

Learning objectives

- *Create 3D scenes*
- *Create a TIN from contours*
- *Drape features onto a TIN*
- *Navigate through scenes*
- *Create a fly-through animation*
- *Create multiple views*
- *Add 3D effects*
- *Edit 3D objects*
- *Perform a line-of-sight analysis*
- *Explore ArcGlobe Web service*

Tutorial 10-1

Create 3D scenes

3D Analyst is one of a collection of extensions to the basic ArcGIS Desktop software package. You must have the extension installed on your computer and then enable it in ArcMap as you will do next.

Add 3D Analyst extension and toolbar

1 Start ArcMap with a new blank map.

2 Click Customize, Extensions.

3 Click the check box beside the 3D Analyst extension and click Close.

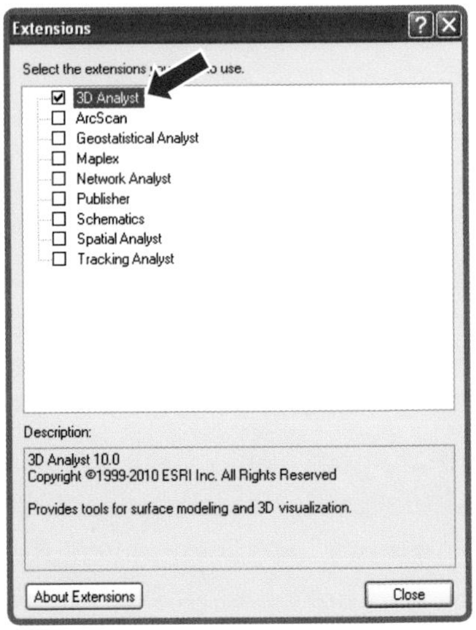

4 **Click Customize, Toolbars, 3D Analyst.** The toolbar for 3D Analyst appears. You can dock it with other toolbars below the main menu.

Launch ArcScene

The ArcMap window is for 2D maps, so to work with 3D maps you need a new window, called ArcScene.

1 **From the 3D Analyst toolbar, click the ArcScene button.**

2 **Click Blank Scene and OK.** A new untitled scene window opens.

Add topo layer

1 **In ArcScene, click the Add Data button.**

2 **In the Add Data browser, browse to \ESRIPress\GIST1\Data\3DAnalyst.gdb\, click Topo, and click Add.** A topography layer of contours near downtown Pittsburgh, Pennsylvania, appears as a 3D view, although the display is 2D at this time. You will make it 3D next.

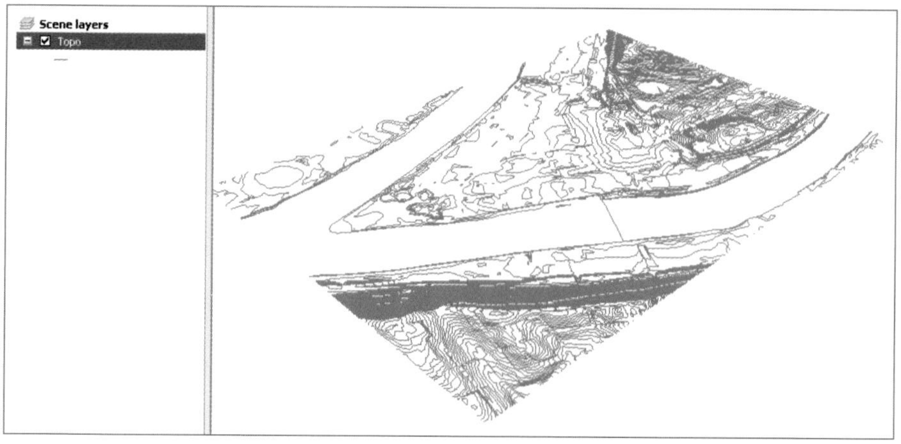

Tutorial 10-2

Create a TIN from contours

3D uses the TIN representation for modeling surfaces. TIN is a vector data model of contiguous, nonoverlapping triangles with vertices created from adjacent sample points of x, y, and z values from 3D space. For example, you will create a TIN from the topo map.

Create a TIN (triangulated irregular network)

1. In the ArcScene window, click Windows, Search.
2. Type TIN in the search text box and press Enter.
3. Click Create TIN from the search results.
4. In the Create TIN from Features window, change the output TIN to **\ESRIPress\GIST1\MyExercises\Chapter10\pgh_tin**.
5. Set the Spatial Reference to NAD_1983_StatePlane_Pennsylvania_South_FIPS_3702_Feet.
6. Select Topo as the input feature class.
7. Change the SF_type to softline.

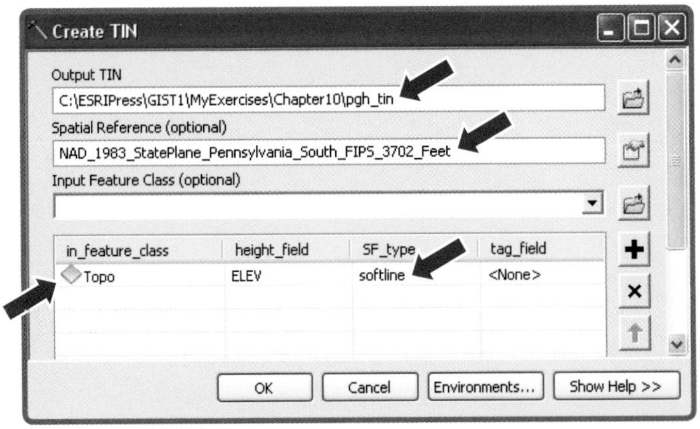

8 **Click OK.** 3D Analyst creates a triangulated irregular network (TIN) from the topography contour lines and adds it as a new layer.

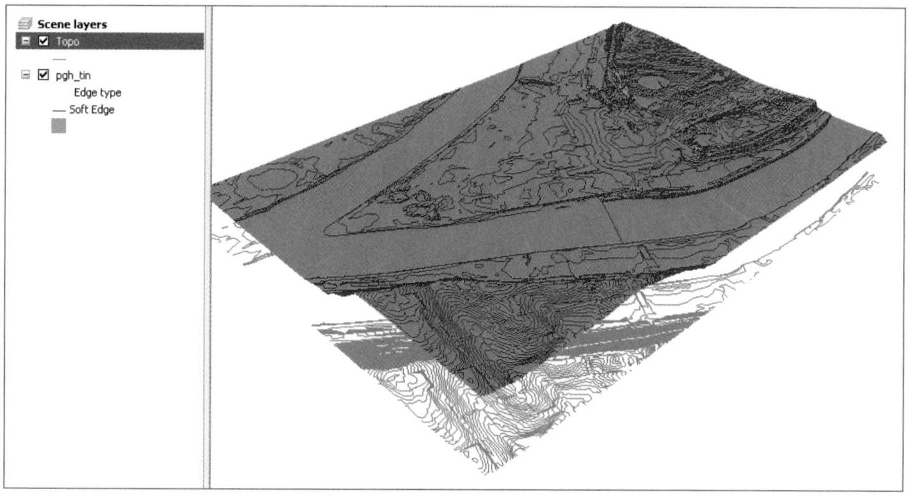

Navigate a 3D view

You do not have a good view of the TIN yet, but you can get one with some navigation.

1 **From the Tools toolbar, click the Navigate button.**

2 **Click and drag the map to view the scene from different angles.** You will see that Pittsburgh is fairly flat where the three rivers converge (this area is known as the "Point") and more hilly in the Mount Washington neighborhood to the right of the Point in the map at right.

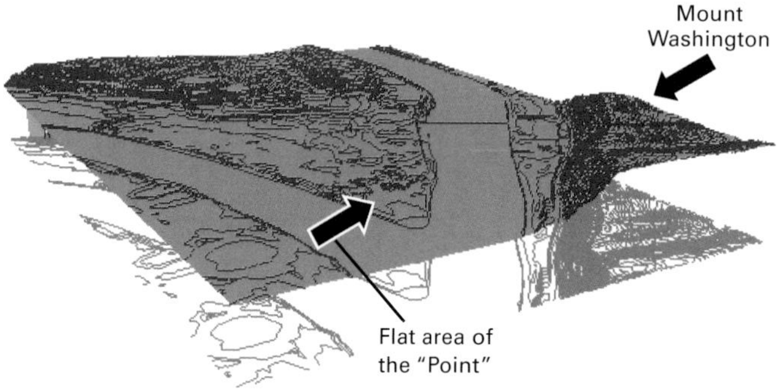

Zoom in to the TIN

Zooming to a small area allows you to see the triangulated irregular network.

1 **Click the Zoom In button .**

2 **Zoom to a small area on Mount Washington.** There are many small triangular facets making up the surface of the pgh_tin. You can see that ArcScene added an artificial light source from the northwest that contributes to the 3D effect.

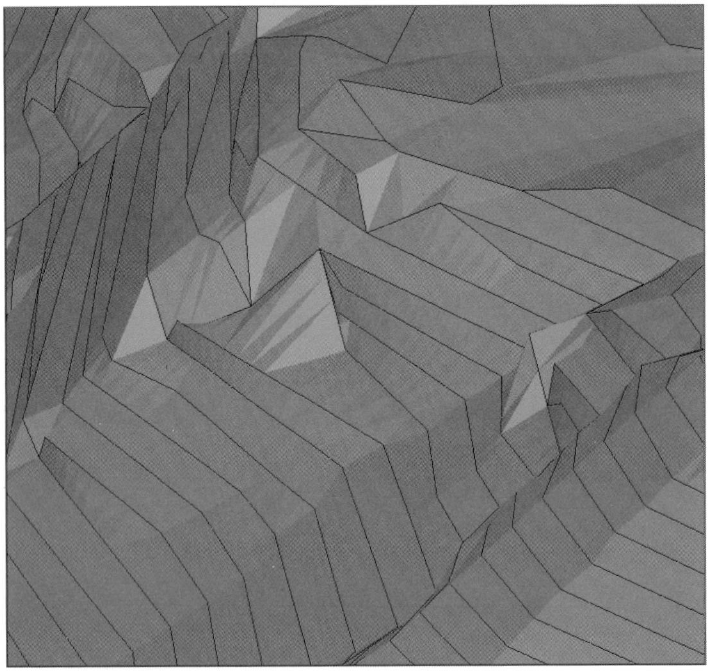

3 **Click the Full Extent button .**

Edit TIN appearance

Changing the appearance of edges on the TIN contours and turning off the topography lines will allow you to see other features (added later) more clearly.

1 **Turn off the original Topo layer.**

2 **In the TOC, double-click the pgh_tin layer's symbol for Soft Edge.**

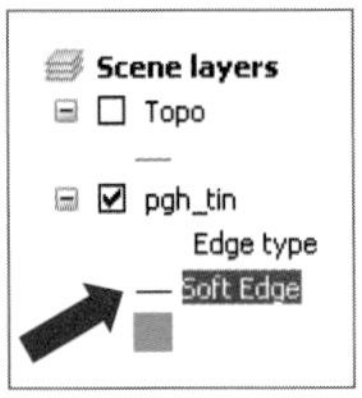

3 **Set the line symbol to No Color and click OK.**

4 **In the TOC, click the green symbol for the pgh_tin layer.**

5 **Set the fill color to Twilight Green.** The resulting scene shows just the pgh_tin layer with a light green color.

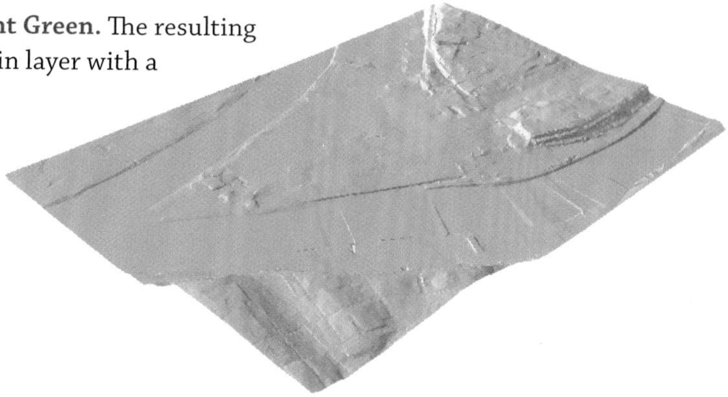

Tutorial 10-3

Drape features onto a TIN

Now that you have a 3D TIN, you can drape 2D map layers on it to show them in 3D.

Drape curbs

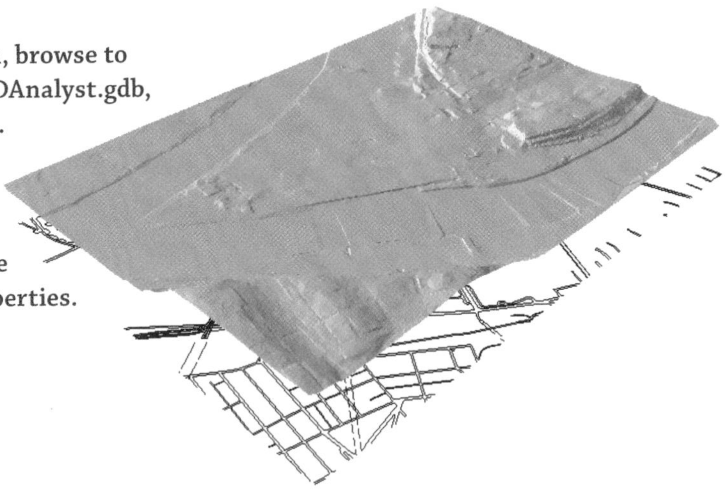

1 Click the Add Data button, browse to \ESRIPress\GIST1\Data\3DAnalyst.gdb, click Curbs, and click Add.
Curbs appear below the TIN contours.

2 In the TOC, right-click the Curbs layer and click Properties.

3 **Click the Base Heights tab, click the radio (option) button beside Floating on a custom surface.** The base height is the elevation at which this flat map layer will appear.

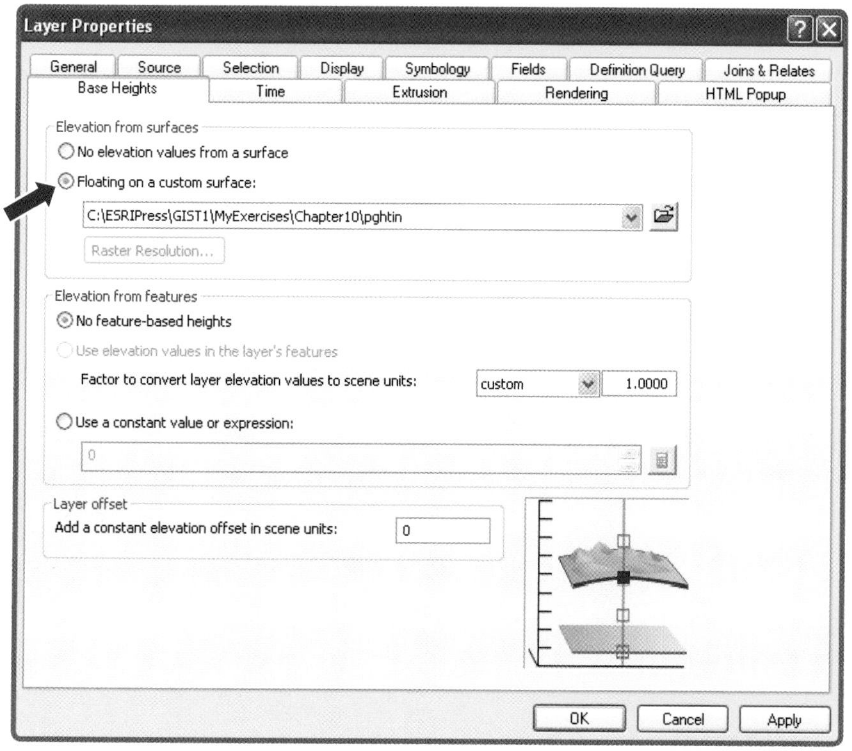

4 **Click OK.** The resulting map displays the curbs draped over contours.

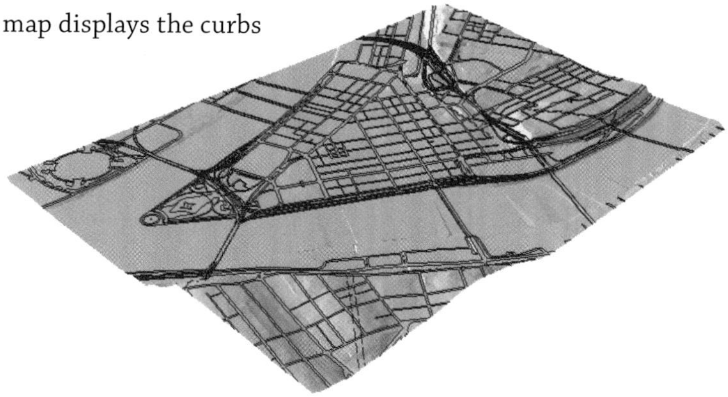

Drape buildings to TIN and extrude buildings

1 **Click the Add Data button, browse to \ESRIPress\GIST1\Data\3DAnalyst.gdb, and click Bldgs and Add.** This action adds the buildings layer with arbitrary heights that ArcScene created by default in the attribute table. Open the Bldgs table to see the new Height attribute.

2 **Right-click the Bldgs layer and click Properties.**

3 **Click the Base Heights tab and click the radio (option) button Floating on a custom surface.**

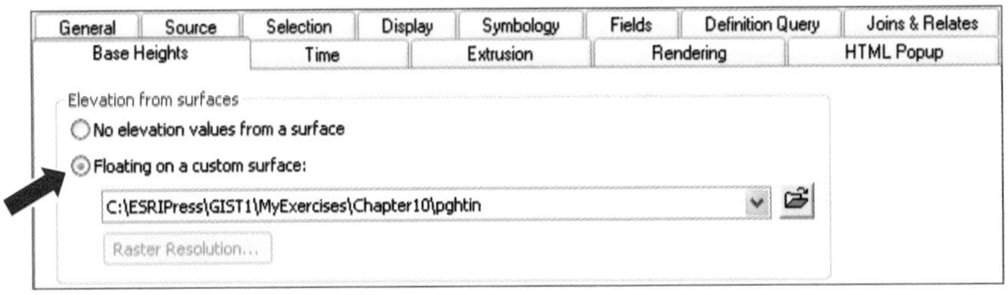

4 **Click the Extrusion tab and make selections as follows:**

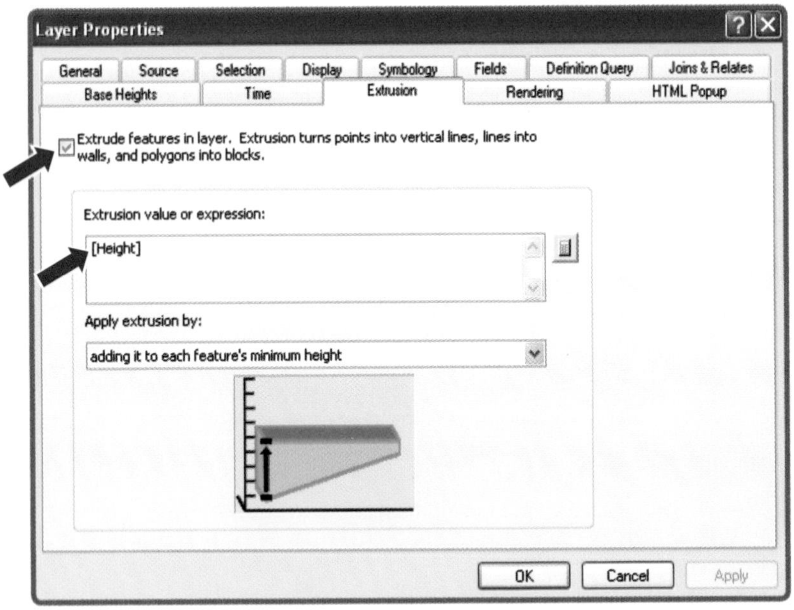

5 Click OK. The resulting view is buildings with various heights all positioned on the TIN. Note that most of the buildings in the Mount Washington neighborhood appear to be residential houses, and the downtown area contains high-rises.

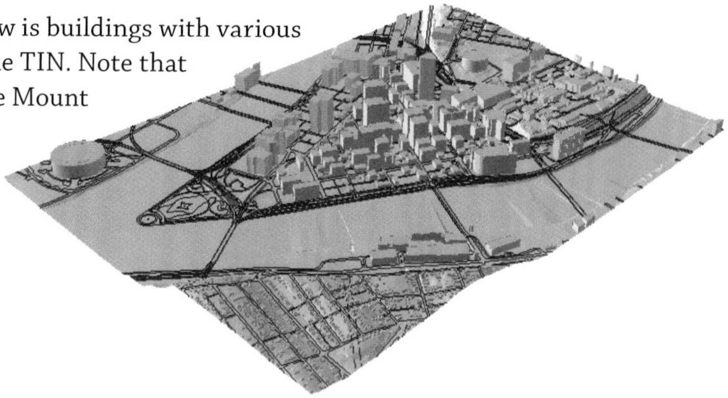

YOUR TURN

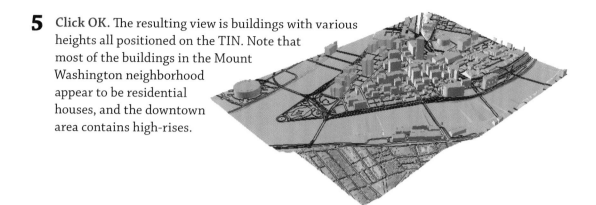

Add rivers polygon features file \ESRIPress\GIST1\Data\AlleghenyCounty.gdb\Rivers and drape it to the TIN. Set the Layer offset to 5. Use the Navigate tool to view the scene from different angles. When finished, zoom to full extent.

Set document properties and save the 3D scene

1 Click File, Scene Document Properties.

2 Click the Pathnames Option button to store relative paths, and click OK.

3 Click File, Save As.

4 Navigate to \ESRIPress\GIST1\MyExercises\Chapter10\ and save the 3D scene as **Tutorial10-1.sxd**.

Tutorial 10-4

Navigate through scenes

Set observer location

1 From the Tools toolbar, click the Set Observer button.

2 Click a location at the point where the three rivers in Pittsburgh meet to set the observer location.

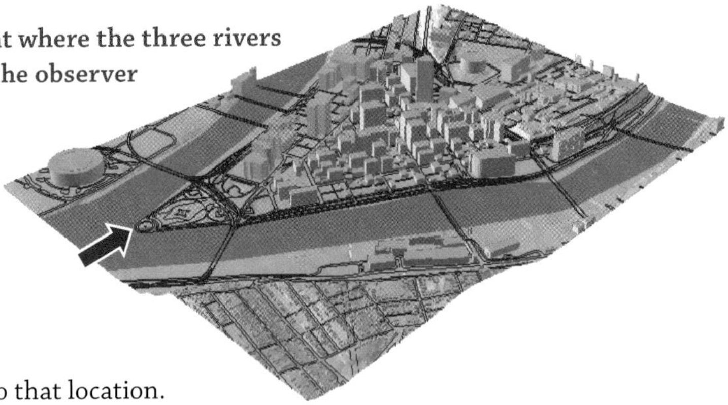

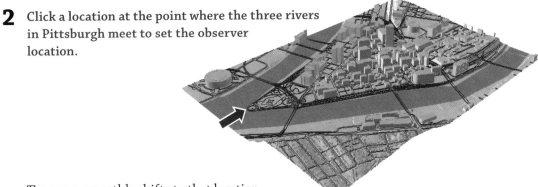

The scene smoothly shifts to that location.

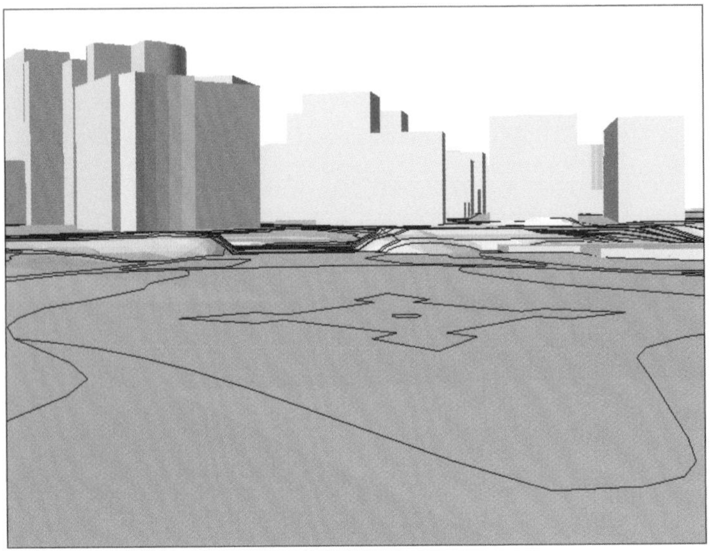

Center view on target location and observer

1 Click the Full Extent button.

2 On the Tools toolbar, click the Center on Target button.

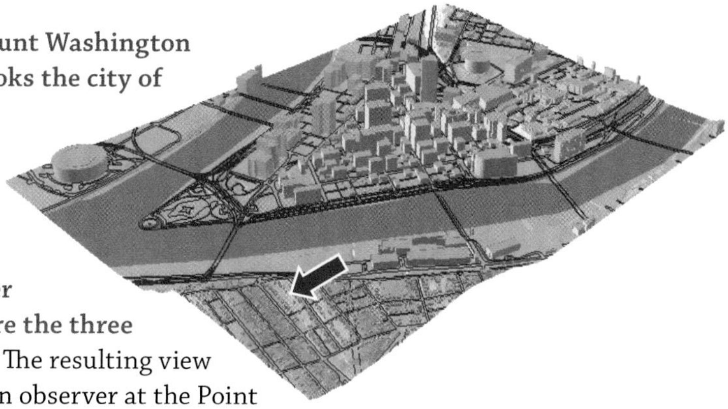

3 Click a location on the Mount Washington neighborhood that overlooks the city of Pittsburgh.

4 Click the Set Observer button.

5 Click the previous observer location at the Point, where the three rivers in Pittsburgh meet. The resulting view is from the perspective of an observer at the Point looking toward the Mount Washington neighborhood.

YOUR TURN

Practice using the Zoom to Target tool, which will zoom to the target location.

Fly through a scene

1 Click the Full Extent button .

2 Set the observer location to the Point.

3 From the Tools toolbar, click the Fly button . Get ready for a wild flight!

4 Click anywhere in the scene with the bird cursor, and click again to start your flight.

5 Slowly move the mouse to the left, right, up, or down. Click the left mouse button to increase your speed, and click the right mouse button to decrease your speed.

6 Press the Escape (Esc) key on the keyboard to stop flight. A new view appears where you stopped the fly-through.

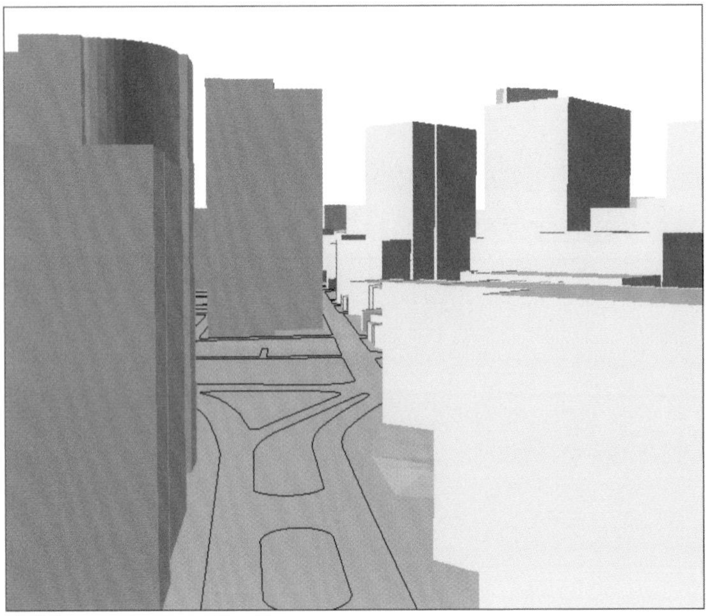

7 Zoom to the full extent.

Create multiple views

1 **From the Standard toolbar, click the Add New Viewer button.**

2 **Click the Navigate and Zoom buttons to change the view.** You can also use the scroll wheel on your mouse to zoom in and out.

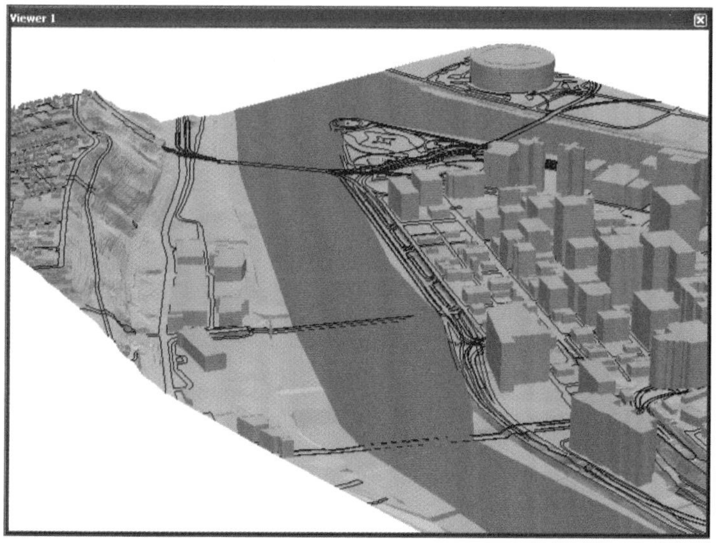

YOUR TURN

Using the Add New Viewer button, create another view of the opposite side of the study area. Close your new views when finished.

Tutorial 10-5

Create a fly-through animation

3D Animations allow you to record movements within your views so that you can save and play them back at a later time.

Add animation toolbar

1. Right-click anywhere in the blank area of a toolbar.
2. Click Animation to display this toolbar. The animation toolbar appears.

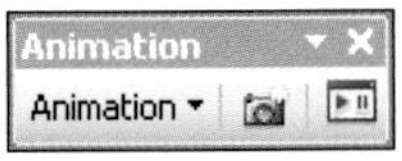

3. Zoom to the full extent.

Record an animation

1. Click the Open Animation Controls button on the Animation toolbar.
2. Click the Record button .
3. Click the Fly button and create a fly-through anywhere in your scene, and then click the Esc key to end your flight.
4. Click the Stop button .

Play an animation

Click the Play button from the Animation Controls toolbar.

YOUR TURN

Practice creating animations by zooming in to a small area first. Explore the Options menu in the Animation Controls toolbar to see animation play and restore options.

Save an animation

1 From the Animation toolbar, click Animation, Save Animation File.

2 Navigate to \ESRIPress\GIST1\MyExercises\Chapter10\ and save the animation as **Animation.asa**.

Export an animation to video

1 From the Animation toolbar, click Animation, ExportAnimation.

2 Navigate to \ESRIPress\GIST1\MyExercises\Chapter10\, save your animation as **Animation.avi**, then click Export and OK. Wait until the animation is fully exported before opening other windows.

Load an animation

1 From the Animation toolbar, click Animation, Load Animation File.

2 In the Open Animation dialog box, navigate to \ESRIPress\GIST1\MyExercises\Chapter10\, click Animation.asa and click Open.

3 Click the Play button ▶ from the Animation Controls toolbar.

4 Zoom to the full extent.

YOUR TURN

Launch a video player such as Windows Media Player and play the AVI video that you created. If you have trouble loading and playing your video, choose the animation files in \ESRIPress\MyExercises\FinishedExercises\Chapter10\. Close the Animation toolbars and zoom to the full extent of your map when finished.

Tutorial 10-6

Add 3D effects and use 3D symbols

Special effects such as transparencies, lighting, and shading modes can greatly enhance the 3D experience for the viewer. The Layer Face Culling command turns off the display of front or back faces of an aerial feature or graphic. Layer Lighting turns lighting on or off for the selected layer. Shading Mode allows you to define the type of shading (smooth or flat) to use for the layer selected. Depth Priority allows you to define which 3D layer should be given higher priority. This is useful when you have two 3D polygon layers that share the same location and might obstruct each other (e.g., land parcels and buildings).

Create transparency effect

1 Right-click anywhere in the blank area of a toolbar and click 3D Effects to display this toolbar.

2 On the 3D Effects toolbar, click the Layer drop-down, and click Bldgs.

3 Click the Layer Transparency button and change the layer's transparency to 50%.

4 Zoom to the center of the city.

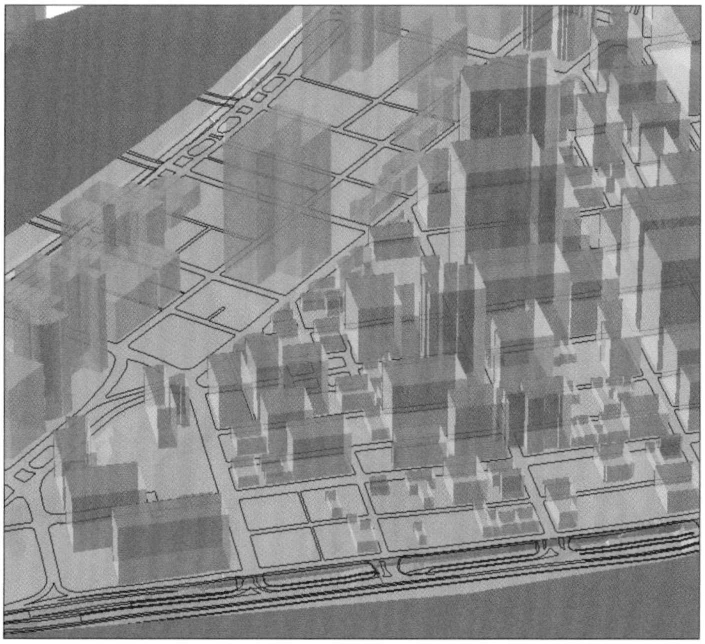

YOUR TURN

Experiment with changing other effects for your 3D buildings, including Layer Face Culling, Layer Lighting, Shading Mode, and Depth Priority. Change the effects for the Rivers and Curbs layers in your 3D scene.

The 3D Analyst extension comes with many 3D symbols for objects such as trees that you will use next.

Add trees layer

1 Click the Add Data button.

2 In the Add Data browser, browse to \ESRIPress\GIST1\Data\3DAnalyst.gdb, click Trees, and click Add.

Display points as 3D trees

1. In the TOC, right-click the Trees layer and click Properties.
2. Click the Base Heights Tab and click the radio button for Floating on a custom surface.
3. Click the Symbology Tab and click the Symbol button .
4. In the Symbol Selector dialog box, click the Style References button (in the lower right of the window), click 3D Trees, and click OK.
5. Scroll through the symbols until you see the 3D trees, choose Bradford Pear, and click OK twice.
6. In the TOC, right-click the Trees layer and click Zoom to Layer.
7. Click the Navigate, Zoom, and Pan buttons to view the trees from street level. The view at right shows trees along a few streets in Pittsburgh in an area undergoing revitalization known as the "Fifth and Forbes" corridor. A 3D study of how adding these trees to help enhance the streets is important to city planners.

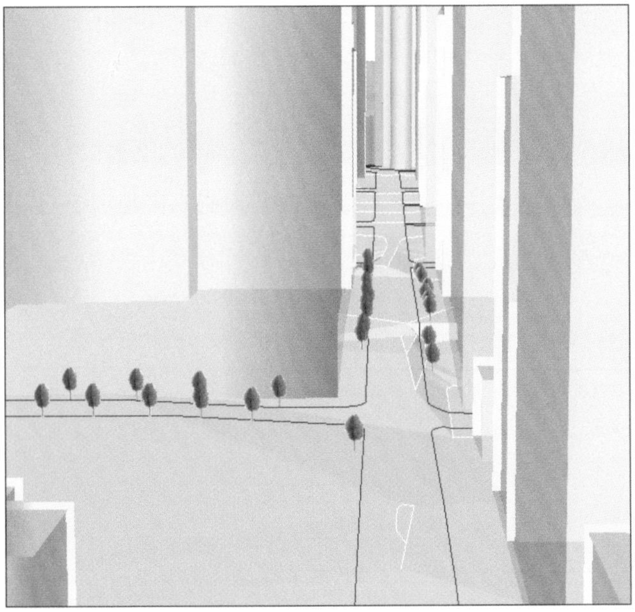

8. Save your ArcScene as **\ESRIPress\GIST1\MyExercises\Chapter10\Tutorial10-2.sxd**.

YOUR TURN

Add the point feature class called \ESRIPress\GIST1\Data\3DAnalyst.gdb\Vehicles and display the vehicles as 3D vehicle symbols using Unique Values based on the "Type" value field from the shapefile's attribute table. Set the base height for the surface to pgh_tin, and zoom to the layer. Explore the other 3D symbols that come standard with ArcEditor, including 3D Basic, 3D Billboards, 3D Buildings, 3D Industrial, 3D Residential, and 3D Street Furniture. Save the scene when you are finished.

Tutorial 10-7

Edit 3D objects

You can edit 3D objects using the 3D Editor toolbar. Edits include changing 3D heights, moving 3D objects, or creating new features.

10-1
10-2
10-3
10-4
10-5
10-6
10-7
10-8
10-9
A10-1
A10-2

Open 3D scene and prepare shapefile for edits

1 On ArcScene's main menu, click File and Open, browse to \ESRIPress\GIST1\MyExercises\Chapter10\, and open Tutorial10-1.sxd.

2 In the TOC, right-click the Bldgs layer, click Data, and click Export.

3 Click "Use the same coordinate system as the dataframe" and save this as a shapefile to **\ESRIPress\GIST1\MyExercises\Chapter10\3DBldgs.shp**.

4 Add the new shapefile to the scene and drape it to the TIN.

5 Remove the original Bldgs layer from the TOC.

YOUR TURN

Extrude the buildings using the Height field. Zoom to the scene shown in the image at right with Pittsburgh's tallest building, the U.S. Steel Tower (formerly the USX Tower).

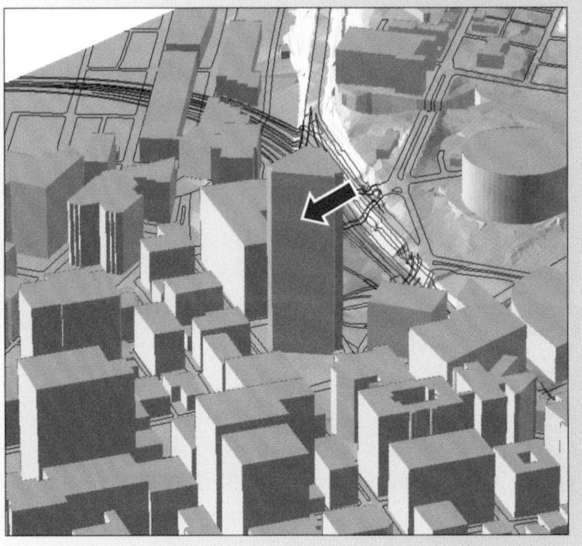

Use the 3D Editor toolbar

1. Click Customize, Toolbars, 3D Editor.
2. On the 3D Editor toolbar, click 3D Editor, Start Editing, 3DBldgs, OK.

Edit 3D building height

1. On the 3D Editor toolbar, click the Edit Vertex tool.
2. Click the U.S. Steel Tower to select it.
3. On the 3D Editor toolbar, click the Attributes button.
4. From the Attributes window, change the building height to 1,000.

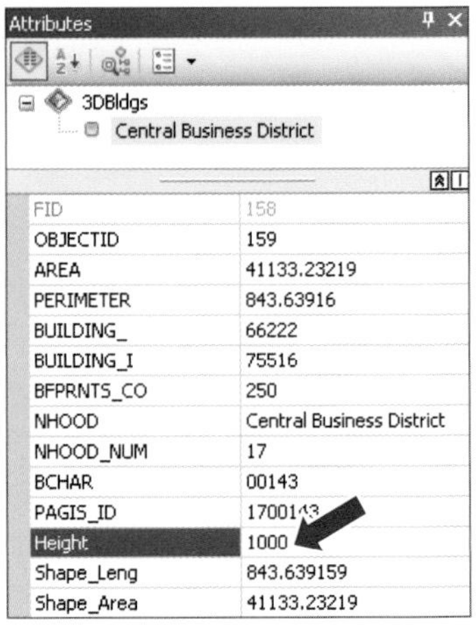

5 **Press the Tab key.** ArcScene uses the new height for display.

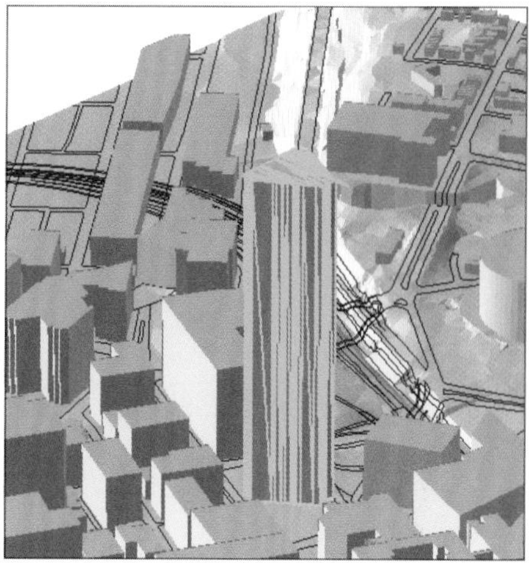

Move a 3D building

1 **Pan the map to the left until you see the circular building to the right of the U.S. Steel Tower.** This is a sports and entertainment center.

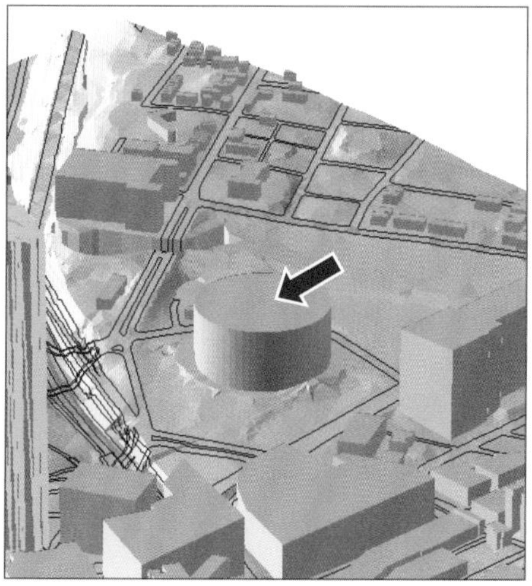

2 **On the 3D Editor toolbar, click the Edit Placement tool.**

3 **Click the arena building to select it, drag the building to a new location on the map, and release.** The sports arena building is in the new location.

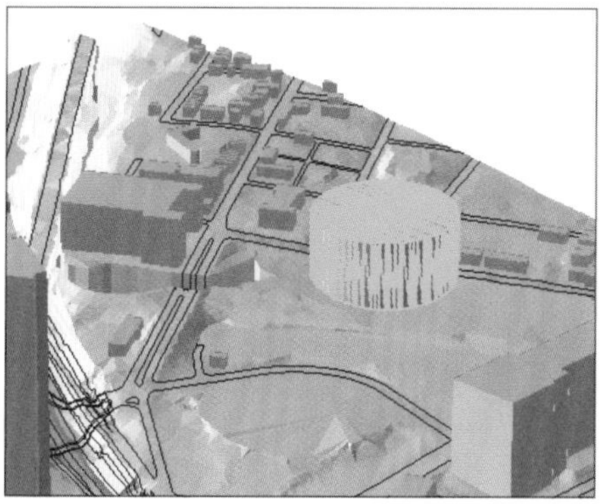

YOUR TURN

Practice moving and editing other building heights. When finished, click 3D Editor, Stop Editing, and Save Edits.

4 Click File, Save As. Navigate to \ESRIPress\GIST1\MyExercises\Chapter10\ and save the 3D scene as **Tutorial10-3.sxd**.

Tutorial 10-8

Perform a line-of-sight analysis

In this tutorial you will explore how to create a line-of-sight analysis using a 2D TIN in ArcMap. A line-of-sight analysis creates a graphic line between two points showing where the view is obstructed between those points.

Start a map document

1 Start ArcMap and open a new empty map. Make sure the 3D Analyst toolbar is turned on.

2 Click the Add Data button, navigate to \ESRIPress\GIST1\MyExercises\Chapter10, click pgh_tin, click Add, and OK in the message box.

Create a line of sight

1 **On the 3D Analyst toolbar, click the Create Line of Sight button .**

2 **In the Line Of Sight window, enter observer and target offsets of 6.** The offset is the number of feet above the TIN for the observer and target. If you set the observer and target heights to zero then typically you will have a view with more obstructions than one with a height greater than one.

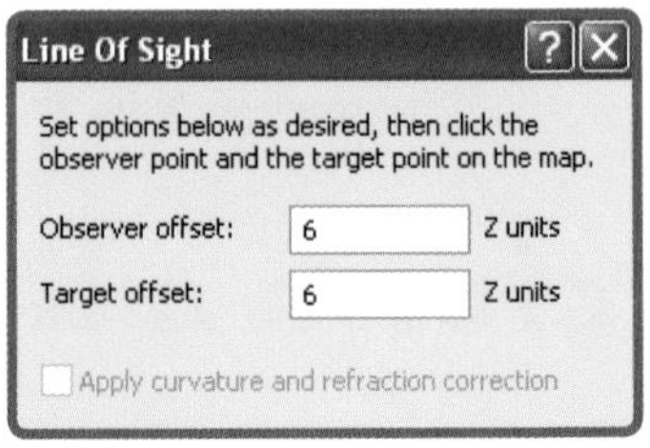

3 **Click a point near Pittsburgh's "Point" where the three rivers meet.**

4 **Click a point on Mount Washington where the elevation is above 910 feet (gray area on the map).** The resulting map shows red along the line where the observer's line of sight is obstructed and green along the line where the view is not obstructed.

YOUR TURN

Choose additional observer and target points to see line-of-sight visibility. Change the observer and target heights to see if the visibility changes. Close ArcMap without saving the document.

Tutorial 10-9

Explore ArcGlobe Web service

ArcGlobe provides a seamless basemap infrastructure in an ArcScene-like interface for the entire world with imagery, elevation, political boundaries, and highways, You can add your own layers to ArcGlobe and quickly have an impressive GIS application. You need a broadband Internet connection to use the Web service that provides the basemaps.

10-1
10-2
10-3
10-4
10-5
10-6
10-7
10-8
10-9
A10-1
A10-2

Launch ArcGlobe

1 **On your desktop, click Start, All Programs, ArcGIS, ArcGlobe 10, and click OK.** ArcGlobe opens showing default layers that are provided by ESRI.

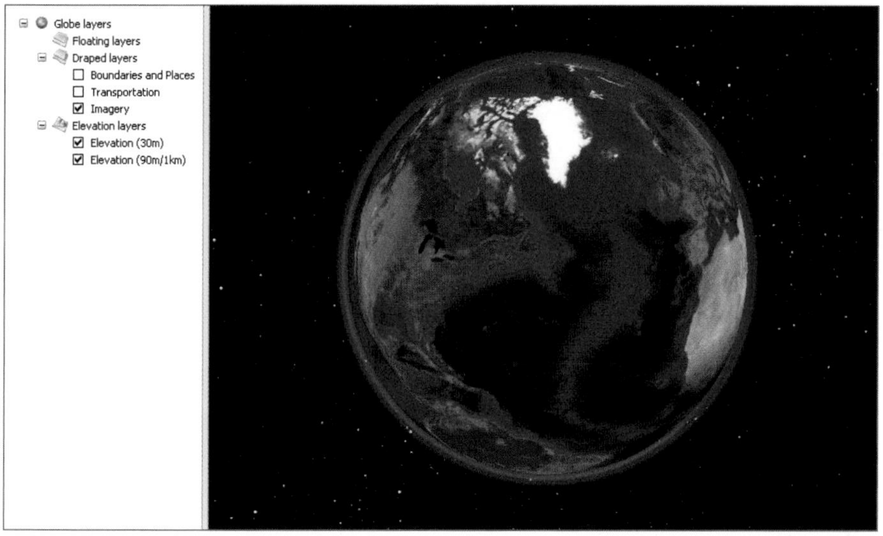

2 **Turn on the Boundaries and Places and Transportation layers in the TOC.** These layers will not display until you zoom in.

Explore ArcGlobe

By default ArcGlobe opened with the Navigate tool selected, ready for you to use next.

1 **Click, hold, and drag the display to the right so that the west coast of North America is at the center of the map.**

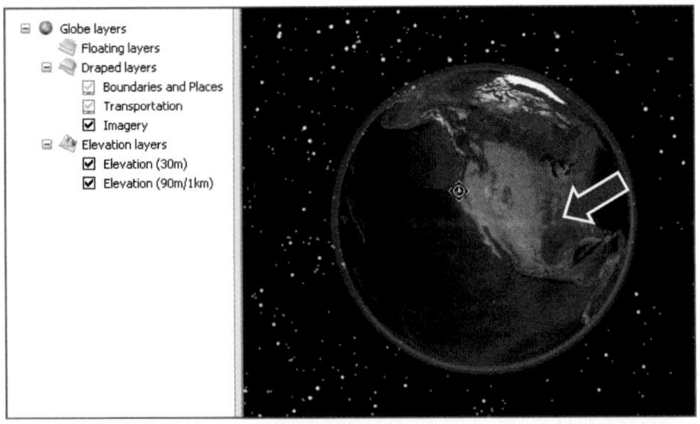

2 Place your cursor as seen at right.

3 **Right-click and drag downward until the label for San Francisco appears, then recenter the map on San Francisco using your left mouse button.**

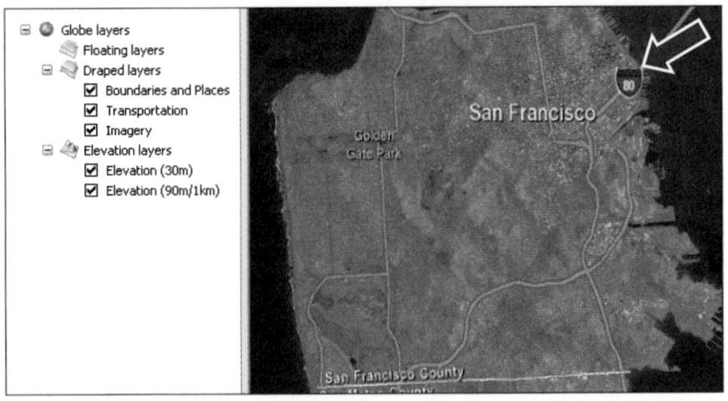

4 **Keep this process up until you can see San Francisco and the Bay Bridge.** If you need to zoom out at some point, right-click the map and drag upwards. You can also use your mouse wheel to zoom in and out.

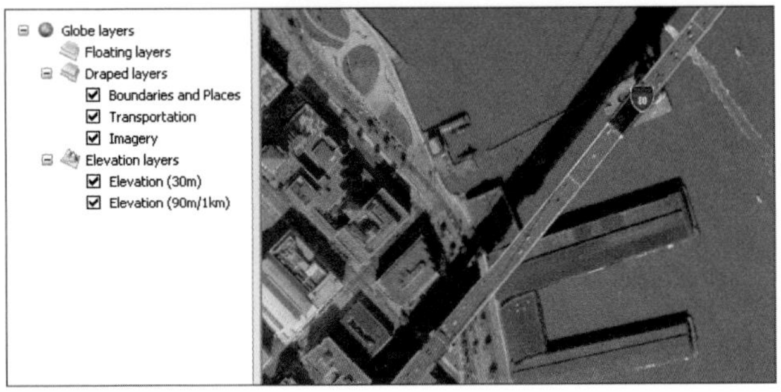

5 **Zoom in even farther as indicated at right.**

6 **Zoom to the full extent.**

YOUR TURN

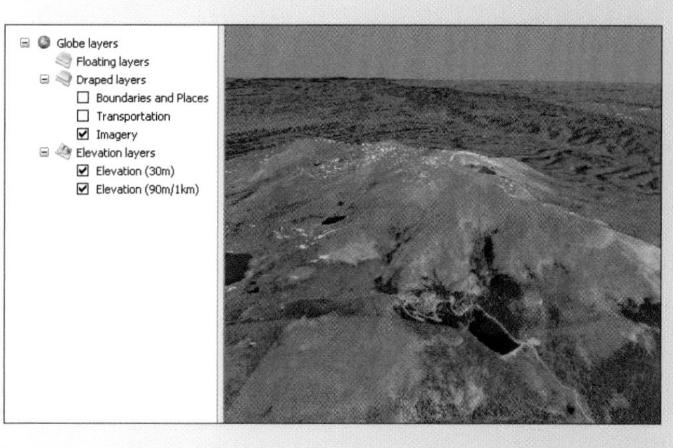

Zoom in to a mountainous area of the globe and use the Navigation Mode button to see elevation. You can view mountains in 3D. You can find particular mountains using Find with the Places tab, zooming to and then zooming out. Then, to view elevation, click the Navigation Mode button . For instance, above is a view of Pikes Peak.

Add and display large-scale vector data

You can add and display map layers for anywhere in the world. Next, you will add two layers for Allegheny County.

1 **Click the Add Data button; navigate to \ESRIPress\GIST1\Data\AlleghenyCounty.gdb; hold down your Ctrl key; click CountySchools, Tracts; release the Ctrl key; and click Add, Finish, and Close.**

2 **Click the List By Type button at the top of the TOC and drag Tracts up in the TOC above Boundaries and Places.**

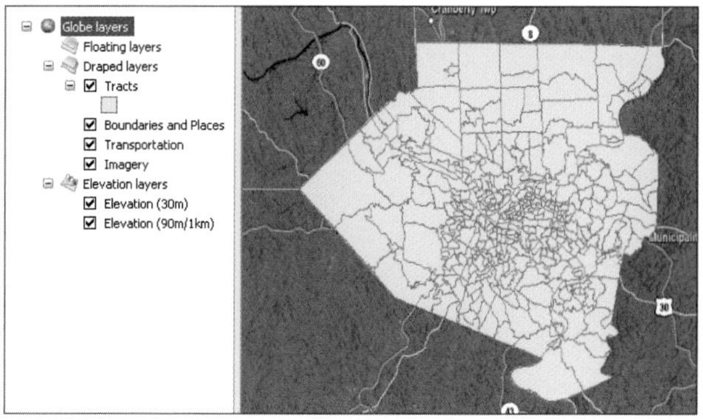

3 **Right-click Tracts and click Zoom to Layer.** You could symbolize the tracts using attributes in any way you wish at this point.

4 **At the top of the TOC, click the List By Source button , right-click County Schools, and click Display XY Data.**

5 **Select POINT_X for the X Field; select POINT_Y for the Y Field; and select Projected Coordinates, State Plane, NAD 1983 (Feet), NAD 1983 StatePlane, Pennsylvania South FIPS 3702 (Feet).prj for the projection. Click OK, Finish.**

6 Turn off Tracts in the TOC, click the Find button and type or make selections as shown in the image at the right.

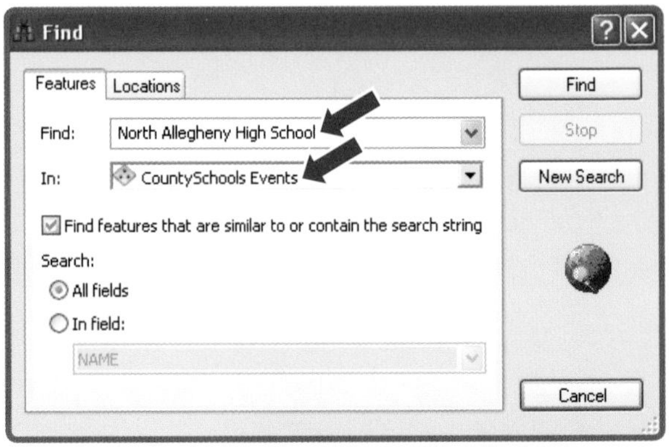

7 Click Find and, in the resulting bottom panel, right-click North Allegheny High School, click Zoom To, and close the Find window.

8 Zoom in until you see the label for North Allegheny High School. You can see that ArcGlobe has its own label for the school as a place and even has local streets for display.

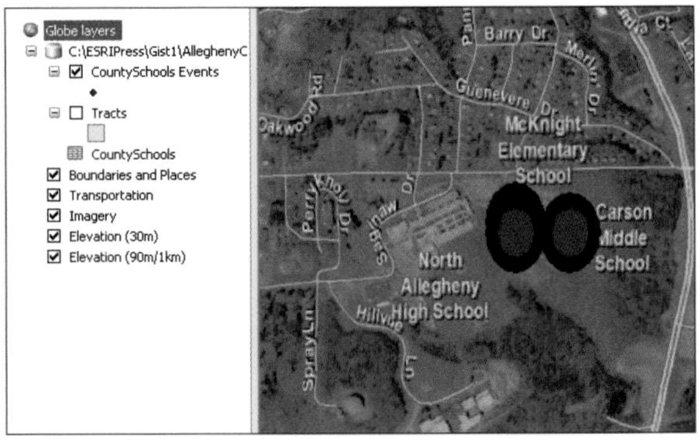

9 Save your ArcGlobe document as **Tutorial10-1.3dd** in \ESRIPress\GIST1\MyExercises\Chapter10\.

YOUR TURN

This Your Turn exercise has you use ArcGlobe for small-scale mapping. Start a new ArcGlobe document by adding \ESRIPress\GIST1\Data\World.gdb\Country. Symbolize Country with Quantities, Graduated colors using POP2006 normalized with SQMI (yields persons per square mile), and quantile for classification method. When finished, do not save your work. Close ArcGlobe. Be sure to move your Country layer above the Imagery layer.

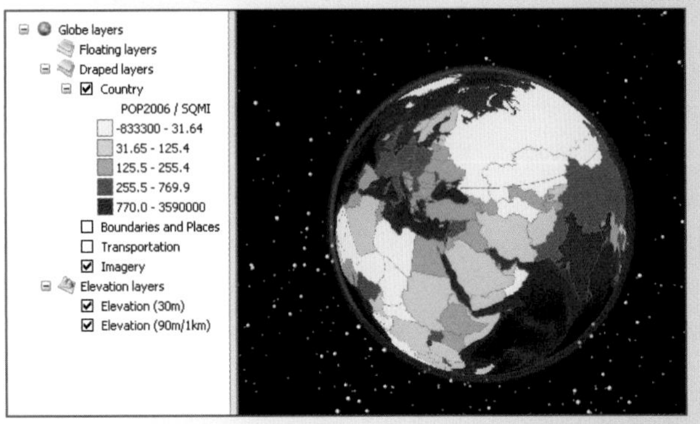

Assignment 10-1

Develop a 3D presentation for downtown historic site evaluation

Many U.S. cities, including the city of Pittsburgh, are experiencing a surge of downtown revitalization. In Pittsburgh, new condominium and apartment projects are in progress, and the city planning department wants to verify that this new development does not interfere with existing historic sites. In this assignment, you will help the city planning department raise the awareness of historic sites in downtown Pittsburgh by developing a 3D model and animation of these areas.

Start with the following:

- \ESRIPress\GIST1\Data\AlleghenyCounty.gdb\Parks—polygon layer of Allegheny County Parks
- \ESRIPress\GIST1\Data\AlleghenyCounty.gdb\Rivers—polygon layer of Allegheny County rivers
- \ESRIPress\GIST1\Data\Pittsburgh\CentralBusinessDistric.gdb\Histsite—polygon layer of historic sites in the city of Pittsburgh's central business district
- \ESRIPress\GIST1\Data\3DAnalyst.gdb\Bldgs—polygon layer of buildings in downtown Pittsburgh
- \ESRIPress\GIST1\3DAnalyst.gdb\Curbs—polyline layer of curbs (sidewalks) in downtown Pittsburgh
- \ESRIPress\GIST1\Data\3DAnalyst.gdb\Topo—polyline layer of topography contours in downtown Pittsburgh

Create a 3D map and animation of historic sites

Create a new ArcMap document called **\ESRIPress\GIST1\MyAssignments\Chapter10\Assignment10-1Yourname.mxd** and add the feature classes listed above. Symbolize the features to your liking and zoom to the Bldgs layer. Create a new feature class of buildings that have their centroid in historic sites and another of buildings whose centroids are not within historic sites (**Hint:** Use switch selection in the attribute table to select the nonhistoric site buildings). Save the new features in a new file geodatabase called **\ESRIPress\GIST1\MyAssignments\Chapter10\Assignment10-1YourName.gdb\HistoricSiteBldgs** and **\ESRIPress\GIST1\MyAssignments\Chapter10\Assignment10-1YourName.gdb\NonHistoricSiteBldgs**. Remove the original Bldgs layer.

In ArcCatalog, create two new point features for trees and street furniture (e.g., benches, signs, trash cans, streetlights, etc.) in historic sites called **\ESRIPress\GIST1\MyAssignments\Chapter10\Assignment10-1YourName.gdb\HistoricSiteTrees** and **\ESRIPress\GIST1\MyAssignments\Chapter10\Assignment10-1YourName.gdb\HistoricSiteFurniture**. Assign them the same spatial reference as the historic sites (NAD_1983_StatePlane_Pennsylvania_South_FIPS_3702_Feet). In ArcMap, digitize points representing trees and street furniture anywhere in the historic site locations.

Create a new 3D ArcScene file called **\ESRIPress\GIST1\MyAssignments\Chapter10\Assignment10-1YourName.sxd** and add all of above features (original and new) except 3DAnalyst.gdb\Bldgs. Create a new TIN from the Topo layer called **\ESRIPress\GIST1\MyAssignments\Chapter10\HistoricSiteTIN** and assign it spatial reference NAD_1983_StatePlane_Pennsylvania_South_FIPS_3702_Feet. Remove the original Topo layer. Drape the new features for historic and nonhistoric buildings, street furniture, and trees as well as existing curbs, historic sites, parks, and rivers to the TIN. Extrude the buildings using the height field. Show the nonhistoric site buildings using a transparency effect of 60% and the historic site buildings as opaque (0% transparency). Swap the 2D points for trees and street furniture with 3D symbols. Symbolize all layers to your liking.

Create a fly-through animation focusing on the historic sites called **\ESRIPress\GIST1\MyAssignments\Chapter10\Assignment10-1YourName.avi**.

WHAT TO TURN IN

If your work is to be graded, turn in the following files:

File geodatabase: \ESRIPress\GIST1\MyAssignments\Chapter10\Assignment10-1YourName.gdb

ArcMap document: \ESRIPress\GIST1\MyAssignments\Chapter10\Assignment10-1YourName.mxd

ArcScene document: \ESRIPress\GIST1\MyAssignments\Chapter10\Assignment10-1YourName.sxd

TIN: \ESRIPress\GIST1\MyAssignments\Chapter10\HistoricSiteTIN

Animation: \ESRIPress\GIST1\MyAssignments\Chapter10\Assignment10-1YourName.avi

If instructed to do so, instead of the above individual files, turn in a compressed file, **Assignment10-1YourName.zip**, with all files included. Do not include path information in the compressed file.

Assignment 10-2

Perform a 3D analysis of conservatory building addition

Pittsburgh's local conservatory, Phipps Conservatory (`http://phipps.conservatory.org`), was built in 1893 by Henry Phipps as a gift to the city of Pittsburgh. Phipps Conservatory recently underwent a major renovation with the addition of a 10,885-square-foot green-engineered welcome center.

Future additions will include state-of-art production greenhouses and a one-of-a-kind tropical forest. The 3D Analyst extension is very useful for envisioning the expansion of the conservatory and also for viewing the entire topography of the study area where the conservatory is located. In this assignment, you will create a 3D TIN; perform a line-of-sight analysis; drape features and an aerial photo to the TIN; and create new 3D features.

Start with the following:

- \ESRIPress\GIST1\Data\Pittsburgh\Phipps.gdb\Bldgs—polygon layer buildings in the Phipps Conservatory study area
- \ESRIPress\GIST1\Data\Pittsburgh\Phipps.gdb\Curbs—polyline layer sidewalk curbs in the Phipps Conservatory study area
- \ESRIPress\GIST1\Data\Pittsburgh\Phipps.gdb\Topo—polyline layer topography contours in the Phipps Conservatory study area
- \ESRIPress\GIST1\CMUCampus\25_45.tif—digital orthographic map

10-1 10-2 10-3 10-4 10-5 10-6 10-7 10-8 10-9 A10-1 A10-2

Create line of sight and perform 3D analysis

In ArcCatalog, create a new file geodatabase called **\ESRIPress\GIST1\MyAssignments\Chapter10\Assignment10-2YourName.gdb**. Create a new polygon feature class in the geodatabase called **PhippsAddition** whose spatial reference system is NAD_1983_StatePlane_Pennsylvania_South_FIPS_3702_Feet.

Create a new ArcMap document called **\ESRIPress\GIST1\MyAssignments\Chapter10\Assignment10-2YourName.mxd** and add the features from Phipps.gdb and PhippsAddition from your new geodatabase. In the PhippsAddition feature class, digitize a simple 20-foot-tall polygon building in the front of Phipps Conservatory (Phipps Conservatory is labeled as PAGIS_ID 770002 in the Bldgs feature class). Create a new TIN from the Topo layer called **\ESRIPress\GIST1\MyAssignments\Chapter10\PhippsTIN** and assign it spatial reference NAD_1983_StatePlane_Pennsylvania_South_FIPS_3702_Feet. Create a line-of-sight analysis from your new building addition to the Carnegie Museum of Pittsburgh (PAGIS_ID 5600589) and to the Café Phipps (PAGIS_ID 7700004). Export your line-of-sight analysis for both buildings to a JPEG file called **\ESRIPress\GIST1\MyAssignments\Chapter10\Assignment10-2YourName.jpg**.

Create a new 3D ArcScene file called **\ESRIPress\GIST1\MyAssignments\Chapter10\Assignment10-2YourName.sxd** with the Bldgs, Curbs, and PhippsAddition features added. Add

PhippsTIN and the aerial image 25_45.tif. Drape the curbs and the aerial photo to PhippsTIN with an offset of one foot (to avoid bleeding into the contours). Be sure to apply the aerial photo to the base height of the new TIN and not the aerial photo itself. Drape both buildings' features (again with an offset of one foot) and display using the building height field. Use 3D effects where you think appropriate to focus attention on the new addition.

Create a PowerPoint presentation called **\ESRIPress\GIST1\MyAssignments\Chapter10\Assignment10-2YourName.ppt** and insert 3D images of views from various angles showing some with the ortho image and some without the ortho image. Focus on the area around Phipps Conservatory and the new addition. **Hint:** Use File, Export Scene, 2D to create images of your views.

WHAT TO TURN IN

If your work is to be graded, turn in the following files:

File geodatabase: \ESRIPress\GIST1\MyAssignments\Chapter10\Assignment10-2YourName.gdb

ArcMap document: \ESRIPress\GIST1\MyAssignments\Chapter10\Assignment10-2YourName.mxd

ArcScene document: \ESRIPress\GIST1\MyAssignments\Chapter10\Assignment10-2YourName.sxd

TIN: \ESRIPress\GIST1\MyAssignments\Chapter10\PhippsTIN

Image file: \ESRIPress\GIST1\MyAssignments\Chapter10\Assignment10-2YourName.jpg

PowerPoint: \ESRIPress\GIST1\MyAssignments\Chapter10\Assignment10-2YourName.ppt

If instructed to do so, instead of the above individual files, turn in a compressed file, **Assignment10-2YourName.zip**, with all files included. Do not include path information in the compressed file.

11

ArcGIS Spatial Analyst

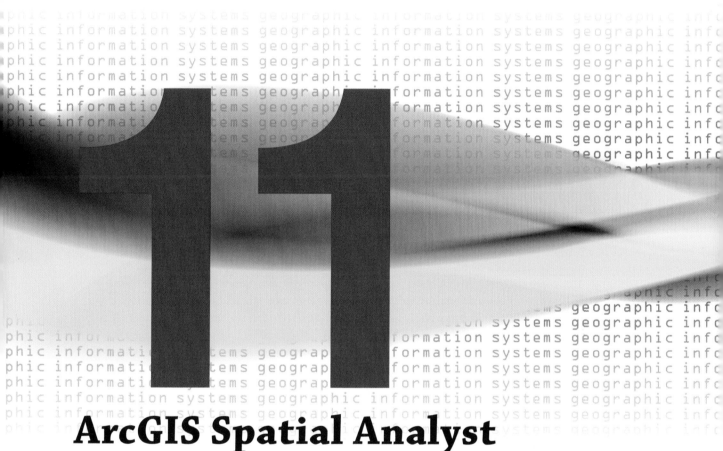

This chapter is an introduction to ArcGIS Spatial Analyst, an extension of ArcGIS Desktop. Spatial Analyst uses or creates raster datasets composed of grid cells to display data that is distributed continuously over space as a surface. In this chapter you will prepare and analyze a demand surface map for the location of heart defibrillators in the city of Pittsburgh with demand based on the number of out-of-hospital cardiac arrests with potential bystander help. You will also learn how to use Spatial Analyst to create a poverty index surface that combines several census data measures from block and block group polygon layers.

Learning objectives

- *Process raster map layers*
- *Create a hillshade raster layer*
- *Make a kernel density map*
- *Extract raster value points*
- *Conduct a raster-based site suitability study*
- *Use ModelBuilder for a risk index*

Tutorial 11-1

Process raster map layers

The map document that you will open has map layers including raster maps from U.S. Geological Survey Web sites—`http://seamless.usgs.gov/website/seamless/viewer.htm` *for digital elevation (NED shaded relief, 1/3 arc second) and* `http://gisdata.usgs.net/website/MRLC/viewer.php` *for land use (NLCD 2001). All raster maps are rectangular in their coordinate systems, but you will use the Pittsburgh boundary as a mask so that cells outside the boundary will have no color and the cells inside will have their assigned colors. In addition, you will use the DEM (digital elevation model) layer to create a hillshade which has a 3D appearance of topography illuminated by the sun. Placing the hillshade under the land-use layer and giving the land-use layer some transparency makes an attractive and informative display.*

Examine raster map layer properties

1 **Open ArcMap and the map document \ESRIPress\GIST1\Maps\Tutorial11-1.mxd.** The vector map layer called OHCA (out-of-hospital cardiac arrests) is the number of heart attacks over a five-year period per census block that occurred outside of hospitals where bystander help was possible because of location. As expected and by definition, you will see that that these heart attacks appear in developed areas. Next, you will examine properties of the raster layers.

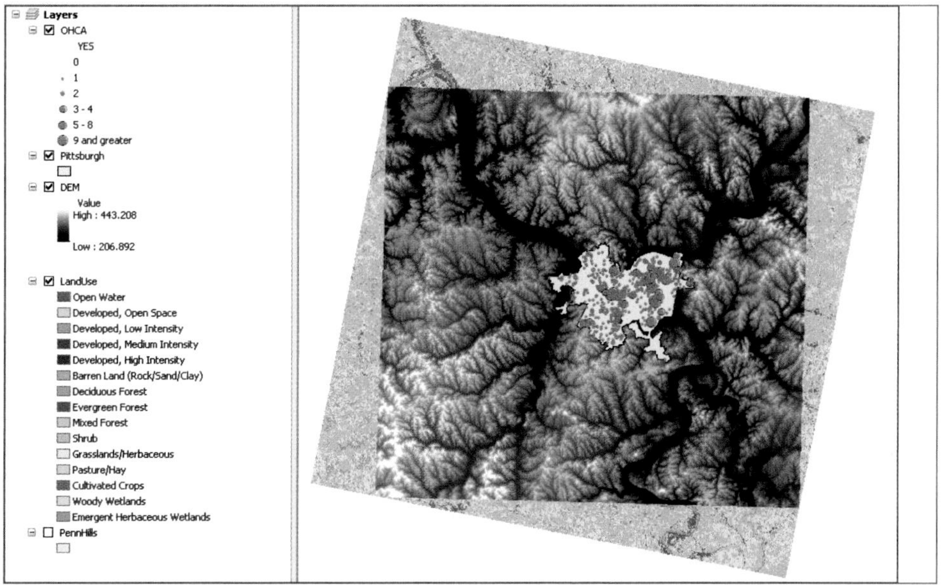

2 **In the TOC, right-click DEM and click Properties and click the Source tab.** All raster maps are rectangular in their coordinate system. This one has 2,106 columns and 1,984 rows with square cells of 90.70912488 decimal degrees on a side.

3 **Scroll down until you see the Extent information.** Here you can see familiar-looking decimal degree values for the extent, so this layer is in geographic coordinates. ArcGIS projected it to the data frame's projection, State Plane for Southern Pennsylvania.

4 **Scroll down further until you see the Statistics information.** Each cell or pixel has a single value—elevation in meters—which is stored as a floating point number. The statistics for elevation over the extent includes a mean elevation above sea level of 323.7 meters and maximum of 443.2 meters.

5 **Close the Properties window.**

6 **Click File and Save As, browse to \ESRIPress\GIST1\MyExercises\Chapter11\, and click Save.**

YOUR TURN

Examine the properties of the land-use layer. Notice that this is a projected layer that uses a projection for the continental United States (and the reason why the layer tilts when ArcGIS reprojects it to the local state plane projection). Also notice that the cell size is larger than that of the DEM, 30 meters on a side, and that the values are integers corresponding to land-use categories.

Set raster environment

Next, you need to set the environment for using the Spatial Analyst tools. Each time you use one of the tools, ArcMap will automatically use the environment settings, thereby saving you time in many instances.

1 **Click Customize and Extensions, check Spatial Analyst, and click Close.** This loads the Spatial Analyst extension, making its functionality available.

2 **On the main menu, click Geoprocessing, Environments, Raster Analysis, and type or make selections as shown in the image.**

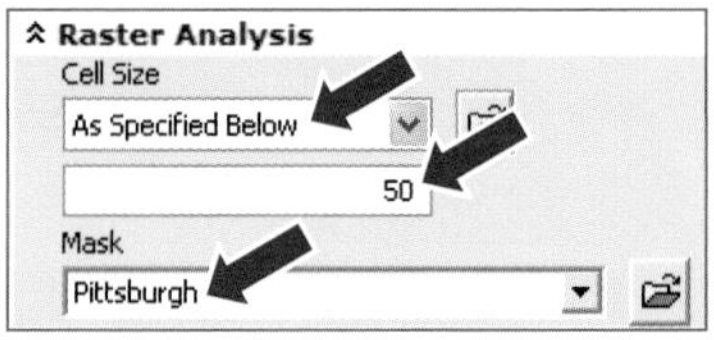

3 Click OK.

Extract land use employing a mask

ArcGIS can display a great many raster or image file formats. The land-use layer in your map document is a TIFF file format image as downloaded from the USGS site. To process the layer, it is necessary to convert it to an ESRI format. You will convert a TIFF file to an ESRI format by saving it in a file geodatabase. At the same time, you will use the Pittsburgh boundary as a mask to clip the original layer to Pittsburgh's rectangular extent and only display cells with Pittsburgh's boundary.

1 **On the main menu, click Windows, Search, Tools; type extract in the search text box, and click Extract by Mask.**

2 **Type or make selections as shown in the image.**

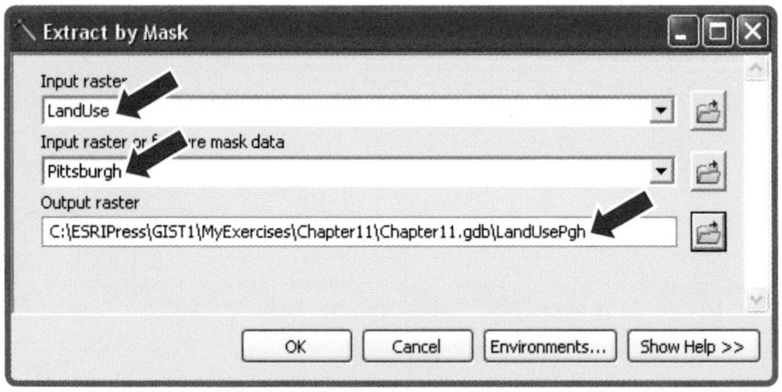

3 Click OK.

4 **In the TOC, turn off all layers except OHCA and LandUsePgh.**

5 **Right-click LandUsePgh and click Zoom to Layer.** ArcMap gave LandUsePgh an arbitrary color ramp (which is unattractive), but next you will add a layer file, created from LandUse, to correctly symbolize the new raster map.

6 **Right-click LandUsePgh, click Properties, click the Symbology tab and Import, browse to \ESRIPress\GIST1\Data\SpatialAnalyst\, click LandUse.lyr, click Add, and click OK twice.** The resulting map is informative and attractive; for example, you can see high-density development along Pittsburgh's rivers, and you can see that the clusters of heart-attack locations are in developed areas. In the next section, you will make the map even better by giving it a 3D appearance, using hillshade based on the DEM layer.

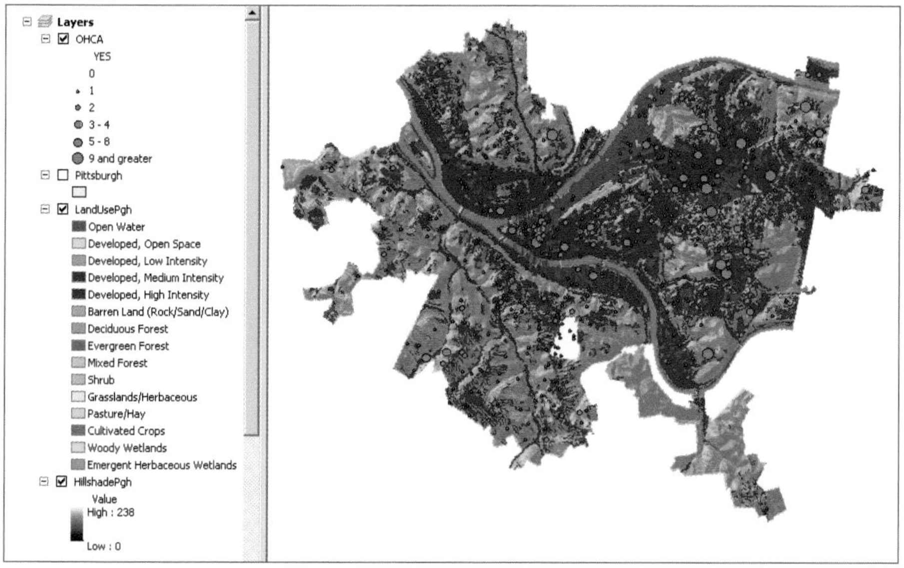

7 **Save your map document.**

YOUR TURN

Extract LandUsePennH from LandUseGrid using the PennHills layer as a mask and saving it as **LandUsePH** in \ESRIPress\GIST1\MyExercises\Chapter11\Chapter11.gdb\. Symbolize the new layer using LandUse.lyr. When finished, turn off all of the Penn Hills layers and zoom back to the Pittsburgh layer if necessary.

Tutorial 11-2

Create a hillshade raster layer

The hillshade function simulates illumination of a surface from an artificial light source representing the sun. Two parameters of this function are the altitude of the light source above the surface's horizon in degrees and its angle (azimuth) relative to true north. The effect of hillshade to a surface, such as elevation above sea level, is striking, giving a 3D appearance due to light and shadow. You can enhance the display of another raster layer, such as land use, by making it partially transparent and placing hillshade beneath it. That is the objective of this tutorial.

Create hillshade for elevation

You will use the default values of the hillshade tool for azimuth and altitude. The sun for your map will be in the west (315°) at an elevation of 45° above the horizon.

1 Type **hillshade** in the search text box, press Enter, and click Hillshade.

2 Type or make selections as shown in the image at the right.

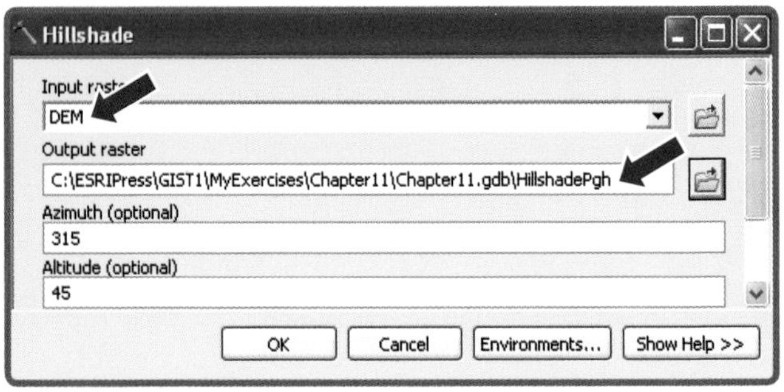

3 Click OK.

4 Move HillshadePgh to just below LandUsePgh in the TOC.

5 Right-click HillshadePgh, click Properties, and click Symbology.

6 Make sure that Stretched is selected in the Show panel, select Standard Deviations in the Stretch panel, and click OK.

7 Right-click LandUsePgh in the TOC, and click Properties and the Display tab.

8 **Type 35 in the Transparency field and click OK.** That's the finished product. Heart-attack locations are in some developed areas, but not all developed areas. Next, you will do additional spatial analysis on population statistics to see if you can determine a major factor affecting the incidence of heart attacks.

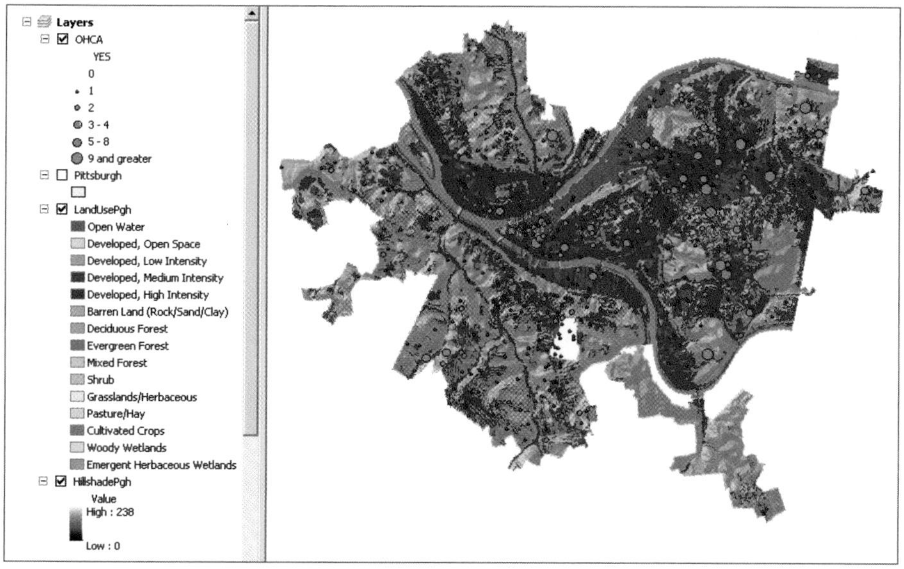

9 Save your map document.

YOUR TURN

Create PennHshade and display it under a 35 percent transparent LandUsePennH.

Tutorial 11-3

Make a kernel density map

The incidence of myocardial infarction (heart attacks) outside of hospitals in the United States for ages 35 to 74 is approximately 5.6 per thousand males per year and 4.2 per thousand females per year[1]. You will use a point feature class of census block centroids in Allegheny County to analyze heart attack incidence as input to an estimation method called kernel density smoothing. This method estimates incidence as heart attacks per unit area (density) and has two parameters, cell size and search radius. There is no "science" of how to set these parameters, but the larger the search radius, the smoother the estimated distribution. When smoothing a particular cell, the farther away, the less influence that other points have. Read ArcGIS Desktop Help on kernel density smoothing to learn more about this method.

Assign environmental settings and get statistics

The map document that you will open shows the observed locations of heart attacks (outside of hospitals and with the potential of bystander assistance) and block centroids symbolized with a color gradient for heart-attack incidence, as well as other supporting layers. The attribute table of block centroids has the incidence attribute, Inc = 0.0042 × [Fem35T74] + 0.0056 × [Male35T74] where Fem35T74 is the population by block of females of age 34 to 74, and Male35T74 is the corresponding population for males. The question is whether incidence does a good job of estimating the observed heart attacks in the OHCA point file.

1 **In ArcMap, open Tutorial11-3.mxd from the \ESRIPress\GIST1\Maps\ folder.** The map display for estimated incidence using block centroids with point markers is as good as vector graphics allow but is difficult to interpret. You will create an alternative representation of incidence by estimating the smoothed mean of the spatial distribution using kernel density smoothing.

1 Rosamond, W. D., L. E. Chambless, A. R. Folsom, L. S. Cooper, D. E. Conwill, L. Clegg, C. H. Wang, and G. Heiss, 1998, "Trends in the incidence of myocardial infarction and in mortality due to coronary heart disease, 1987 to 1994," *New England Journal of Medicine*, Vol. 339 (1998): 861–867.

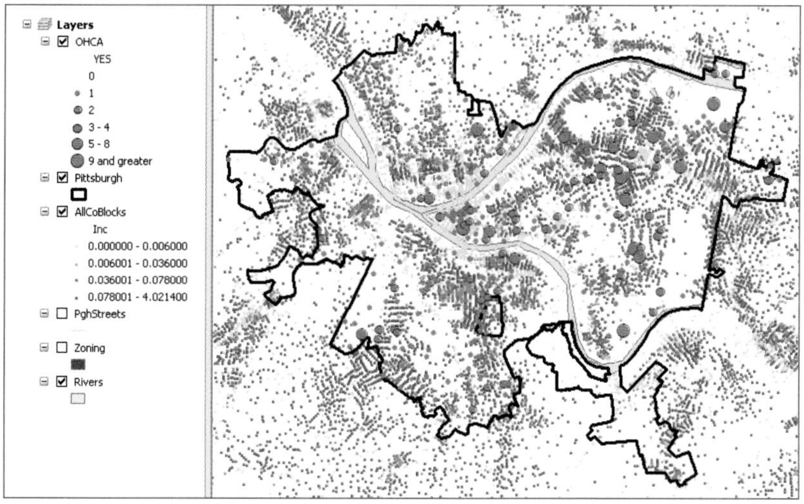

2 Click File and Save As, browse to \ESRIPress\GIST1\MyExercises\Chapter11\, and click Save.

3 On the main menu, click Geoprocessing, Environments, Raster Analysis, and type or make selections as shown in the image on the right.

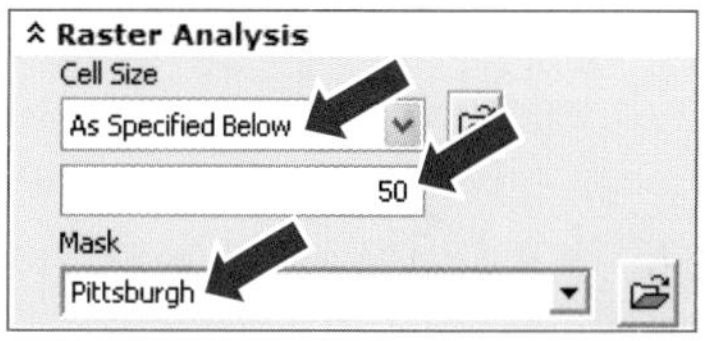

4 On the main menu, click Selection, Select By Location, and make selections as shown in the image on the right.

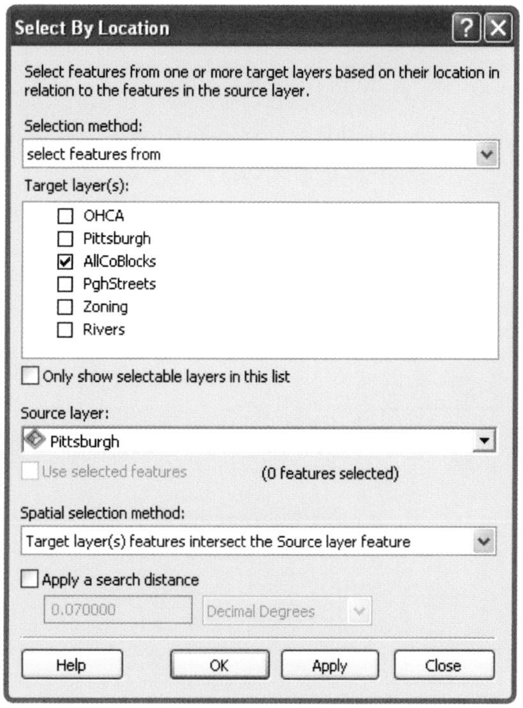

5 Click OK.

6 Right-click AllCoBlocks in the TOC, click Open Attribute Table, right-click the column heading for Inc, and click Statistics. Note the sum of Inc, 684 for Pittsburgh, the expected annual number of heart attacks in Pittsburgh outside of hospitals. Below you will verify that density smoothing preserves this sum in any surface it estimates. Kernel density smoothing simply spreads the total around on a smooth surface, preserving the input total number of heart attacks.

7 Close the Selection Statistics window and the table, and clear the selection.

Make a density map for heart-attack incidence

The OHCA map layer shows heart attacks per census block in Pittsburgh. Blocks in Pittsburgh average a little less than 300 feet per side in length. Suppose that policy analysts estimate that a defibrillator with public access can be made known to residents and retrieved for use as far away as 2.5 blocks from the location. They thus recommend looking at areas that are five blocks by five blocks in size, or 1,500 feet on a side, with defibrillators located in the center. Therefore, you will use a 150-foot cell and 1,500-foot search radius. The 150-foot cell approximates the middle of a street segment, the average location of a heart attack.

1 Type **kernel density** in the search text box, press Enter, and click Kernel Density.

2 Type or make selections as shown in the image.

3 Click OK. The resulting surface does not appear useful at this point, but it will after you symbolize it better next.

4 Right-click Kernel1500; click Properties, the Symbology tab, and the Classify button; and select Standard Deviation for Classification Method. Standard Deviation is a good option for showing variation in raster grids because it yields a central category and an equal number of categories on each side of the center. That makes dichromatic color scales, such as you will use next, more meaningful and easier to interpret. You control the number of categories in the next step by choosing the fraction of standard deviation for which to create break points, every 1, 1/3, 1/4, etc., standard deviations.

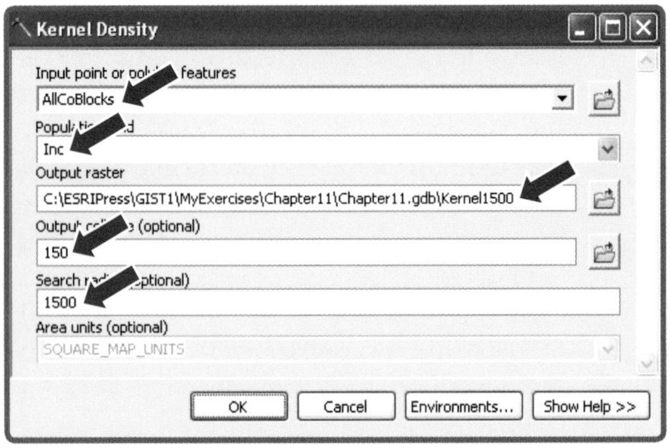

5 Select 1/3 Std Dev for Interval Size and click OK.

6 Select the color ramp that runs from green to yellow to red, and click OK.

7 Turn off AllCoBlocks, PghStreets, and Zoning, and turn on all other layers in the TOC. Incidence matches clusters of the OHCA heart-attack data in many, but not all areas. For example, there is a cluster in Pittsburgh's central business district (triangle just to the right of where the three rivers join), but estimated incidence is low there. The problem is that the density map, based on population data, shows expected heart attacks per square foot in reference to where people live, not necessarily where they have heart attacks. Many people shop or work in the central business district and unfortunately have heart attacks there, but few live there.

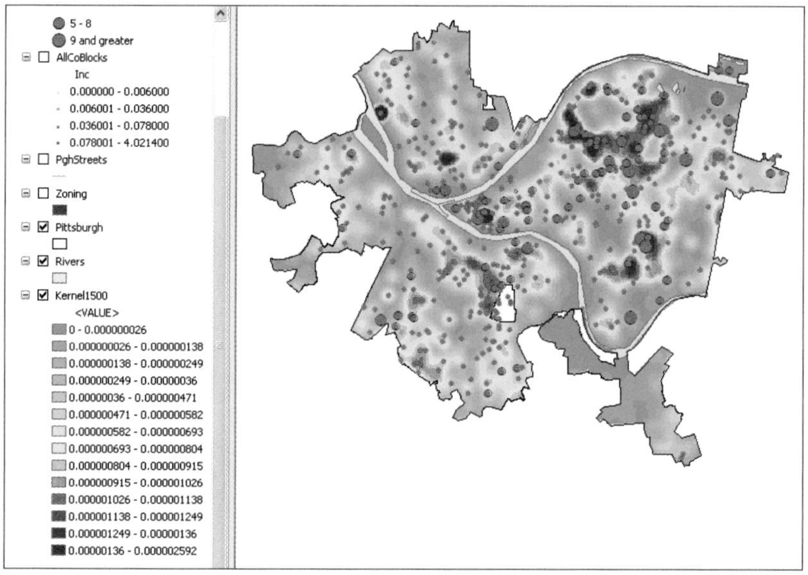

Check the density surface, to see if it preserves the total number of heart attacks. The estimated incidence that you found using block centroids was 684. Open the properties for the density surface, click the Symbology tab, and click Classify. There you will find useful statistics: 72,315 cells with a mean of 0.000000415 heart attacks per square foot. Remember that each cell is 150 feet by 150 feet. Therefore, 150 × 150 × 72,315 × 0.000000415 = 675 heart attacks, which is close to 684. So what kernel density smoothing did here was to move the input number of heart attacks around and to distribute them smoothly. The kernel density map is a better estimate of incidence than raw data, because smoothing averages out randomness and provides an estimate of the mean or average surface.

YOUR TURN

Create a second kernel density surface for incidence, called Kernel3000, with all inputs and outputs the same except you will use a search radius of 3,000 instead of 1,500. Symbolize the output the same as Density1500. While keeping Kernel1500 turned on, turn Kernel3000 on and off to see the differences in the two layers. Kernel3000 is more spread out and smoother, but it has the same corresponding number of estimated heart attacks: close to 684.

Tutorial 11-4

Extract raster value points

While the estimated densities appear to match the actual heart-attack data in OHCA, the match may or may not stand up to closer investigation. ArcMap has a tool that will extract point estimates from the raster surface for each point in OHCA. Then you can use the extracted densities multiplied by block areas to estimate number of heart attacks. If there is a strong correlation between the estimated and actual heart attacks, there would be evidence that population alone is a good predictor of heart attacks.

1 Type **extract values to points** in the search text box, press Enter, and click Extract Values to Points.

2 Type or make selections as shown in the image.

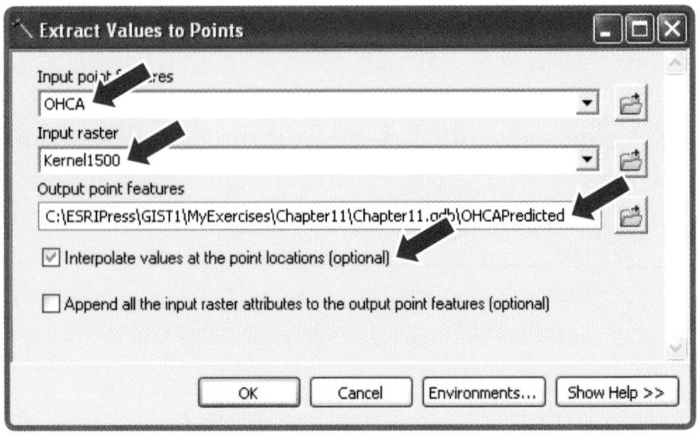

3 Click OK. The resultant layer, OHCAPredicted, has an attribute, RASTERVALU, which is an estimate of heart attack density, or heart-attacks per square foot, in the vicinity of each block.

Calculate predicted heart attacks

You can expect that the resulting estimate will be larger than the actual number of heart attacks in OHCA's YES attribute, which is just a subset of all heart attacks (those in which bystander help was possible, given the location).

1 Right-click OHCAPredicted and open its attribute table.

2 Click Options, Add Field, and add a field called **Predicted** that will contain floating point values.

3 Right-click the Predicted column heading and click Field Calculator.

4 Create the expression **5 × [RASTERVALU] × [Area]** and click OK. OHCA data is a five-year sample for heart attacks, thus the expression includes the multiple 5.

5 Close the attribute table. A few of the points in OHCA have no raster values near them, so ArcGIS assigns the value –9999 to them to signify missing values. Before looking at a scatter plot of predicted and actual values, you will first select only OHCA points with positive predicted values.

6 Click Selection, Select by Attributes.

7 For the OHCAPredicted layer, create the expression "**Predicted**" >= **0** and click OK.

8 Right-click OHCAPredicted, and click Data and Export Data.

9 Export selected features to **\ESRIPress\GIST1\MyExercises\Chapter11\OHCAPredicted2.shp** and click Yes to add the shapefile to the map.

10 Clear the selected features and turn off the OHCA_Predicted layer.

11-1
11-2
11-3
11-4
11-5
11-6
A11-1
A11-2

Create scatter plot of actual versus predicted heart attacks

Note: While Pittsburgh has a total of 7,466 blocks, only 1,509 blocks had heart attacks. The scatter plot that you will eventually construct includes data only for the 1,509 blocks, but should ideally include the balance of the total blocks, which had actual values of zero but predicted values sometimes much larger than zero. Nevertheless, you will be able to get an indication of the correlation between predicted and actual heart attacks. Adding the balance of blocks would only make the correlation worse, but the correlation is actually already very low, as you will see next.

1 Click View, Graphs, Create.

2 Type or make selections as follows:

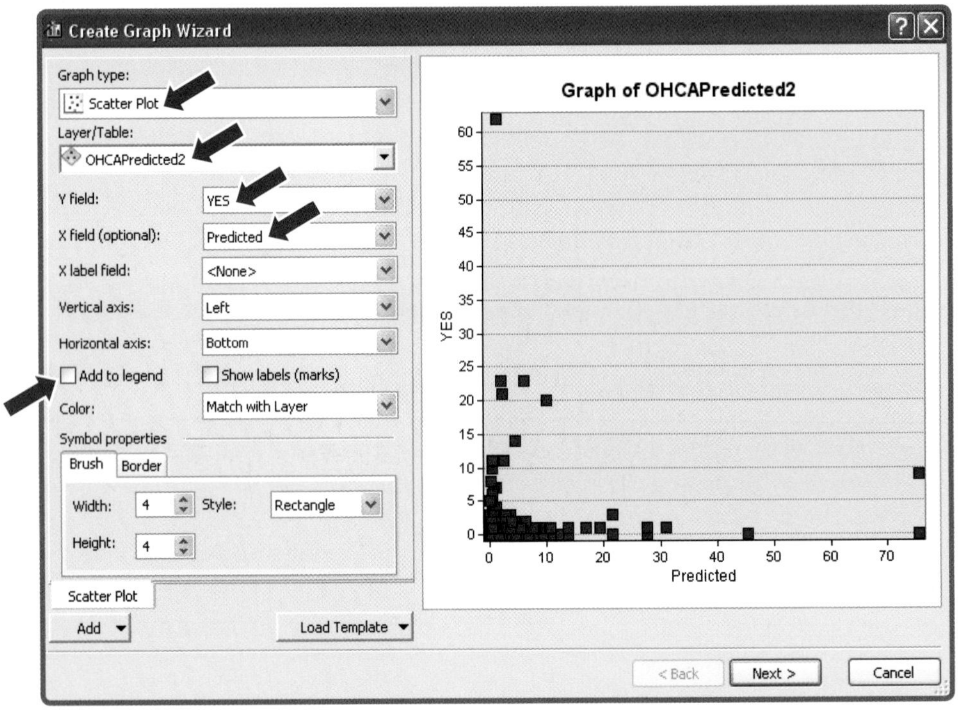

3 Click Next, Finish. At the scale of blocks, the predicted values seem to correlate poorly with the actual values. A good correlation would have a graph with Actual (YES attribute) and predicted values scattering around a 45-degree slope line. This scatter plots shows no correlation at all. If you export the corresponding data to a statistical package or Excel, you would find that the correlation coefficient between predicted and actual values is only 0.0899, which is very low. Evidently, factors other than where the population resides affect the locations and clustering of heart attacks occurring outside of hospitals.

4 Save your map document.

Tutorial 11-5

Conduct a raster-based site suitability study

The objective is to find locations that have high heart-attack rates and that have heart defibrillators accessible to the public. The approach includes using kernel density smoothing on the available heart-attack data to remove randomness from the spatial distribution. This provides a more reliable estimate of demand. An assumption is that any locations within commercial areas provide needed public accessibility.

Open a map document

A vector map layer is available for commercial area boundaries. To conduct a raster-based analysis, you will have to convert this map layer into a raster layer. This is the first task that you will undertake.

1 **In ArcMap, open Tutorial11-5.mxd from the \ESRIPress\GIST1\Maps\ folder.** The map document shows the observed locations of heart attacks (outside of hospitals and with the potential of bystander assistance), a 600-foot buffer of commercially zoned areas in Pittsburgh, and other supporting layers. The 600-foot (or two-block) buffer of commercial areas includes adjacent noncommercial areas that have sufficient access to defibrillators.

2 **Click File and Save As, browse to \ESRIPress\GIST1\MyExercises\Chapter11\, and click Save.**

3 **On the main menu, click Geoprocessing, Environments, Raster Analysis, and type or make selections as shown in the image at the right.**

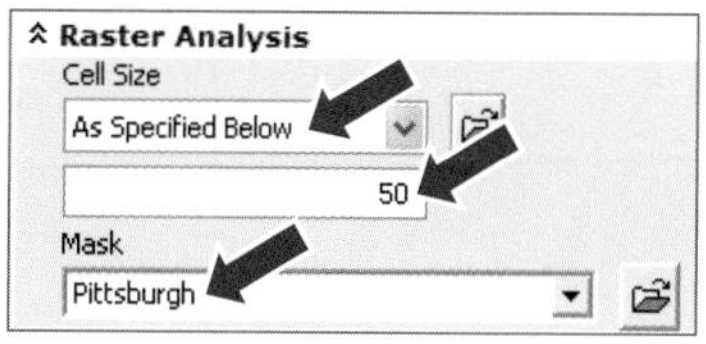

4 Click OK.

11-1
11-2
11-3
11-4
11-5
11-6
A11-1
A11-2

Convert feature buffer to a raster dataset

The ZoningCommercialBuffer layer has two polygons and corresponding records with a single attribute: Commercial. The Commercial value of 1 corresponds to commercial land use or land within 600 feet of commercial land use. The other value, 0, represents the balance of Pittsburgh and includes all other zoned land uses. You will convert this vector layer into a raster dataset using a conversion tool. First, however, you need to select both records in the vector file in order for them to convert.

1 **Right-click the ZoningCommercialBuffer layer and click Open Attribute table.**

2 **Select both records by clicking the row selector of the first row, drag the mouse to select both rows, and close the table.**

3 **Type feature to raster in the search text box, press Enter, and click Feature to Raster.**

4 **Type or make selections as shown in the image at the right.**

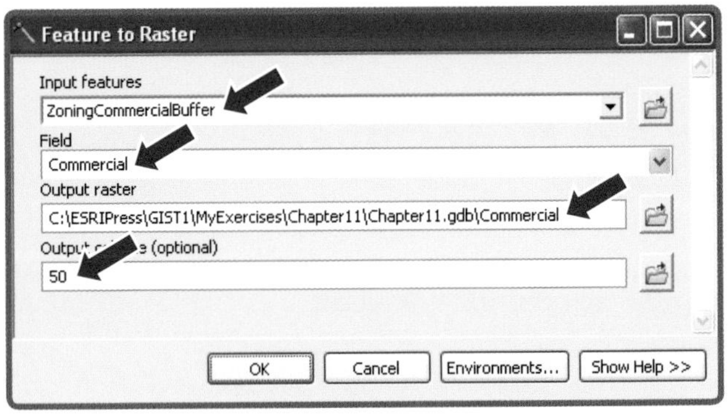

5 **Click OK.**

6 **Remove the ZoningCommercialBuffer layer and turn off the OHCA layer.**

7 **Right-click Commercial, click Properties, click the Symbology tab, click Unique Values in the Show panel, and resymbolize the new Commercial area to have two colors: white for noncommercial and gray for commercial.**

YOUR TURN

Create a kernel density map based on the YES attribute of the OHCA point layer that has 150-foot cells, a search radius of 1,500 feet, and area units of SQUARE_FEET. Call the new raster layer **HeartAttack** and save it in \ESRIPress\GIST1\MyExercises\Chapter11\Chapter11.gdb. Symbolize the layer using the standard deviation method with interval size 1/3 Std Dev. Use the green-to-yellow-to-red color ramp. Try turning the OHCA layer on and off to see how well the density surface represents heart attacks, then remove the OHCA layer. The resulting raster map is below.

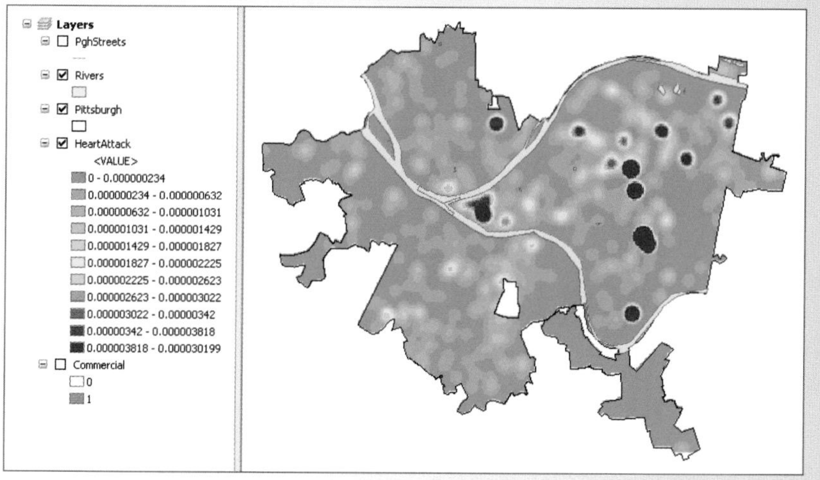

Query a raster dataset with a single criterion using reclassify

First, you will reclassify your kernel density map, HeartAttack, for areas that have sufficiently high heart-attack density to merit a defibrillator. Suppose that policy makers seek 25-block areas, roughly five blocks on a side, that would have 10 or more heart attacks every five years in locations where bystander help is possible. A square 25-block area is 5 × 300 feet = 1,500 feet on a side with 1,500 feet × 1,500 feet = 2.25×10^6 square feet of area.

Thus the heart-attack density sought is 10 heart attacks / 2.25×10^6 square feet = 0.000004444 heart attacks per square foot or higher. While the density map you just created has a continuous range of values, next you will reclassify values into just two values: 0 for cells with density less than 0.000004444, and 1 for cells with density greater than or equal to 0.000004444.

11-1 11-2 11-3 11-4 11-5 11-6 A11-1 A11-2

1 Type **Reclassify** in the search text box, press Enter, and click Reclassify.

2 Select HeartAttack for the Input raster and click Classify.

3 Select 2 for Classes, select Manual for Method, type **0.000004444** to replace 0.000006 in the Break Values panel, and click OK. The Old values column shows values with only 6 decimal places, but Spatial Analyst has all 9 decimal places in memory.

4 For New values, replace the 1 with **0** and replace the 2 with **1**.

5 Finish filling in the form by typing or making selections as shown in the image at the right.

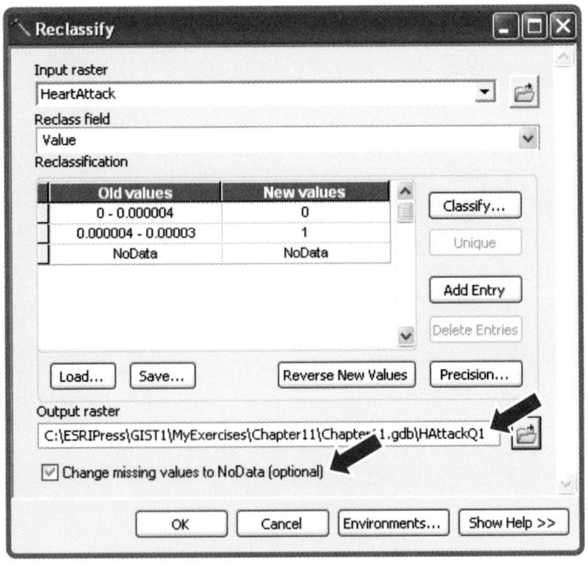

6 Click OK.

7 Resymbolize Calculation so that 0 has no color and 1 is dark blue, and make sure that HeartAttack is turned on and below HAttackQ1. You can see that relatively few peak areas, eight, have sufficiently high heart-attack density. Some of them are likely too small, but you will not make that determination until you consider all query criteria.

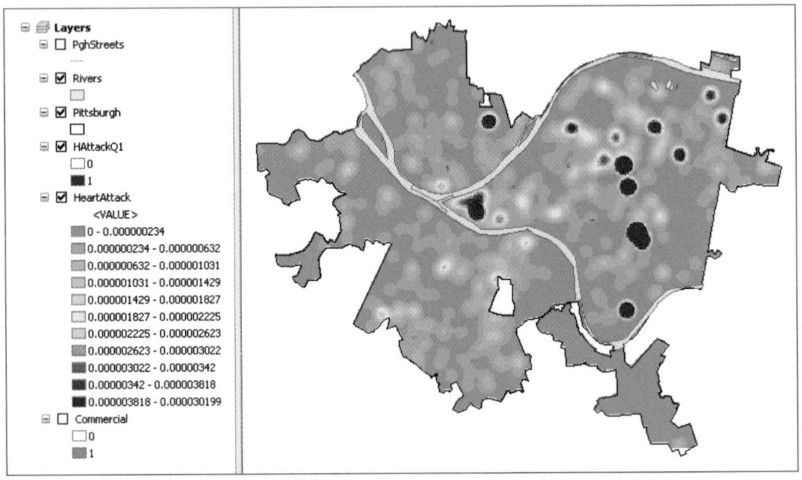

Query a raster dataset with a two criteria

Next, you will include a second criterion in the query— locations within the commercial buffer—for suitable defibrillator sites. The Boolean And tool combines two raster datasets by giving all cells the value 0 except where both input cells are 1, in which case the output cell gets the value 1. In this case, the resulting areas defined by cells with value 1 are both in the commercial buffer and the sufficiently high heart-attack area of HAttackQ1.

1 **Type Boolean And in the search text box, press Enter, and click Boolean And.**

2 **Type or make selections as shown in the image at the right.**

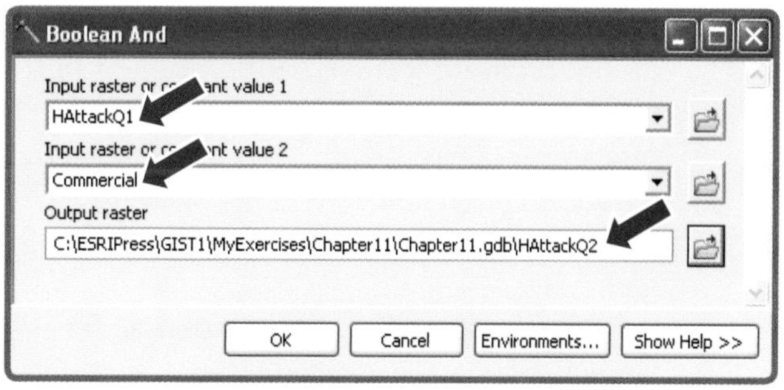

3 **Click OK.**

4 **Resymbolize HAttackq2 so that 0 has no color and 1 is Tourmaline Green (eighth column, third row of the color chips array), and make sure that HAttackQ1 and HeartAttack are turned on and below HAttack2.** As you would expect, adding a second criterion with the AND connection has reduced the size of areas meeting criteria.

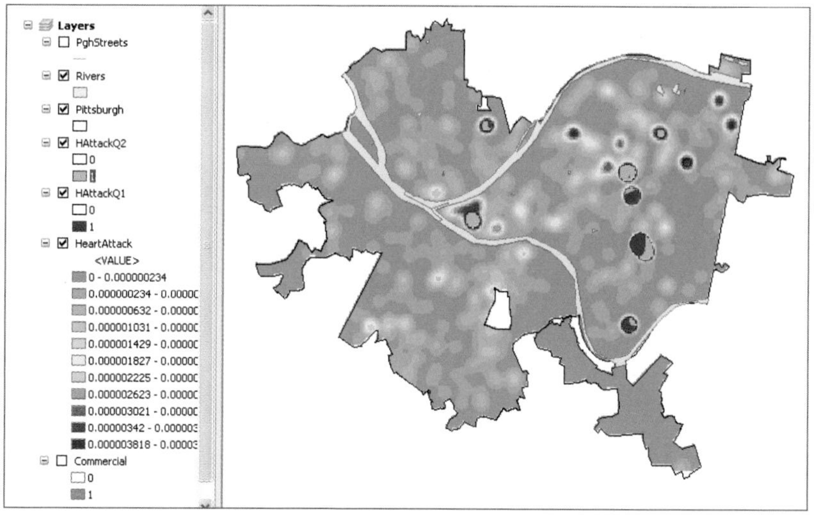

Three of the formerly promising areas are significantly reduced.

YOUR TURN

Turn on the Streets layer and zoom in to each feasible area to check the third criterion that there be at least 25 blocks, or roughly 2.25 million square feet, in a square area. Use the Measure tool on the toolbar to measure feasible areas. Which areas remain feasible? What would you report back to policy makers? Save the map document.

Tutorial 11-6

Use ModelBuilder for a risk index

People who live in poverty often have poor health care, unhealthy diets, and unhealthy habits such as smoking—all factors contributing to heart attacks. In this tutorial, you will create an index for identifying poverty areas by combining four poverty indicators[2]: population (1) below the poverty income line, (2) of female-headed households with children, (3) with less than a high school education, and (4) of workforce males who are unemployed.

Robyn Dawes provides a simple method for combining such measures into a poverty index[3]. If you have a reasonably good theory that several variables are indicative or predictive of a dependent variable of interest (and whether the dependent variable is observable or not), then Dawes makes a good case that all you need to do is to remove scale from each input, so each has the same weight, and then average the scaled inputs to create a predictive index. A good way to remove scale from a variable is to calculate z-scores, subtracting the mean and then dividing by the standard deviation for each variable.

You can see in the following table that if you simply averaged the four variables, then Male unemployed arbitrarily gets the highest weight while female-headed households would have practically no weight, given the means of the variables. Z-scores for all four variables, however, all have means of zero and standard deviations of one, so when averaged they each will have equal weight.

Indicator variable	Mean	Standard Deviation
Female-headed households with children	1.422	4.431
Less than high school education	110.060	80.812
Male unemployed	154.500	124.804
Poverty income	126.021	147.188

There are three parts to creating the poverty index. First you will calculate the z-scores for each of the four indicators. The map layers for the indicators are centroids of blocks for the population of female-headed households with children and centroids of block groups

2 W. O'Hare and M. Mather, "The growing number of kids in severely distressed neighborhoods: Evidence from the 2000 census," *Kids Count* (2003). *Kids Count* is a publication of the Annie E. Casey Foundation and the Population Reference Bureau, and is available from `http://www.aecf.org/upload/publicationfiles/da3622h1280.pdf`.

3 R. M. Dawes, "The robust beauty of improper linear models in decision making," *American Psychologist* 34 (1979): 571–582.

for the other three indicators (which are not available at the more desirable, smaller block level). Thus to make these layers comparable for combining them into an index, you will transform them into kernel density maps, all with the same 150-foot-square grid cells. So the second part is to create kernel density maps for all four input variables. The third part is to use a Spatial Analyst tool to add the surfaces, weighted by 0.25, to average them. You will carry out parts 2 and 3 using ModelBuilder to document the work and provide a reusable tool for creating an index. Note that you can work through the following exercises successfully even if you did not complete the introduction to ModelBuilder in tutorial 8-7. Tutorial 8-7 has a more complete introduction to ModelBuilder.

Set geoprocessing environment

The map document you will open has inputs for preparing the poverty index: AllCoBlkGrps, which has block group centroids and needed attributes (NoHighSch2 = population with less than high school education, Male16Unem = males in the workforce who are unemployed, and Poverty = population below poverty income), and AllCoBlocks, which has block centroids and the attribute FHHChld = female-headed households with children.

1 **In ArcMap open Tutorial11-6.mxd from the \ESRIPress\GIST1\Maps\ folder.** Shown are the block group centroids and block centroids, each displaying one of the four poverty indicators via a color ramp. You can see that it is difficult to represent the spatial patterns effectively using vector graphics, plus it is difficult to integrate the information from just two spatial distributions out of the four needed for the poverty index. The raster poverty index that you will create will do a better job on both issues.

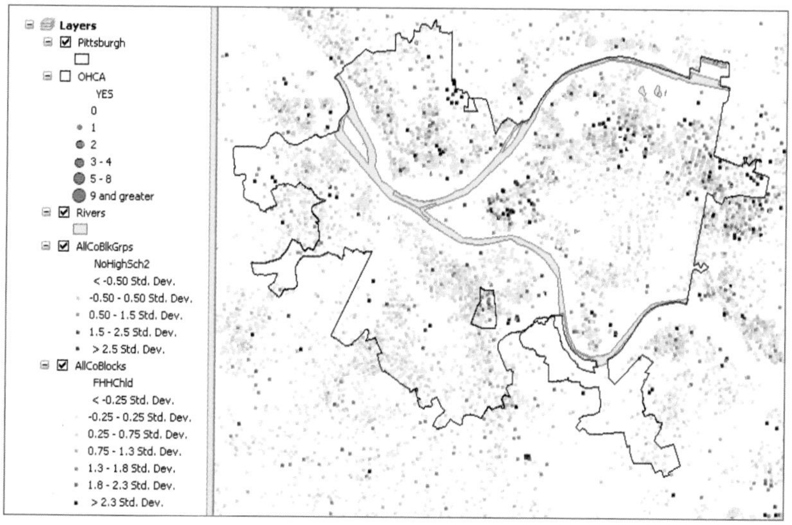

2 On the main menu, click Geoprocessing, Geoprocessing Options, and make selections as shown in the image at the right.

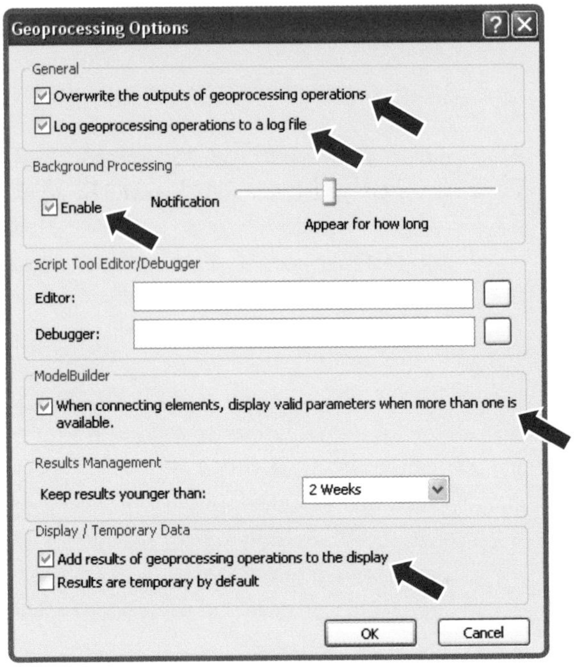

3 Click OK.

4 On the main menu, click Geoprocessing, Environments, Raster Analysis, and Select As Specified Below for Cell Size. Type **150** for the specification, select Pittsburgh for the Mask, and click OK.

5 Save the map document in \ESRIPress\GIST1\MyExercises\Chapter11\.

Standardize input variables

Here you will calculate the z-score in an attribute table of one of the input feature classes. To save time, the other three variables already have z-scores ready for use.

1 Right-click AllCoBlocks in the TOC and click Open Attribute Table.

2 Scroll to the right, right-click FHHChld, and click Statistics. It is convenient to copy and paste the statistics to Notepad and then copy and paste them later to the field calculator that you will use.

3 On your desktop, click Start, All Programs, Accessories, Notepad.

4 Select all of the statistics in the Statistics of AllCoBlocks window, pres Ctrl+C, click inside the Notepad window, and press Ctrl+V.

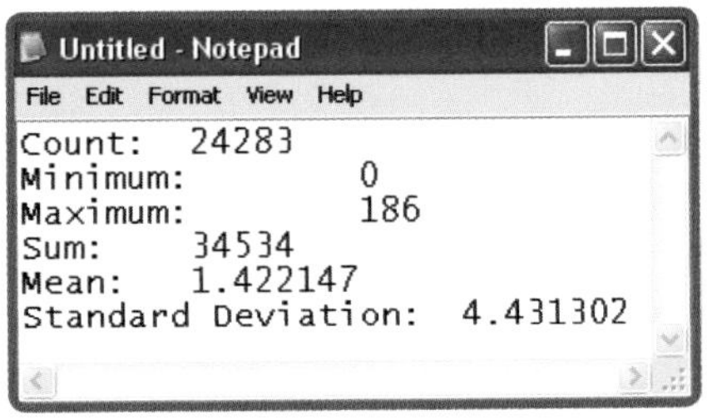

5 Close the Statistics of AllCoBlocks window, click the Table Options button in the Table window, click Add Field, type **ZFHHChld** for Name, select Float for Type, and click OK.

6 Right-click the column heading for ZFHHChld, click Field Calculator, and create the following expression in the bottom panel of the Field Calculator window by copying and pasting from your Notepad window,:

([FHHChld] – 1.422147) / 4.431302

7 Click OK. To the right are the first six values for the calculated z-scores.

FHHChld	ZFHHChld
0	-0.320932
0	-0.320932
0	-0.320932
2	0.130403
4	0.581737
1	-0.095265

8 Close the table and Notepad.

Create a new toolbox and model

1 Click Windows and Catalog, and expand Home – Chapter 11 in the folder/file tree.

2 Right-click Home – Chapter 11, click New and Toolbox, and rename the new toolbox **UnweightedIndices.tbx**.

3 Right-click UnweightedIndices.tbx and click New, Model.

4 In the Model window, click Model, Model Properties.

5 On the General tab, for Name type **PovertyIndex** (no spaces allowed), for Label type **Poverty Index**, click OK, and hide the Catalog window.

Create a kernel density layer for an input

The next task is to create kernel density layers for the four inputs using the z-scores. After you create model elements for one kernel density layer, you can easily copy it and make adjustments for the remaining three.

1 Click Windows, Search.

2 In the Search window, type **Kernel Density** in the search text box, press Enter, and drag Kernel Density to the Poverty Index model window and drop it in.

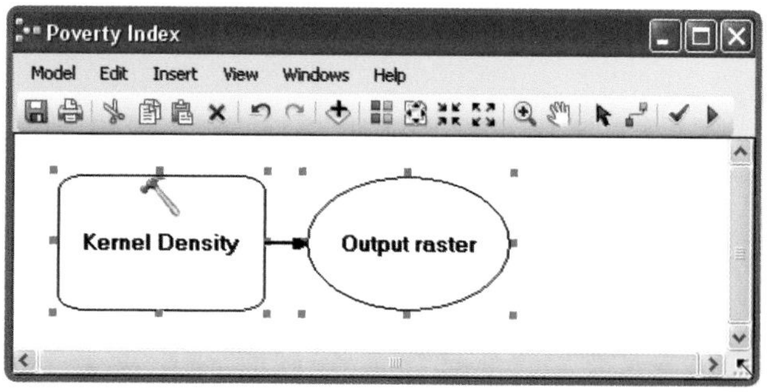

3 In the model, right-click Kernel Density and click Open.

4 Type or make the selections as shown in the image at the right.

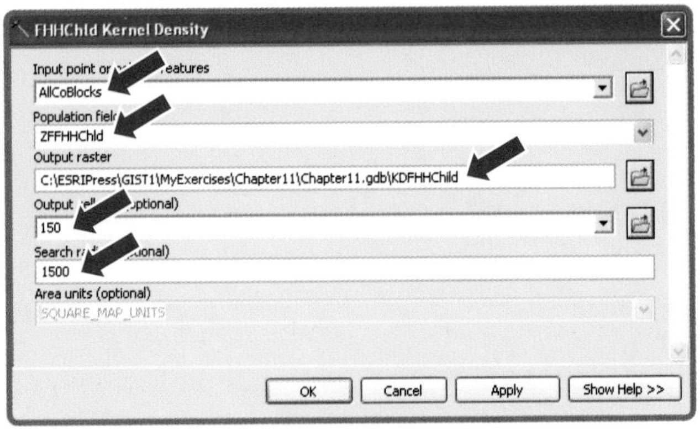

5 Click OK.

6 Right-click the Kernel Density tool element, click Rename, and change the name to **FHHChld Kernel Density**.

7 Right-click FHHChld Kernel Density and click Run.

8 Right-click the KDFHHChld and click Add to Display.

YOUR TURN

Resymbolize the new layer using the Classified method with 1/4 standard deviations and the color ramp that runs from blue to yellow to red. Turn off the point feature layers. The result is as follows:

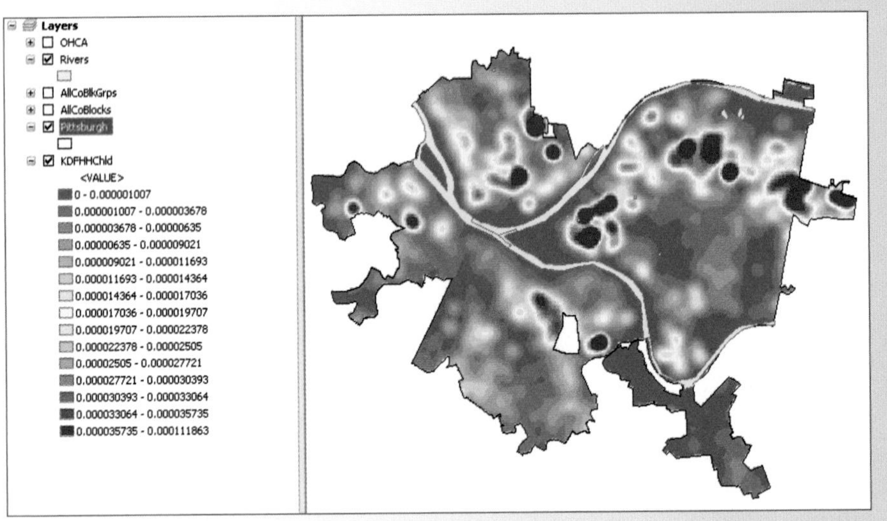

Create a kernel density layer for a second input

You can reuse the model elements you just built. While blocks work very well with a search radius of 1,500 feet, there are fewer block groups (the remaining three poverty inputs are at the block group level), so you need a larger search radius of 3,000 feet for them.

1. In the Model window, right-click FHHChld Kernel Density and click Copy.
2. Click Edit, Paste.
3. Right-click the new FHHChld Kernel Density 2 model element and rename it **NoHighSchKernelDensity**.
4. Right-click NoHighSch Kernel Density and click Open. Ignore the error messages. You will make changes that eliminate them.
5. Type or make the selections as shown in the image.
6. Click OK.
7. Right-click KDFHHChld(2) and rename it **NoHighSch**.
8. Right-click NoHighSch Kernel Density, click Run, and resymbolize as you like.

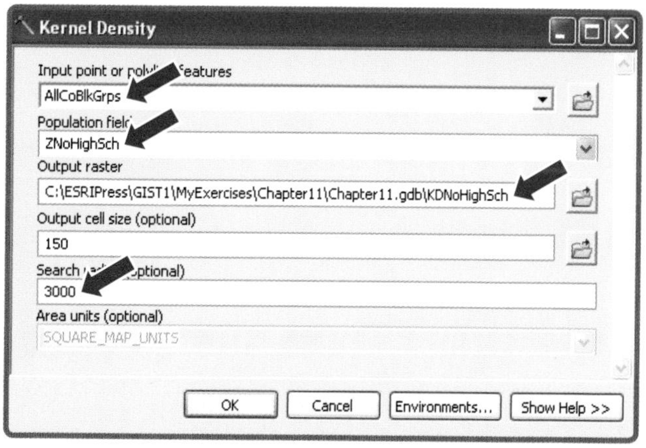

YOUR TURN

Copy and paste the NoHighSch Kernel Density model element two times to use block group attributes ZMaleUnem and ZPoverty to create two new raster layers. See the resulting partial model at right for element names that you need to use. Then run each of the two new model elements and resymbolize resulting map layers. Examine each of the four raster maps. You will see that they have overlapping but different patterns. The index will combine these patterns into a single, overall pattern. Resize and rearrange model elements. Run the three processes you just created. Resymbolize each new layer using standard deviations and color ramps of your choice. Notice that the process model elements acquire drop shadows in the model window after you run them. To reset the model so that you can run it again, if needed, click Model, Validate Entire Model. ModelBuilder removes the drop shadows. Save your model.

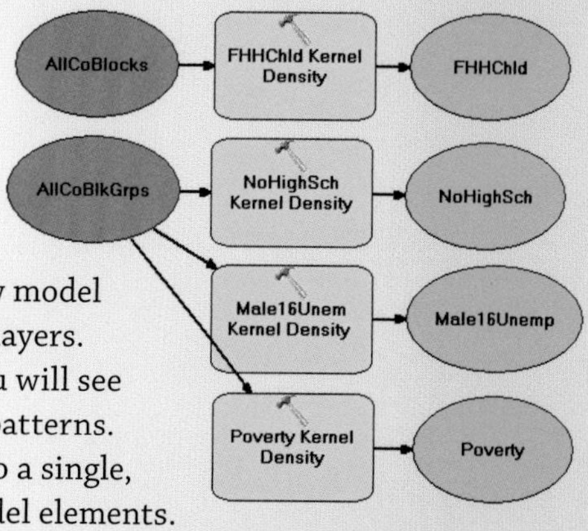

Average kernel density maps

1. Type **Weighted Sum** in the search text box and press Enter.
2. Drag the Weighted Sum link to your model, to the right of the kernel density outputs, and drop it in.
3. Right-click Weighted Sum, click Open, and type or make selections as shown in the image to the right.

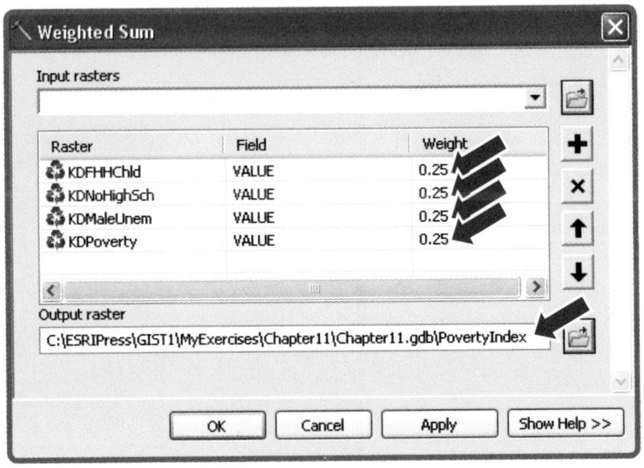

4. Click OK and use the following model to complete renaming model elements.

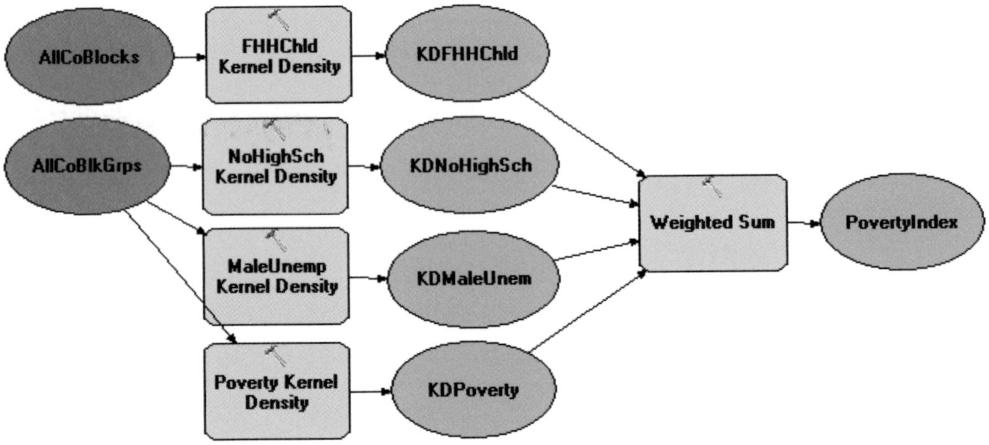

5 Run the Weighted Sum process and resymbolize the resulting PovertyIndex using standard deviations and the green-to-yellow-to-red color ramp.

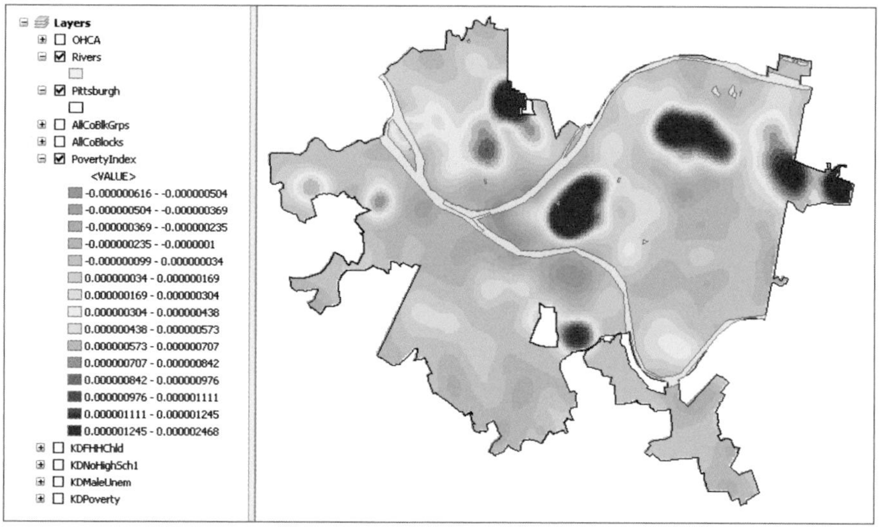

6 Save your model.

Create poverty contour

Suppose that after consideration, policy analysts wish to use the poverty index density of 0.0000009 or higher to define poverty. Next, you will create a feature class that has the contour line for that index "elevation."

1 In the search text box, type **Contour List** and click the Contour List link.

2 Type or make the selections as shown.

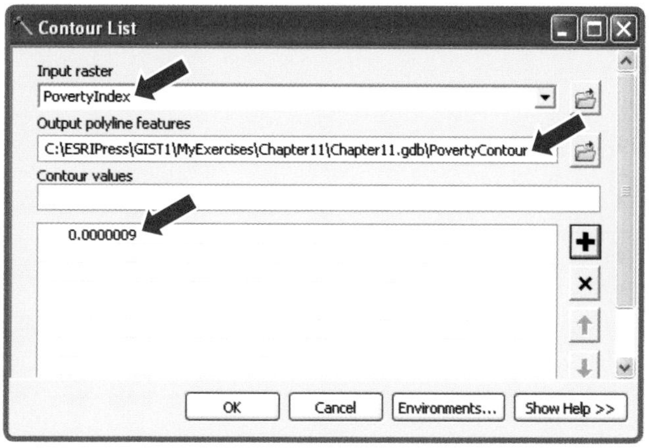

3 **Click OK.** You now have a set of polygons, shown with thick black outlines, that explicitly define poverty areas and can be used for many policy purposes.

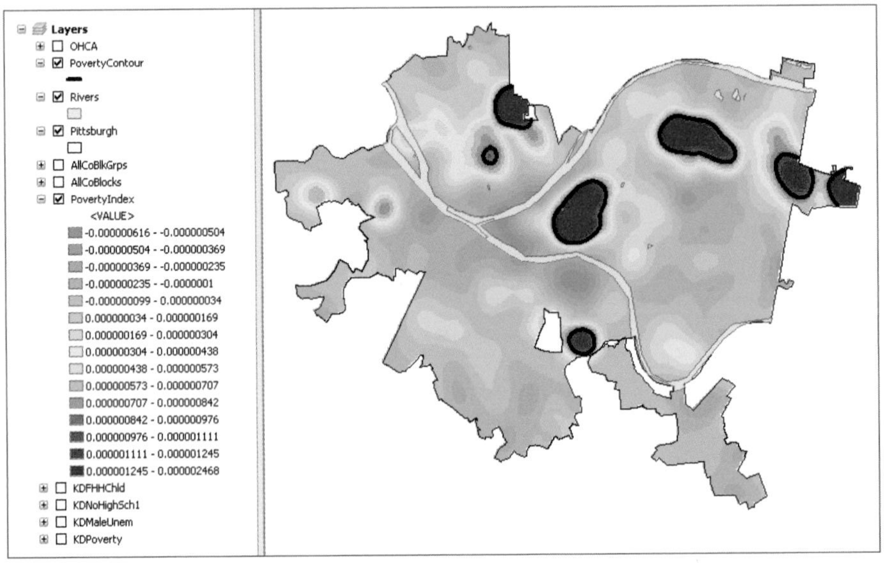

4 **Save the map document and close ArcMap.**

Assignment 11-1

Create hillshade for suburbs

The first ring of suburbs around urban areas are good areas for revitalization in the future as suburban homeowners attempt to downsize houses and relocate closer to work. The houses in these areas tend to be relatively small but well constructed, though in need of renovation.

This assignment has you choose a subset of municipalities in Allegheny County comprising the first ring of suburbs around Pittsburgh to display land use with hillshade. You will also display school locations for the suburbs. The resulting map document and layers provide a good starting point for redevelopment work.

Start with the following:

- \ESRIPress\GIST1\Data\AlleghenyCounty.gdb\Munic—polygon layer for municipalities in Allegheny County
- \ESRIPress\GIST1\Data\AlleghenyCounty.gdb\CountySchools—XY data file that has names of schools and (x,y) point coordinates in Pennsylvania South State Plane 1983 projection
- \ESRIPress\GIST1\Data\SpatialAnalyst\SpatialAnalyst.gdb\Pittsburgh—boundary polygon for Pittsburgh
- \ESRIPress\GIST1\Data\SpatialAnalyst\LandUse\28910720—land use for Allegheny County
- \ESRIPress\GIST1\Data\SpatialAnalyst\SpatialAnalyst.gdb\DEM—digital elevation model for Allegheny County
- \ESRIPress\GIST1\Data\SpatialAnalyst\LandUse.lyr—layer file for rendering land-use raster dataset

Preprocess vector layers

Create a map document called **\ESRIPress\GIST1\MyAssignments\Chapter11\Assignment 11-1YourName.mxd** with each of the above layers added. Add the municipalities first, so that your data frame inherits that layer's projection, which is the local 1983 State Plane projection. Add county schools as a XY layer. Turn off Pittsburgh to simplify the next step, in which you will create a ring of suburbs. Create a new file geodatabase called **\ESRIPress\GIST1\MyAssignments \Chapter11\Assignment11-1YourName.gdb** and add all new layers that you create to it.

Define the first ring of suburbs as those within one mile of Pittsburgh, but not including Pittsburgh in Munic. Start by making Munic the only selectable layer. Then use Selection, Select By Location, and select municipalities that are within a distance of one mile of Pittsburgh. Then use the Select Features tool, hold down your Shift key, and click inside the Pittsburgh polygon in the Munic layer to deselect it. Finally, right-click Munic; click Data, Export data to create Suburbs; and add it to your map document. Now select schools that intersect with suburbs and create **SuburbanSchools**.

11-1 11-2 11-3 11-4 11-5 11-6 A11-1 A11-2

Process raster layers

Using Suburbs as the mask and cell size of 50, extract a raster from LandUse called LandUseSub and import LandUse.lyr for symbolization. Create a hillshade from DEM called **HillshadeSub**. Make LandUseSub transparent, move it above the hillshade, turn off unneeded layers, and display suburban schools with the shaded land-use layer. Housing will be in the red, developed areas.

WHAT TO TURN IN

If your work is to be graded, turn in the following files:

ArcMap document: \ESRIPress\GIST1\MyAssignments\Chapter11 \Assignment11-1YourName.mxd

File geodatabase: \ESRIPress\GIST1\MyAssignments\Chapter11 \Assignment11-1YourName.gdb

If instructed to do so, instead of the above individual files, turn in a compressed file, **Assignment11-1YourName.zip**, with all files included. Do not include path information in the compressed file.

Assignment 11-2

Determine heart attack fatalities outside of hospitals in Mount Lebanon by gender

Unfortunately, females have more fatal heart attacks outside of hospitals than males, perhaps because symptoms of heart attacks in females are less well known than those for males. Heart attacks outside of hospitals are roughly 1.5 per thousand for males aged 35 to 74 and 2.3 per thousand for females in the same age range. In this assignment you will create two density map layers—one for males and one for females—using these incidence rates for the municipality of Mount Lebanon in Allegheny County. You will do all raster processing using Spatial Analysis tools in a model.

Start with the following:

- \ESRIPress\GIST1\Data\AlleghenyCounty.gdb\Munic—polygon layer for municipalities in Allegheny County
- \ESRIPress\GIST1\Data\SpatialAnalyst\SpatialAnalyst.gdb\AllCoBlocks—point layer for census block centroids in Allegheny County

In ArcMap create a map document called **\ESRIPress\GIST1\MyAssignments\Chapter11 \Assignment11-2YourName.mxd** with the above layers added. Create a file geodatabase called **\ESRIPress\GIST1\MyAssignments\Chapter11\Assignment11-2YourName.gdb** and save all new layers and other files that you create.

Select the Mount Lebanon polygon from Munic and export it as **MtLebanon**. Extract the Mount Lebanon blocks from AllCoBlocks and save them as **MtLebBlocks**. Add floating point fields to the attribute table for MtLebBlocks: MMortinc = 0.0015 × [Male35T74] for the annual number of heart-attack fatalities for males aged 35 to 74 and FMortInc = 0.0023 × [Fem35T74] for females aged 35 to 74.

Create kernel density map layers for MMortinc and FMortinc using MtLebanon as the mask and with a cell size of 100 and search radius of 1,500 square feet. Give the outputs descriptive names, add them to the map, and apply the same symbology scheme to both. Symbolize the kernel density map for females first and then import that symbolization for the male map.

WHAT TO TURN IN

If your work is to be graded, turn in the following files:

ArcMap document: \ESRIPress\GIST1\MyAssigments\Chapter11 \Assignment11-2YourName.mxd

File geodatabase: \ESRIPress\GIST1\MyAssigments\Chapter11 \Assignment11-2YourName.gdb

If instructed to do so, instead of the above individual files, turn in a compressed file, **Assignment11-2YourName.zip**, with all files included. Do not include path information in the compressed file.

Appendix A

Task index

Software tool/concept, **tutorial(s) in which it appears**

Appendix B

Data source credits

Chapter 1 data sources include

\ESRIPress\GIST1\Data\UnitedStates.gdb\USStates, from ESRI Data & Maps, 2007, courtesy of ArcUSA, U.S. Census, ESRI(Pop2005 field).

\ESRIPress\GIST1\Data\UnitedStates.gdb\USCities, from ESRI Data & Maps, 2007, courtesy of U.S. Census.

\ESRIPress\GIST1\Data\UnitedStates.gdb\COCounties, ESRI Data & Maps, 2007, courtesy of ArcUSA, U.S. Census, ESRI(Pop2005 field).

\ESRIPress\GIST1\Data\UnitedStates.gdb\COStreets, ESRI Data & Maps, 2007, courtesy of ArcUSA, U.S. Census, ESRI(Pop2005 field).

\ESRIPress\GIST1\Data\Pittsburgh\MidHill.gdb\Streets, courtesy of the City of Pittsburgh, Department of City Planning.

\ESRIPress\GIST1\Data\Pittsburgh\MidHill.gdb\Curbs, courtesy of the City of Pittsburgh, Department of City Planning.

\ESRIPress\GIST1\Data\Pittsburgh\MidHill.gdb\Buildings, courtesy of the City of Pittsburgh, Department of City Planning.

\ESRIPress\GIST1\Data\Pittsburgh\MidHill.gdb\CADCalls, courtesy of the City of Pittsburgh, Department of City Planning.

Chapter 2 data sources include

\ESRIPress\GIST1\Data\UnitedStates.gdb\USStates, from ESRI Data & Maps, 2007, courtesy of ArcUSA, U.S. Census, ESRI(Pop2005 field).

\ESRIPress\GIST1\Data\UnitedStates.gdb\USCounties, from ESRI Data & Maps, 2007, courtesy of ArcUSA, U.S. Census, ESRI(Pop2005 field).

\ESRIPress\GIST1\Data\UnitedStates.gdb\UTTracts, from ESRI Data & Maps, 2007, courtesy of Tele Atlas, U.S. Census, ESRI(Pop2005 field).

\ESRIPress\GIST1\Data\UnitedStates.gdb\NVTracts, from ESRI Data & Maps, 2007, courtesy of Tele Atlas, U.S. Census, ESRI(Pop2005 field).

\ESRIPress\GIST1\Data\UnitedStates.gdb\USCities, from ESRI Data & Maps, 2007, courtesy of U.S. Census.

\ESRIPress\GIST1\Data\UnitedStates.gdb\PACounties, from ESRI Data & Maps, 2007, courtesy of ArcUSA, U.S. Census, ESRI(Pop2005 field).

\ESRIPress\GIST1\Data\UnitedStates.gdb\PACities, from ESRI Data & Maps, 2007, courtesy of National Atlas of the United States.

\ESRIPress\GIST1\Data\Pittsburgh\City.gdb\Neighborhoods, courtesy of the City of Pittsburgh, Department of City Planning.

\ESRIPress\GIST1\Data\Pittsburgh\City.gdb\Schools, courtesy of the City of Pittsburgh, Department of City Planning.

\ESRIPress\GIST1\Data\UnitedStates.gdb\PATracts, from ESRI Data & Maps, 2007, courtesy of Tele Atlas, U.S. Census, ESRI(Pop2005 field).

\ESRIPress\GIST1\Data\Pittsburgh\City.gdb\BlockGroups, from ESR I Data & Maps, 2007, courtesy of U.S. Census.

Chapter 3 data sources include

\ESRIPress\GIST1\Data\UnitedStates.gdb\USStates, from ESRI Data & Maps, 2007, courtesy of ArcUSA, U.S. Census, ESRI(Pop2005 field).

\ESRIPress\GIST1\Data\UnitedStates.gdb\USCities, from ESRI Data & Maps, 2007, courtesy of U.S. Census.

\ESRIPress\GIST1\Data\AlleghenyCounty.gdb\Parks, courtesy of Southwestern Pennsylvania Commission.

\ESRIPress\GIST1\Data\AlleghenyCounty.gdb\Munic, courtesy of Southwestern Pennsylvania Commission.

\ESRIPress\GIST1\Data\AlleghenyCounty.gdb\Rivers, courtesy of Southwestern Pennsylvania Commission.

\ESRIPress\GIST1\Data\AlleghenyCounty.gdb\CountySchools, from ESR I Data & Maps, 2007, courtesy of U.S. Census.

\ESRIPress\GIST1\Data\Pittsburgh\MidHill.gdb\Curbs, courtesy of the City of Pittsburgh, Department of City Planning.

\ESRIPress\GIST1\Data\Pittsburgh\MidHill.gdb\Streets, courtesy of the City of Pittsburgh, Department of City Planning.

\ESRIPress\GIST1\Data\Pittsburgh\MidHill.gdb\AutoTheftCrimeSeries, courtesy of the City of Pittsburgh, Department of City Planning.

\ESRIPress\GIST1\Data\Pittsburgh\MidHill.gdb\MiddleHill, courtesy of the City of Pittsburgh, Department of City Planning.

\ESRIPress\GIST1\Data\Pittsburgh\MidHill.gdb\CADCalls, courtesy of the City of Pittsburgh, Department of City Planning.

\ESRIPress\GIST1\Data\Pittsburgh\CentralBusinessDistrict.gdb\CBDOutline, courtesy of the City of Pittsburgh, Department of City Planning.

\ESRIPress\GIST1\Data\Pittsburgh\CentralBusinessDistrict.gdb\CBDBLDG, courtesy of the City of Pittsburgh, Department of City Planning.

\ESRIPress\GIST1\Data\Pittsburgh\CentralBusinessDistrict.gdb\CBDStreets, courtesy of the City of Pittsburgh, Department of City Planning.

\ESRIPress\GIST1\Data\Pittsburgh\CentralBusinessDistrict.gdb\Histpnts, courtesy of the City of Pittsburgh, Department of City Planning.

\ESRIPress\GIST1\Data\Pittsburgh\ CentralBusinessDistrict.gdb\Histsite, courtesy of the City of Pittsburgh, Department of City Planning.

\ESRIPress\GIST1\Data\UnitedStates.gdb\CAOrangeCountyTracts, from ESRI Data & Maps, 2007, courtesy of Tele Atlas, U.S. Census, ESRI(Pop2005 field).

Chapter 4 data sources include

\ESRIPress\GIST1\Data\MaricopaCounty\tgr04013ccd00.shp, courtesy of the U.S. Census Bureau TIGER.

\ESRIPress\GIST1\Data\MaricopaCounty\tgr04013trt00.shp, courtesy of the U.S. Census Bureau TIGER.

\ESRIPress\GIST1\Data\MaricopaCounty\CensusDat.dbf, courtesy of the U.S. Census Bureau.

\ESRIPress\GIST1\Data\RochesterNY \RochesterPolice.gdb\carbeats, courtesy of the Rochester, New York, Police Department.

\ESRIPress\GIST1\Data\RochesterNY \RochesterPolice.gdb\business, courtesy of the Rochester, New York, Police Department.

\ESRIPress\GIST1\Data\AlleghenyCounty.gdb\Munic, courtesy of Southwestern Pennsylvania Commission.

\ESRIPress\GIST1\Data\AlleghenyCounty.gdb\Rivers, courtesy of Southwestern Pennsylvania Commission.

\ESRIPress\GIST1\Data\Pittsburgh\City.gdb\PghTracts, courtesy of Tele Atlas, U.S. Census, ESRI(Pop2005 field).

\ESRIPress\GIST1\Data\Pittsburgh\City.gdb\Schools, courtesy of the City of Pittsburgh, Department of City Planning.

Chapter 5 data sources include

Screen capture of `www.esri.com` home page, from ESRI Data & Maps, 2008, courtesy of the U.S. Census.

Screen capture of `www.esri.com/data/download/census2000_tigerline/index.html`, from ESRI Data & Maps, 2000, courtesy of U.S. Census.

Screen captures of `www.census.gov`, courtesy of the U.S. Census. All U.S. Census Bureau materials, regardless of the media, are entirely in the public domain. There are no user fees, site licenses, or any special agreements, etc., for the public or private use, and/or reuse of any census title. As a tax-funded product, it is all in the public record.

Screen captures of `http://seamless.usgs.gov`, courtesy of the USGS, regardless of the media, are entirely in the public domain. There are no user fees, site licenses, or any special agreements, etc., for the public or private use, and/or reuse of any census title. As a tax-funded product, it is all in the public record.

\ESRIPress\GIST1\Data\AlleghenyCounty.gdb\Tracts, from ESRI Data & Maps, 2007, courtesy of U.S. Census.

\ESRIPress\GIST1\Data\AlleghenyCounty.gdb\Munic, from ESRI Data & Maps, 2007, courtesy of Southwestern Pennsylvania Commission.

\ESRIPress\GIST1\Data\World.gdb\Country, from ESRI Data & Maps, 2004, courtesy of ArcWorld Supplement.

\ESRIPress\GIST1\Data\World.gdb\Ocean, from ESRI Data & Maps, courtesy of ESRI.

\ESRIPress\GIST1\Data\UnitedStates.gdb\USStates, from ESRI Data & Maps, 2007, courtesy of ArcUSA, U.S. Census, ESRI(Pop2005 field).

\ESRIPress\GIST1\Data\Pittsburgh\EastLiberty\Building, courtesy of the City of Pittsburgh, Department of City Planning.

\ESRIPress\GIST1\Data\CMUCampus\CampusMap.dwg, courtesy of the Carnegie Mellon University.

\ESRIPress\GIST1\Data\Datafiles\Earthquakes.dbf, from ESRI Data & Maps, 2007, courtesy of National Atlas of the United States, USGS.

\ESRIPress\GIST1\Data\UnitedStates.gdb\CACounties, from ESRI Data & Maps, 2007, courtesy of ArcUSA, U.S. Census, ESRI(Pop2005 field).

\ESRIPress\GIST1\Data\UnitedStates.gdb\NYManhattanCounty, from ESR I Data & Maps, 2007, courtesy of U.S. Census.

\ESRIPress\GIST1\Data\MaricopaCounty\CountySchools.dbf, from ESR I Data & Maps, 2007, courtesy of U.S. Census.

Chapter 6 data sources include

\ESRIPress\GIST1\Data\Pittsburgh\MidHill.gdb\MiddleHill, courtesy of the City of Pittsburgh, Department of City Planning.

\ESRIPress\GIST1\Data\Pittsburgh\MidHill.gdb\Streets, courtesy of the City of Pittsburgh, Department of City Planning.

\ESRIPress\GIST1\Data\Pittsburgh\MidHill.gdb\CommercialProperties, courtesy of the City of Pittsburgh, Department of City Planning.

\ESRIPress\GIST1\Data\Pittsburgh\MidHill.gdb\Buildings, courtesy of the City of Pittsburgh, Department of City Planning.

\ESRIPress\GIST1\Data\Pittsburgh\Zone2.gdb\streets, courtesy of the City of Pittsburgh, Department of City Planning.

\ESRIPress\GIST1\Data\Pittsburgh\Zone2.gdb\zone2, courtesy of the City of Pittsburgh, Department of City Planning.

\ESRIPress\GIST1\Data\CMUCampus\Hbh.shp, courtesy of Carnegie Mellon University.

\ESRIPress\GIST1\Data\CMUCampus\25_45.tif, courtesy of the Southwestern Pennsylvania Commission.

\ESRIPress\GIST1\Data\CMUCampus\26_45.tif, courtesy of the Southwestern Pennsylvania Commission.

\ESRIPress\GIST1\Data\CMUCampus\CampusMap.dwg, courtesy of the Carnegie Mellon University.

Chapter 7 data sources include

\ESRIPress\GIST1\Data\UnitedStates.gdb\PAZip, courtesy of Tele Atlas, ESRI(Pop2005 field).

\ESRIPress\GIST1\Data\Flux\FLUXEvent.mdb\tAttendees, courtesy of FLUX

\ESRIPress\GIST1\Data\Pittsburgh\City.gdb\Neighborhoods, courtesy of the City of Pittsburgh, Department of City Planning.

\ESRIPress\GIST1\Data\Pittsburgh\City.gdb\PghStreets, courtesy of the City of Pittsburgh, Department of City Planning.

\ESRIPress\GIST1\Data\Pittsburgh\CentralBusinessDistrict.gdb\CBDStreets, courtesy of the City of Pittsburgh, Department of City Planning.

\ESRIPress\GIST1\Data\Pittsburgh\CentralBusinessDistrict.gdb\CBDOutline, courtesy of the City of Pittsburgh, Department of City Planning.

\ESRIPress\GIST1\Data\Pittsburgh\CentralBusinessDistrict.gdb\Clients.dbf, courtesy of Kristen Kurland, Carnegie Mellon University.

\ESRIPress\GIST1\Data\UnitedStates.gdb\HHWZipCodes, courtesy of the Pennsylvania Resources Council.

\ESRIPress\GIST1\Data\Pittsburgh\CentralBusinessDistrict.gdb\BldgAliasNames, courtesy of Wil Gorr, Carnegie Mellon University.

\ESRIPress\GIST1\Data\UnitedStates.gdb\PAZip, courtesy of Tele Atlas, ESRI(Pop2005 field).

\ESRIPress\GIST1\Data\UnitedStates.gdb\PACounties, from ESRI Data & Maps, 2007, courtesy of ArcUSA, U.S. Census, ESRI(Pop2005 field).

\ESRIPress\GIST1\Data\Pittsburgh\ForeignBusinesses.dbf, courtesy of Kristen Kurland, Carnegie Mellon University

Chapter 8 data sources include

\ESRIPress\GIST1\Data\UnitedStates.gdb\NYBoroughs, from ESRI Data & Maps, 2007, courtesy of U.S. Census.

\ESRIPress\GIST1\Data\UnitedStates.gdb\NYMetroRoads, from ESRI Data & Maps, 2004, courtesy of U.S. Census.

\ESRIPress\GIST1\Data\UnitedStates.gdb\NYMetroZIP, from ESRI Data & Maps, 2004, courtesy of GDT, ESRI BIS(Pop2003 field).

\ESRIPress\GIST1\Data\UnitedStates.gdb\NYBronxCountyWater, courtesy of the U.S. Census Bureau TIGER.

\ESRIPress\GIST1\Data\UnitedStates.gdb\NYKingsCountyWater, courtesy of the U.S. Census Bureau TIGER.

\ESRIPress\GIST1\Data\UnitedStates.gdb\NYNewYorkCountyWater, courtesy of the U.S. Census Bureau TIGER.

\ESRIPress\GIST1\Data\UnitedStates.gdb\NYQueensCountyWater, courtesy of the U.S. Census Bureau TIGER.

\ESRIPress\GIST1\Data\UnitedStates.gdb\NYRichmondCountyWater, courtesy of the U.S. Census Bureau TIGER.

\ESRIPress\GIST1\Data\UnitedStates.gdb\NYWater, courtesy of the U.S. Census Bureau TIGER.

\ESRIPress\GIST1\Data\UnitedStates.gdb\NYManhattanZipCodes, from ESRI Data & Maps, 2004, courtesy of GDT, ESRI BIS(Pop2003 field).

\ESRIPress\GIST1\Data\UnitedStates.gdb\NYManhattanTracts, from ESRI Data & Maps, 2004, courtesy of GDT, ESRI BIS(Pop2003 field).

\ESRIPress\GIST1\Data\AlleghenyCounty.gdb\Munic, courtesy of Southwestern Pennsylvania Commission.

\ESRIPress\GIST1\Data\AlleghenyCounty.gdb\Tracts, courtesy of the U.S. Census Bureau TIGER.

\ESRIPress\GIST1\Data\Pittsburgh\City.gdb\Neighborhoods, courtesy of the City of Pittsburgh, Department of City Planning.

\ESRIPress\GIST1\Data\UnitedStates.gdb\COCounties, from ESRI Data & Maps, 2007, courtesy of ArcUSA, U.S. Census, ESRI(Pop2005 field).

\ESRIPress\GIST1\Data\UnitedStates.gdb\COStreets, from ESRI Data & Maps, 2007, courtesy of ArcUSA, U.S. Census, ESRI(Pop2005 field).

\ESRIPress\GIST1\Data\UnitedStates.gdb\COStreets2, from ESRI Data & Maps, 2007, courtesy of ArcUSA, U.S. Census, ESRI(Pop2005 field).

\ESRIPress\GIST1\Data\UnitedStates.gdb\COUrban, from ESRI Data & Maps, 2007, courtesy of ArcUSA, U.S. Census, ESRI(Pop2005 field).

\ESRIPress\GIST1\Data\UnitedStates.gdb\COUrban2, from ESRI Data & Maps, 2007, courtesy of ArcUSA, U.S. Census, ESRI(Pop2005 field).

\ESRIPress\GIST1\Data\UnitedStates.gdb\USCities_dtl, from ESRI Data & Maps, 2007, courtesy of U.S. Census.

\ESRIPress\GIST1\Data\Pittsburgh\EastLiberty\Parcel, courtesy of the City of Pittsburgh, Department of City Planning.

\ESRIPress\GIST1\Data\Pittsburgh\EastLiberty\EastLib, courtesy of the City of Pittsburgh, Department of City Planning.

\ESRIPress\GIST1\Data\Pittsburgh\City.gdb\Pittsburgh, courtesy of the Southwestern Pennsylvania Commission.

Chapter 9 data sources include

\ESRIPress\GIST1\Data\RochesterNY\LakePrecinct.gdb\lakebars, courtesy of the Rochester, New York, Police Department.

\ESRIPress\GIST1\Data\RochesterNY\LakePrecinct.gdb\lakeassualts, courtesy of the Rochester, New York, Police Department.

\ESRIPress\GIST1\Data\RochesterNY\LakePrecinct.gdb\LakeBlockCentroids, from ESRI Data & Maps, 2004, courtesy of U.S. Census.

\ESRIPress\GIST1\Data\RochesterNY\LakePrecinct.gdb\LakeBusinesses, courtesy of InfoUSA.

\ESRIPress\GIST1\Data\RochesterNY\LakePrecinct.gdb\lakecarbeats, courtesy of the Rochester, New York, Police Department.

\ESRIPress\GIST1\Data\RochesterNY\LakePrecinct.gdb\lakeprecinct, courtesy of the Rochester, New York, Police Department.

\ESRIPress\GIST1\Data\RochesterNY\LakePrecinct.gdb\lakestreets, courtesy of the Rochester, New York, Police Department.

\ESRIPress\GIST1\Data\RochesterNY\LakePrecinct.gdb\lakestrct2000, courtesy of the U.S. Census Bureau TIGER.

\ESRIPress\GIST1\Data\UnitedStates.gdb\CACounties, from ESRI Data & Maps, 2007, courtesy of ArcUSA, U.S. Census, ESRI(Pop2005 field).

\ESRIPress\GIST1\Data\Datafiles\Earthquakes.dbf, from ESRI Data & Maps, 2004, courtesy of National Atlas of the United States, USGS.

\ESRIPress\GIST1\Data\UnitedStates.gdb\USCities_dtl, from ESRI Data & Maps, 2007, courtesy of U.S. Census.

\ESRIPress\GIST1\Data\Pittsburgh\15222.gdb\Streets, courtesy of the City of Pittsburgh, Department of City Planning.

ESRIPress\GIST1\Data\Pittsburgh\15222.gdb\Curbs, courtesy of the City of Pittsburgh, Department of City Planning.

\ESRIPress\GIST1\Data\Pittsburgh\15222.gdb\Restaurants, courtesy of Wil Gorr, Carnegie Mellon University.

\ESRIPress\GIST1\Data\AlleghenyCounty.gdb\Rivers, courtesy of Southwestern Pennsylvania Commission.

\ESRIPress\GIST1\Data\RochesterNY\HouseholdIncome.xls, courtesy of the U.S. Census Bureau American Factfinder.

\ESRIPress\GIST1\Data\RochesterNY\LakePrecinct.gdb\LakeBlockGroupCentroids, from ESRI Data & Maps, 2004, courtesy of U.S. Census.

\ESRIPress\GIST1\Data\RochesterNY\LakePrecinct.gdb\LakeBlockGroups, from ESRI Data & Maps, 2004, courtesy of U.S. Census.

Chapter 10 data sources include

\ESRIPress\GIST1\Data\3DAnalyst.gdb\Bldgs, courtesy of the City of Pittsburgh, Department of City Planning.

\ESRIPress\GIST1\Data\3DAnalyst.gdb\Curbs, courtesy of the City of Pittsburgh, Department of City Planning.

\ESRIPress\GIST1\Data\3DAnalyst.gdb\Topo, courtesy of the City of Pittsburgh, Department of City Planning.

\ESRIPress\GIST1\Data\3DAnalyst.gdb\Trees, courtesy of Kristen Kurland, Carnegie Mellon University.

\ESRIPress\GIST1\Data\3DAnalyst.gdb\Vehicles, courtesy of Kristen Kurland, Carnegie Mellon University

\ESRIPress\GIST1\Data\AlleghenyCounty.gdb\Parks, courtesy of Southwestern Pennsylvania Commission.

\ESRIPress\GIST1\Data\AlleghenyCounty.gdb\Rivers, courtesy of Southwestern Pennsylvania Commission.

\ESRIPress\GIST1\Data\Pittsburgh\Phipps.gdb\Bldgs, courtesy of the City of Pittsburgh, Department of City Planning.

\ESRIPress\GIST1\Data\Pittsburgh\Phipps.gdb\Curbs, courtesy of the City of Pittsburgh, Department of City Planning.

\ESRIPress\GIST1\Data\Pittsburgh\Phipps.gdb\Topo, courtesy of the City of Pittsburgh, Department of City Planning.

\ESRIPress\GIST1\Data\CMUCampus\25_45.tif, courtesy of the Southwestern Pennsylvania Commission.

\ESRIPress\GIST1\Data\CMUCampus\26_45.tif, courtesy of the Southwestern Pennsylvania Commission.

\ESRIPress\GIST1\Data\AlleghenyCounty.gdb\Tracts, from ESR I Data & Maps, 2007, courtesy of U.S. Census.

\ESRIPress\GIST1\Data\AlleghenyCounty.gdb\CountySchools, from ESR I Data & Maps, 2007, courtesy of U.S. Census.

\ESRIPress\GIST1\Data\World.gdb\Country, from ESR I Data & Maps, 2004, courtesy of ArcWorld Supplement.

\ESRIPress\GIST1\Data\AlleghenyCounty.gdb\Parks, courtesy of Southwestern Pennsylvania Commission.

\ESRIPress\GIST1\Data\AlleghenyCounty.gdb\Rivers, courtesy of Southwestern Pennsylvania Commission.

\ESRIPress\GIST1\Data\Pittsburgh\ CentralBusinessDistrict.gdb\Histsite, courtesy of the City of Pittsburgh, Department of City Planning.

Screen captures of ArcGlobe,

Elevation (30m) – Source: USGS. The data is from the National Elevation Dataset (NED) produced by the United States Geological Survey (USGS).

Elevation (90m/1km) – Source: NASA, NGA, USGS. The data is from the National Elevation Dataset (NED) produced by the United States Geological Survey (USGS). Copyright:© 2009 ESRI, i-cubed, GeoEye elevation data includes 90m SRTM elevation data from NASA and NGA where it is available and 1km GTOPO30 data from the USGS elsewhere.

Imagery – Copyright:© 2009 ESRI, i-cubed, GeoEye. This globe presents low-resolution imagery for the world and high-resolution imagery for the United States and other metropolitan areas around the world. The globe includes NASA Blue Marble: Next Generation 500m resolution imagery at small scales (above 1:1,000,000), i-cubed 15m eSAT imagery at medium-to-large scales (down to 1:70,000) for the world, and USGS 15m Landsat imagery for Antarctica. It also includes 1m i-cubed Nationwide Select imagery for the continental United States, and GeoEye IKONOS 1m resolution imagery for Hawaii, parts of Alaska, and several hundred metropolitan areas around the world.

Boundaries and Places – Copyright:© 2009 ESRI, AND, TANA. The map was developed by ESRI using administrative and cities data from ESRI and AND Mapping for the world and Tele Atlas administrative, cities, and landmark data for North America and Europe.

Transportation – Copyright:© 2009 ESRI, AND, TANA. The map was developed by ESRI using ESRI highway data, National Geospatial-Intelligence Agency (NGA) airport data, AND road and railroad data for the world, and Tele Atlas Dynamap® and Multinet® street data for North America and Europe.

Chapter 11 data sources include

\ESRIPress\GIST1\Data\SpatialAnalyst\SpatialAnalyst.gdb\Zoning, courtesy of the City of Pittsburgh, Department of City Planning.

\ESRIPress\GIST1\Data\SpatialAnalyst\SpatialAnalyst.gdb\AllCoBlkGrps, courtesy of the U.S. Census TIGER.

\ESRIPress\GIST1\Data\SpatialAnalyst\SpatialAnalyst.gdb\AllCoBlks, courtesy of the U.S. Census Bureau.

\ESRIPress\GIST1\Data\SpatialAnalyst\SpatialAnalyst.gdb\OHCA, courtesy of Children's Hospital of Pittsburgh.

\ESRIPress\GIST1\Data\SpatialAnalyst\SpatialAnalyst.gdb\PennHills, courtesy of the Southwestern Pennsylvania Commission.

\ESRIPress\GIST1\Data\SpatialAnalyst\SpatialAnalyst.gdb\Pittsburgh, courtesy of the Southwestern Pennsylvania Commission.

\ESRIPress\GIST1\Data\SpatialAnalyst\SpatialAnalyst.gdb\Rivers, courtesy of the U.S. Census Bureau.

\ESRIPress\GIST1\Data\Pittsburgh\City.gdb\PghStreets, courtesy of the City of Pittsburgh, Department of City Planning.

\ESRIPress\GIST1\Data\SpatialAnalyst\SpatialAnalyst.gdb\ZoningCommercialBuffer, courtesy of the City of Pittsburgh, Department of City Planning.

\ESRIPress\GIST1\Data\SpatialAnalyst\SpatialAnalyst.gdb\DEM, courtesy of U.S. Geological Survey, Department of the Interior/USGS.

\ESRIPress\GIST1\Data\SpatialAnalyst\SpatialAnalyst.gdb\LandUse, image courtesy of U.S. Geological Survey, Department of the Interior/USGS.

\ESRIPress\GIST1\Data\AlleghenyCounty.gdb\Munic, courtesy of Southwestern Pennsylvania Commission.

\ESRIPress\GIST1\Data\AlleghenyCounty.gdb\CountySchools, from ESR I Data & Maps, 2007, courtesy of U.S. Census.

Appendix C

Data license agreement

Important:
Read carefully before opening the sealed media package

Environmental Systems Research Institute Inc. (ESRI) is willing to license the enclosed data and related materials to you only upon the condition that you accept all of the terms and conditions contained in this license agreement. Please read the terms and conditions carefully before opening the sealed media package. By opening the sealed media package, you are indicating your acceptance of the ESRI License Agreement. If you do not agree to the terms and conditions as stated, then ESRI is unwilling to license the data and related materials to you. In such event, you should return the media package with the seal unbroken and all other components to ESRI.

ESRI license agreement

This is a license agreement, and not an agreement for sale, between you (Licensee) and Environmental Systems Research Institute Inc. (ESRI). This ESRI License Agreement (Agreement) gives Licensee certain limited rights to use the data and related materials (Data and Related Materials). All rights not specifically granted in this Agreement are reserved to ESRI and its Licensors.

Reservation of Ownership and Grant of License: ESRI and its Licensors retain exclusive rights, title, and ownership to the copy of the Data and Related Materials licensed under this Agreement and, hereby, grant to Licensee a personal, nonexclusive, nontransferable, royalty-free, worldwide license to use the Data and Related Materials based on the terms and conditions of this Agreement. Licensee agrees to use reasonable effort to protect the Data and Related Materials from unauthorized use, reproduction, distribution, or publication.

Proprietary Rights and Copyright: Licensee acknowledges that the Data and Related Materials are proprietary and confidential property of ESRI and its Licensors and are protected by United States copyright laws and applicable international copyright treaties and/or conventions.

Permitted Uses: Licensee may install the Data and Related Materials onto permanent storage device(s) for Licensee's own internal use.

Licensee may make only one (1) copy of the original Data and Related Materials for archival purposes during the term of this Agreement unless the right to make additional copies is granted to Licensee in writing by ESRI.

Licensee may internally use the Data and Related Materials provided by ESRI for the stated purpose of GIS training and education.

Uses Not Permitted: Licensee shall not sell, rent, lease, sublicense, lend, assign, time-share, or transfer, in whole or in part, or provide unlicensed Third Parties access to the Data and Related Materials or portions of the Data and Related Materials, any updates, or Licensee's rights under this Agreement.

Licensee shall not remove or obscure any copyright or trademark notices of ESRI or its Licensors.

Term and Termination: The license granted to Licensee by this Agreement shall commence upon the acceptance of this Agreement and shall continue until such time that Licensee elects in writing to discontinue use of the Data or Related Materials and terminates this Agreement. The Agreement shall automatically terminate without notice if Licensee fails to comply with any provision of this Agreement. Licensee shall then return to ESRI the Data and Related Materials. The parties hereby agree that all provisions that operate to protect the rights of ESRI and its Licensors shall remain in force should breach occur.

Disclaimer of Warranty: The Data and Related Materials contained herein are provided "as-is," without warranty of any kind, either express or implied, including, but not limited to, the implied warranties of merchantability, fitness for a particular purpose, or noninfringement. ESRI does not warrant that the Data and Related Materials will meet Licensee's needs or expectations, that the use of the Data and Related Materials will be uninterrupted, or that all nonconformities, defects, or errors can or will be corrected. ESRI is not inviting reliance on the Data or Related Materials for commercial planning or analysis purposes, and Licensee should always check actual data.

Data Disclaimer: The Data used herein has been derived from actual spatial or tabular information. In some cases, ESRI has manipulated and applied certain assumptions, analyses, and opinions to the Data solely for educational training purposes. Assumptions, analyses, opinions applied, and actual outcomes may vary. Again, ESRI is not inviting reliance on this Data, and the Licensee should always verify actual Data and exercise their own professional judgment when interpreting any outcomes.

LIMITATION OF LIABILITY: ESRI SHALL NOT BE LIABLE FOR DIRECT, INDIRECT, SPECIAL, INCIDENTAL, OR CONSEQUENTIAL DAMAGES RELATED TO LICENSEE'S USE OF THE DATA AND RELATED MATERIALS, EVEN IF ESRI IS ADVISED OF THE POSSIBILITY OF SUCH DAMAGE.

No Implied Waivers: No failure or delay by ESRI or its Licensors in enforcing any right or remedy under this Agreement shall be construed as a waiver of any future or other exercise of such right or remedy by ESRI or its Licensors.

Order for Precedence: Any conflict between the terms of this Agreement and any FAR, DFAR, purchase order, or other terms shall be resolved in favor of the terms expressed in this Agreement, subject to the government's minimum rights unless agreed otherwise.

Export Regulation: Licensee acknowledges that this Agreement and the performance thereof are subject to compliance with any and all applicable United States laws, regulations, or orders relating to the export of data thereto. Licensee agrees to comply with all laws, regulations, and orders of the United States in regard to any export of such technical data.

Severability: If any provision(s) of this Agreement shall be held to be invalid, illegal, or unenforceable by a court or other tribunal of competent jurisdiction, the validity, legality, and enforceability of the remaining provisions shall not in any way be affected or impaired thereby.

Governing Law: This Agreement, entered into in the County of San Bernardino, shall be construed and enforced in accordance with and be governed by the laws of the United States of America and the State of California without reference to conflict of laws principles. The parties hereby consent to the personal jurisdiction of the courts of this county and waive their rights to change venue.

Entire Agreement: The parties agree that this Agreement constitutes the sole and entire agreement of the parties as to the matter set forth herein and supersedes any previous agreements, understandings, and arrangements between the parties relating hereto.

Appendix D

Installing the data and software

GIS Tutorial 1: Basic Workbook includes a DVD containing exercise and assignment data and a DVD containing a fully functioning 180-day trial version of ArcGIS Desktop 10 software (ArcEditor license level). If you already have a licensed copy of ArcGIS Desktop 10 installed on your computer (or have access to the software through a network), do not install the trial software. Use your licensed software to do the exercises in this book. If you have an older version of ArcGIS installed on your computer, you must uninstall it before you can install the software that is provided with this book.

.NET Framework 3.5 SP1 must be installed on your computer before you install ArcGIS Desktop 10. Some features of ArcGIS Desktop 10 require Microsoft Internet Explorer version 8.0. If you do not have Microsoft Internet Explorer version 8.0, you must install it before installing ArcGIS Desktop 10.

Installing the exercise data

Follow the steps below to install the exercise data.

1 Put the data DVD in your computer's DVD drive. A splash screen will appear.

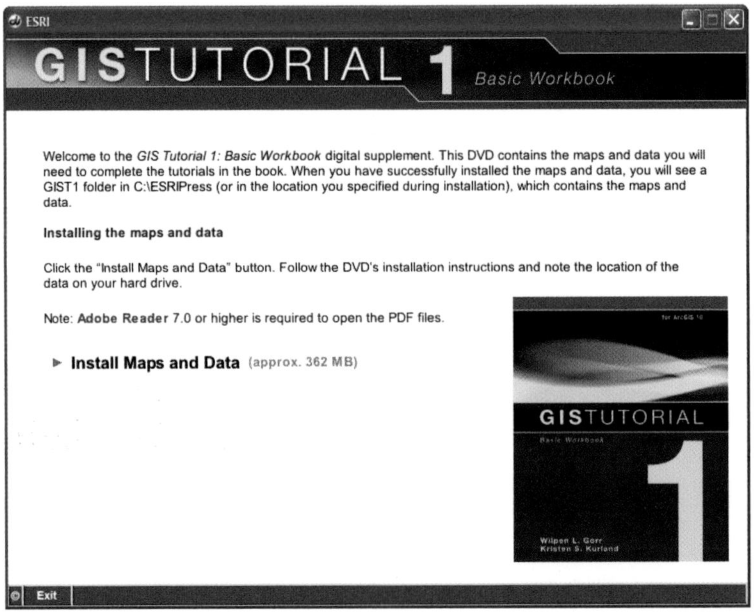

2 Read the welcome, then click the Install Exercise Data link. This launches the InstallShield Wizard.

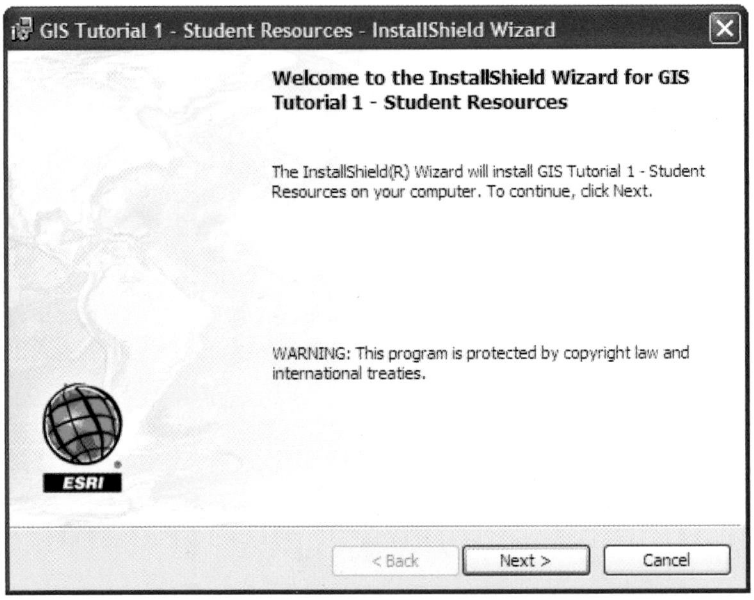

3 Click Next. Read and accept the license agreement terms, then click Next.